Biomechanics: Concepts and Computation

This quantitative approach integrates the classical concepts of mechanics and computational modelling techniques, in a logical progression through a wide range of fundamental biomechanics principles. Online MATLAB-based software, along with examples and problems using biomedical applications, will motivate undergraduate biomedical engineering students to practise and test their skills. The book covers topics such as kinematics, equilibrium, stresses and strains, and also focuses on large deformations and rotations and non-linear constitutive equations, including visco-elastic behaviour and the behaviour of long slender fibre-like structures. This is the first textbook that integrates both general and specific topics, theoretical background and biomedical engineering applications, as well as analytical and numerical approaches. This is the definitive textbook for students.

Cees Oomens is Associate Professor in Biomechanics and Continuum Mechanics at the Eindhoven University of Technology, the Netherlands. He has lectured many different courses ranging from basic courses in continuum mechanics at bachelor level, to courses on mechanical properties of materials and advanced courses in computational modelling at masters and postgraduate level. His current research focuses on damage and adaptation of soft biological tissues, with emphasis on skeletal muscle tissue and skin.

Marcel Brekelmans is Associate Professor in Continuum Mechanics at the Eindhoven University of Technology. Since 1998 he has also lectured in the Biomedical Engineering Faculty at the University; here his teaching addresses continuum mechanics, basic level and numerical analysis. He has published a considerable number of papers in well-known journals, and his research interests in continuum mechanics include the modelling of history-dependent material behaviour (plasticity, damage and fracture) in forming processes.

Frank Baaijens is Full Professor in Soft Tissue Biomechanics and Tissue Engineering at the Eindhoven University of Technology, where he has also been a part-time Professor in the Polymer Group of the Division of Computational and Experimental Mechanics since 1990. He is currently Scientific Director of the national research program on BioMedical Materials (BMM), and his research focuses on soft tissue biomechanics and tissue engineering.

CAMBRIDGE TEXTS IN BIOMEDICAL ENGINEERING

Cambridge Texts in Biomedical Engineering provides a forum for high-quality accessible textbooks targeted at undergraduate and graduate courses in biomedical engineering. It will cover a broad range of biomedical engineering topics from introductory texts to advanced topics including, but not limited to, biomechanics, physiology, biomedical instrumentation, imaging, signals and systems, cell engineering, and bioinformatics. The series blends theory and practice, aimed primarily at biomedical engineering students, it also suits broader courses in engineering, the life sciences and medicine.

Biomechanics

Concepts and Computation

Cees Oomens, Marcel Brekelmans, Frank Baaijens

Eindhoven University of Technology
Department of Biomedical Engineering
Tissue Biomechanics & Engineering

CAMBRIDGE UNIVERSITY PRESS
Cambridge, New York, Melbourne, Madrid, Cape Town, Singapore,
São Paulo, Delhi, Dubai, Tokyo

Cambridge University Press
The Edinburgh Building, Cambridge CB2 8RU, UK

Published in the United States of America by Cambridge University Press, New York

www.cambridge.org
Information on this title: www.cambridge.org/9780521172967

First published 2009
This digitally printed version 2010

A catalogue record for this publication is available from the British Library

ISBN 978-0-521-87558-5 Hardback
ISBN 978-0-521-17296-7 Paperback

Contents

About the cover

The cover contains images reflecting biomechanics research topics at the Eindhoven University of Technology. An important aspect of mechanics is experimental work to determine material properties and to validate models. The application field ranges from microscopic structures at the level of cells to larger organs like the heart. The core of biomechanics is constituted by models formulated in terms of partial differential equations and computer models to derive approximate solutions.

- *Main image*: Myogenic precursor cells have the ability to differentiate and fuse to form multinucleated myotubes. This differentiation process can be influenced by means of mechanical as well as biochemical stimuli. To monitor this process of early differentiation, immunohistochemical analyses are performed to provide information concerning morphology and localization of characteristic structural proteins of muscle cells. In the illustration, the sarcomeric proteins actin (red), and myosin (green) are shown. Nuclei are stained blue. Image courtesy of Mrs Marloes Langelaan.
- *Left top*: To study the effect of a mechanical load on the damage evolution of skeletal tissue an in-vitro model system using tissue engineered muscle was developed. The image shows this muscle construct in a set-up on a confocal microscope. In the device the construct can be mechanically deformed by means of an indentor. Fluorescent identification of both necrotic and apoptotic cells can be established using different staining techniques Image courtesy of Mrs Debby Gawlitta.
- *Left middle*: A three-dimensional finite element mesh of the human heart ventricles is shown. This mesh is used to solve the equations of motion for the beating heart. The model was used to study the effect of depolarization waves and mechanics in the paced heart. Image courtesy of Mr Roy Kerckhoffs.
- *Left bottom*: The equilibrium equations are derived from Newton's laws and describe (quasi-)static force equilibrium in a three-dimensional continuum. Chapter 9 of the present book.

Preface

In September 1997 an educational programme in Biomedical Engineering, unique in the Netherlands, started at the Eindhoven University of Technology, together with the University of Maastricht, as a logical step after almost two decades of research collaboration between both universities. This development culminated in the foundation of the Department of Biomedical Engineering in April 1999 and the creation of a graduate programme (MSc) in Biomedical Engineering in 2000 and Medical Engineering in 2002.

Already at the start of this educational programme, it was decided that a comprehensive course in biomechanics had to be part of the curriculum and that this course had to start right at the beginning of the Bachelor phase. A search for suitable material for this purpose showed that excellent biomechanics textbooks exist. But many of these books are very specialized to certain aspects of biomechanics. The more general textbooks are addressing mechanical or civil engineers or physicists who wish to specialize in biomechanics, so these books include chapters or sections on biology and physiology. Almost all books that were found are at Masters or post-graduate level, requiring basic to sophisticated knowledge of mechanics and mathematics. At a more fundamental level only books could be found that were written for mechanical and civil engineers.

We decided to write our own course material for the basic training in mechanics appropriate for our candidate biomedical engineers at Bachelor level, starting with the basic concepts of mechanics and ending with numerical solution procedures, based on the Finite Element Method. The course material assembled in the current book, comprises three courses for our biomedical engineers curriculum, distributed over the three years of their Bachelor studies. Chapters 1 to 6 mostly treat the basic concepts of forces, moments and equilibrium in a discrete context in the first year. Chapters 7 to 13 in the second year discuss the basis of continuum mechanics and Chapters 14 to 18 in the third year are focussed on solving the field equations of mechanics using the Finite Element Method.

What makes this book different from other basic mechanics or biomechanics treatises? Of course there is the usual attention, as in standard books, focussed on kinematics, equilibrium, stresses and strains. But several topics are discussed that are normally not found in one single textbook or only described briefly.

- Much attention is given to large deformations and rotations and non-linear constitutive equations (see Chapters 4, 9 and 10).
- A separate chapter is devoted to one-dimensional visco-elastic behaviour (Chapter 5).
- There is special attention to long slender fibre-like structures (Chapter 4).
- The similarities and differences in describing the behaviour of solids and fluids and aspects of diffusion and filtration are discussed (Chapters 12 to 16).
- Basic concepts of mechanics and numerical solution strategies for partial differential equations are integrated in one single textbook (Chapters 14 to 18).

Because of the usually rather complex geometries (and non-linear aspects) found in biomechanical problems hardly any relevant analytical solutions can be derived for the field equations and approximate solutions have to be constructed. It is the opinion of the authors that at Bachelor level at least the basis for these numerical techniques has to be addressed.

In Chapters 14 to 18 extensive use is made of a finite element code written in Matlab by one of the authors, which is especially developed as a tool for students. Applying this code requires that the user has a licence for the use of Matlab, which can be obtained via MathWorks (www.mathworks.com). The finite element code, which is a set of Matlab scripts, including manuals, is freely available and can be downloaded from the website: www.mate.tue.nl/biomechanicsbook.

1 Vector calculus

1.1 Introduction

Before we can start with biomechanics it is necessary to introduce some basic mathematical concepts and to introduce the mathematical notation that will be used throughout the book. The present chapter is aimed at understanding some of the basics of vector calculus, which is necessary to elucidate the concepts of force and momentum that will be treated in the next chapter.

1.2 Definition of a vector

A **vector** is a physical entity having both a magnitude (length or size) and a direction. For a vector $\vec{a}$ it holds, see Fig. 1.1:

$$\vec{a} = a\vec{e}. \tag{1.1}$$

The **length** of the vector $\vec{a}$ is denoted by $|\vec{a}|$ and is equal to the length of the arrow. The length is equal to a, when a is positive, and equal to $-a$ when a is negative. The **direction** of $\vec{a}$ is given by the unit vector $\vec{e}$ combined with the sign of a. The unit vector $\vec{e}$ has length 1. The vector $\vec{0}$ has length zero.

1.3 Vector operations

Multiplication of a vector $\vec{a} = a\vec{e}$ by a positive scalar α yields a vector $\vec{b}$ having the same direction as $\vec{a}$ but a different magnitude $\alpha|\vec{a}|$:

$$\vec{b} = \alpha\vec{a} = \alpha a\vec{e}. \tag{1.2}$$

This makes sense: pulling twice as hard on a wire creates a force in the wire having the same orientation (the direction of the wire does not change), but with a magnitude that is twice as large.

Figure 1.1

The vector $\vec{a} = a\vec{e}$ with $a > 0$.

Figure 1.2

Graphical representation of the sum of two vectors: $\vec{c} = \vec{a} + \vec{b}$.

The **sum** of two vectors $\vec{a}$ and $\vec{b}$ is a new vector $\vec{c}$, equal to the diagonal of the parallelogram spanned by $\vec{a}$ and $\vec{b}$, see Fig. 1.2:

$$\vec{c} = \vec{a} + \vec{b}. \tag{1.3}$$

This may be interpreted as follows. Imagine two thin wires which are attached to a point P. The wires are being pulled at in two different directions according to the vectors $\vec{a}$ and $\vec{b}$. The length of each vector represents the magnitude of the pulling force. The net force vector exerted on the attachment point P is the vector sum of the two vectors $\vec{a}$ and $\vec{b}$. If the wires are aligned with each other and the pulling direction is the same, the resulting force direction is clearly coinciding with the direction of the two wires and the length of the resulting force vector is the sum of the two pulling forces. Alternatively, if the two wires are aligned but the pulling forces are in opposite directions and of equal magnitude, the resulting force exerted on point P is the zero vector $\vec{0}$.

The **inner product** or **dot product** of two vectors is a scalar quantity, defined as

$$\vec{a} \cdot \vec{b} = |\vec{a}||\vec{b}| \cos(\phi), \tag{1.4}$$

where ϕ is the smallest angle between $\vec{a}$ and $\vec{b}$, see Fig. 1.3. The inner product is **commutative**, i.e.

$$\vec{a} \cdot \vec{b} = \vec{b} \cdot \vec{a}. \tag{1.5}$$

Figure 1.3

Definition of the angle ϕ.

The inner product can be used to define the length of a vector, since the inner product of a vector with itself yields ($\phi = 0$):

$$\vec{a} \cdot \vec{a} = |\vec{a}||\vec{a}| \cos(0) = |\vec{a}|^2. \tag{1.6}$$

If two vectors are perpendicular to each other the inner product of these two vectors is equal to zero, since in that case $\phi = \frac{\pi}{2}$:

$$\vec{a} \cdot \vec{b} = 0, \text{ if } \phi = \frac{\pi}{2}. \tag{1.7}$$

The **cross product** or **vector product** of two vectors $\vec{a}$ and $\vec{b}$ yields a new vector $\vec{c}$ that is perpendicular to both $\vec{a}$ and $\vec{b}$ such that $\vec{a}$, $\vec{b}$ and $\vec{c}$ form a right-handed system. The vector $\vec{c}$ is denoted as

$$\vec{c} = \vec{a} \times \vec{b}. \tag{1.8}$$

The length of the vector $\vec{c}$ is given by

$$|\vec{c}| = |\vec{a}||\vec{b}| \sin(\phi), \tag{1.9}$$

where ϕ is the smallest angle between $\vec{a}$ and $\vec{b}$. The length of $\vec{c}$ equals the area of the parallelogram spanned by the vectors $\vec{a}$ and $\vec{b}$. The vector system $\vec{a}$, $\vec{b}$ and $\vec{c}$ forms a right-handed system, meaning that if a corkscrew is used rotating from $\vec{a}$ to $\vec{b}$ the corkscrew would move into the direction of $\vec{c}$.

The vector product of a vector $\vec{a}$ with itself yields the zero vector since in that case $\phi = 0$:

$$\vec{a} \times \vec{a} = \vec{0}. \tag{1.10}$$

The vector product is **not** commutative, since the vector product of $\vec{b}$ and $\vec{a}$ yields a vector that has the opposite direction of the vector product of $\vec{a}$ and $\vec{b}$:

$$\vec{a} \times \vec{b} = -\vec{b} \times \vec{a}. \tag{1.11}$$

The **triple product** of three vectors $\vec{a}$, $\vec{b}$ and $\vec{c}$ is a scalar, defined by

$$\vec{a} \times \vec{b} \cdot \vec{c} = (\vec{a} \times \vec{b}) \cdot \vec{c}. \tag{1.12}$$

So, first the vector product of $\vec{a}$ and $\vec{b}$ is determined and subsequently the inner product of the resulting vector with the third vector $\vec{c}$ is taken. If all three vectors $\vec{a}$, $\vec{b}$ and $\vec{c}$ are non-zero vectors, while the triple product is equal to zero then the vector $\vec{c}$ lies in the plane spanned by the vectors $\vec{a}$ and $\vec{b}$. This can be explained by the fact that the vector product of $\vec{a}$ and $\vec{b}$ yields a vector perpendicular to the plane spanned by $\vec{a}$ and $\vec{b}$. Reversely, this implies that if the triple product is non-zero then the three vectors $\vec{a}$, $\vec{b}$ and $\vec{c}$ are not in the same plane. In that case the absolute value of the triple product of the vectors $\vec{a}$, $\vec{b}$ and $\vec{c}$ equals the volume of the parallelepiped spanned by $\vec{a}$, $\vec{b}$ and $\vec{c}$.

The **dyadic** or **tensor product** of two vectors $\vec{a}$ and $\vec{b}$ defines a linear transformation operator called a **dyad** $\vec{a}\vec{b}$. Application of a dyad $\vec{a}\vec{b}$ to a vector $\vec{p}$ yields a vector into the direction of $\vec{a}$, where $\vec{a}$ is multiplied by the inner product of $\vec{b}$ and $\vec{p}$:

$$\vec{a}\vec{b} \cdot \vec{p} = \vec{a}\,(\vec{b} \cdot \vec{p}). \tag{1.13}$$

So, application of a dyad to a vector transforms this vector into another vector. This transformation is linear, as can be seen from

$$\vec{a}\vec{b} \cdot (\alpha\vec{p} + \beta\vec{q}) = \vec{a}\vec{b} \cdot \alpha\vec{p} + \vec{a}\vec{b} \cdot \beta\vec{q} = \alpha\vec{a}\vec{b} \cdot \vec{p} + \beta\vec{a}\vec{b} \cdot \vec{q}. \tag{1.14}$$

The transpose of a dyad $(\vec{a}\vec{b})^{\mathrm{T}}$ is defined by

$$(\vec{a}\vec{b})^{\mathrm{T}} \cdot \vec{p} = \vec{b}\vec{a} \cdot \vec{p}, \tag{1.15}$$

or simply

$$(\vec{a}\vec{b})^{\mathrm{T}} = \vec{b}\vec{a}. \tag{1.16}$$

An operator A that transforms a vector $\vec{a}$ into another vector $\vec{b}$ according to

$$\vec{b} = A \cdot \vec{a}, \tag{1.17}$$

is called a second-order tensor A. This implies that the dyadic product of two vectors is a second-order tensor.

In the three-dimensional space a set of three vectors $\vec{c}_1$, $\vec{c}_2$ and $\vec{c}_3$ is called a **basis** if the triple product of the three vectors is non-zero, hence if all three vectors are non-zero vectors and if they do not lie in the same plane:

$$\vec{c}_1 \times \vec{c}_2 \cdot \vec{c}_3 \neq 0. \tag{1.18}$$

The three vectors $\vec{c}_1$, $\vec{c}_2$ and $\vec{c}_3$, composing the basis, are called basis vectors.

If the basis vectors are mutually perpendicular vectors the basis is called an **orthogonal** basis. If such basis vectors have unit length, then the basis is called **orthonormal**. A **Cartesian basis** is an orthonormal, right-handed basis with basis vectors independent of the location in the three-dimensional space. In the following we will indicate the Cartesian basis vectors with $\vec{e}_x$, $\vec{e}_y$ and $\vec{e}_z$.

1.4 Decomposition of a vector with respect to a basis

As stated above, a Cartesian vector basis is an orthonormal basis. Any vector can be decomposed into the sum of, at most, three vectors parallel to the three basis vectors $\vec{e}_x$, $\vec{e}_y$ and $\vec{e}_z$:

$$\vec{a} = a_x \vec{e}_x + a_y \vec{e}_y + a_z \vec{e}_z. \tag{1.19}$$

The components a_x, a_y and a_z can be found by taking the inner product of the vector $\vec{a}$ with respect to each of the basis vectors:

$$\begin{aligned} a_x &= \vec{a} \cdot \vec{e}_x \\ a_y &= \vec{a} \cdot \vec{e}_y \\ a_z &= \vec{a} \cdot \vec{e}_z, \end{aligned} \tag{1.20}$$

where use is made of the fact that the basis vectors have unit length and are mutually orthogonal, for example:

$$\vec{a} \cdot \vec{e}_x = a_x \vec{e}_x \cdot \vec{e}_x + a_y \vec{e}_y \cdot \vec{e}_x + a_z \vec{e}_z \cdot \vec{e}_x = a_x. \tag{1.21}$$

The components, say a_x, a_y and a_z, of a vector $\vec{a}$ with respect to the Cartesian vector basis, may be collected in a **column**, denoted by $\underset{\sim}{a}$:

$$\underset{\sim}{a} = \begin{bmatrix} a_x \\ a_y \\ a_z \end{bmatrix}. \tag{1.22}$$

So, with respect to a Cartesian vector basis any vector $\vec{a}$ may be decomposed in components that can be collected in a column:

$$\vec{a} \longleftrightarrow \underset{\sim}{a}. \tag{1.23}$$

This 'transformation' is only possible and meaningful if the vector basis with which the components of the column $\underset{\sim}{a}$ are defined has been specified. The choice of a different vector basis leads to a different column representation $\underset{\sim}{a}$ of the vector $\vec{a}$, this is illustrated in Fig. 1.4. The vector $\vec{a}$ has two different column representations, $\underset{\sim}{a}$ and $\underset{\sim}{a}^*$, depending on which vector basis is used. If, in a two-dimensional context $\{\vec{e}_x, \vec{e}_y\}$ is used as a vector basis then

$$\vec{a} \longrightarrow \underset{\sim}{a} = \begin{bmatrix} a_x \\ a_y \end{bmatrix}, \tag{1.24}$$

while, if $\{\vec{e}_x^{\,*}, \vec{e}_y^{\,*}\}$ is used as vector basis:

$$\vec{a} \longrightarrow \underset{\sim}{a}^* = \begin{bmatrix} a_x^* \\ a_y^* \end{bmatrix}. \tag{1.25}$$

Figure 1.4

Vector $\vec{a}$ with respect to vector basis $\{\vec{e}_x, \vec{e}_y\}$ and $\{\vec{e}_x^*, \vec{e}_y^*\}$.

Consequently, with respect to a Cartesian vector basis, vector operations such as multiplication, addition, inner product and dyadic product may be rewritten as 'column' (actually matrix) operations.

Multiplication of a vector $\vec{a} = a_x \vec{e}_x + a_y \vec{e}_y + a_z \vec{e}_z$ with a scalar α yields a new vector, say $\vec{b}$:

$$
\begin{aligned}
\vec{b} = \alpha \vec{a} &= \alpha(a_x \vec{e}_x + a_y \vec{e}_y + a_z \vec{e}_z) \\
&= \alpha a_x \vec{e}_x + \alpha a_y \vec{e}_y + \alpha a_z \vec{e}_z.
\end{aligned}
\tag{1.26}
$$

So

$$
\vec{b} = \alpha \vec{a} \longrightarrow \underset{\sim}{b} = \alpha \underset{\sim}{a}.
\tag{1.27}
$$

The sum of two vectors $\vec{a}$ and $\vec{b}$ leads to

$$
\vec{c} = \vec{a} + \vec{b} \longrightarrow \underset{\sim}{c} = \underset{\sim}{a} + \underset{\sim}{b}.
\tag{1.28}
$$

Using the fact that the Cartesian basis vectors have unit length and are mutually orthogonal, the inner product of two vectors $\vec{a}$ and $\vec{b}$ yields a scalar c according to

$$
\begin{aligned}
c = \vec{a} \cdot \vec{b} &= (a_x \vec{e}_x + a_y \vec{e}_y + a_z \vec{e}_z) \cdot (b_x \vec{e}_x + b_y \vec{e}_y + b_z \vec{e}_z) \\
&= a_x b_x + a_y b_y + a_z b_z.
\end{aligned}
\tag{1.29}
$$

In column notation this result is obtained via

$$
c = \underset{\sim}{a}^{\mathrm{T}} \underset{\sim}{b},
\tag{1.30}
$$

where $\underset{\sim}{a}^{\mathrm{T}}$ denotes the **transpose** of the column $\underset{\sim}{a}$, defined as

$$
\underset{\sim}{a}^{\mathrm{T}} = \begin{bmatrix} a_x & a_y & a_z \end{bmatrix},
\tag{1.31}
$$

such that:

$$
\underset{\sim}{a}^{\mathrm{T}} \underset{\sim}{b} = \begin{bmatrix} a_x & a_y & a_z \end{bmatrix} \begin{bmatrix} b_x \\ b_y \\ b_z \end{bmatrix} = a_x b_x + a_y b_y + a_z b_z.
\tag{1.32}
$$

Using the properties of the basis vectors of the Cartesian vector basis:

$$\vec{e}_x \times \vec{e}_x = \vec{0}$$
$$\vec{e}_x \times \vec{e}_y = \vec{e}_z$$
$$\vec{e}_x \times \vec{e}_z = -\vec{e}_y$$

$$\vec{e}_y \times \vec{e}_x = -\vec{e}_z$$
$$\vec{e}_y \times \vec{e}_y = \vec{0} \tag{1.33}$$
$$\vec{e}_y \times \vec{e}_z = \vec{e}_x$$

$$\vec{e}_z \times \vec{e}_x = \vec{e}_y$$
$$\vec{e}_z \times \vec{e}_y = -\vec{e}_x$$
$$\vec{e}_z \times \vec{e}_z = \vec{0},$$

the vector product of a vector $\vec{a}$ and a vector $\vec{b}$ is directly computed by means of

$$\vec{a} \times \vec{b} = (a_x \vec{e}_x + a_y \vec{e}_y + a_z \vec{e}_z) \times (b_x \vec{e}_x + b_y \vec{e}_y + b_z \vec{e}_z)$$
$$= (a_y b_z - a_z b_y) \vec{e}_x + (a_z b_x - a_x b_z) \vec{e}_y + (a_x b_y - a_y b_x) \vec{e}_z. \tag{1.34}$$

If by definition $\vec{c} = \vec{a} \times \vec{b}$, then the associated column $\underset{\sim}{c}$ can be written as:

$$\underset{\sim}{c} = \begin{bmatrix} a_y b_z - a_z b_y \\ a_z b_x - a_x b_z \\ a_x b_y - a_y b_x \end{bmatrix}. \tag{1.35}$$

The dyadic product $\vec{a}\vec{b}$ transforms another vector $\vec{c}$ into a vector $\vec{d}$, according to the definition

$$\vec{d} = \vec{a}\vec{b} \cdot \vec{c} = A \cdot \vec{c}, \tag{1.36}$$

with A the second-order tensor equal to the dyadic product $\vec{a}\vec{b}$. In column notation this is equivalent to

$$\underset{\sim}{d} = \underset{\sim}{a}(\underset{\sim}{b}^T \underset{\sim}{c}) = (\underset{\sim}{a}\,\underset{\sim}{b}^T)\underset{\sim}{c}, \tag{1.37}$$

with $\underset{\sim}{a}\,\underset{\sim}{b}^T$ a (3×3) matrix given by

$$\underline{A} = \underset{\sim}{a}\,\underset{\sim}{b}^T = \begin{bmatrix} a_x \\ a_y \\ a_z \end{bmatrix} \begin{bmatrix} b_x & b_y & b_z \end{bmatrix} = \begin{bmatrix} a_x b_x & a_x b_y & a_x b_z \\ a_y b_x & a_y b_y & a_y b_z \\ a_z b_x & a_z b_y & a_z b_z \end{bmatrix}, \tag{1.38}$$

or

$$\underset{\sim}{d} = \underline{A}\,\underset{\sim}{c}. \tag{1.39}$$

In this case $\underline{A}$ is called the matrix representation of the second-order tensor A, as the comparison of Eqs. (1.36) and (1.39) reveals.

Exercises

1.1 The basis $\{\vec{e}_x, \vec{e}_y, \vec{e}_z\}$ has a right-handed orientation and is orthonormal.
   (a) Determine $|\vec{e}_i|$ for $i = x, y, z$.
   (b) Determine $\vec{e}_i \cdot \vec{e}_j$ for $i, j = x, y, z$.
   (c) Determine $\vec{e}_x \cdot \vec{e}_y \times \vec{e}_z$.
   (d) Why is: $\vec{e}_x \times \vec{e}_y = \vec{e}_z$?

1.2 Let $\{\vec{e}_x, \vec{e}_y, \vec{e}_z\}$ be an orthonormal vector basis. The force vectors $\vec{F}_x = 3\vec{e}_x + 2\vec{e}_y + \vec{e}_z$ and $\vec{F}_y = -4\vec{e}_x + \vec{e}_y + 4\vec{e}_z$ act on point P. Calculate a vector $\vec{F}_z$ acting on P in such a way that the sum of all force vectors is the zero vector.

1.3 Let $\{\vec{e}_x, \vec{e}_y, \vec{e}_z\}$ be a right-handed and orthonormal vector basis. The following vectors are given: $\vec{a} = 4\vec{e}_z$, $\vec{b} = -3\vec{e}_y + 4\vec{e}_z$ and $\vec{c} = \vec{e}_x + 2\,\vec{e}_z$.
   (a) Write the vectors in column notation.
   (b) Determine $\vec{a} + \vec{b}$ and $3(\vec{a} + \vec{b} + \vec{c})$.
   (c) Determine $\vec{a} \cdot \vec{b}$, $\vec{b} \cdot \vec{a}$, $\vec{a} \times \vec{b}$ and $\vec{b} \times \vec{a}$.
   (d) Determine $|\vec{a}|$, $|\vec{b}|$, $|\vec{a} \times \vec{b}|$ and $|\vec{b} \times \vec{a}|$.
   (e) Determine the smallest angle between $\vec{a}$ and $\vec{b}$.
   (f) Determine a unit normal vector on the plane defined by $\vec{a}$ and $\vec{b}$.
   (g) Determine $\vec{a} \times \vec{b} \cdot \vec{c}$ and $\vec{a} \times \vec{c} \cdot \vec{b}$.
   (h) Determine $\vec{a}\vec{b} \cdot \vec{c}$, $(\vec{a}\vec{b})^{\mathrm{T}} \cdot \vec{c}$ and $\vec{b}\vec{a} \cdot \vec{c}$.
   (i) Do the vectors $\vec{a}$, $\vec{b}$ and $\vec{c}$ form a suitable vector basis? If the answer is yes, do they form an orthogonal basis? If the answer is yes, do they form an orthonormal basis?

1.4 Consider the basis $\{\vec{a}, \vec{b}, \vec{c}\}$ with $\vec{a}, \vec{b}$ and $\vec{c}$ defined as in the previous exercise. The following vectors are given: $\vec{d} = \vec{a} + 2\vec{b}$ and $\vec{e} = 2\vec{a} - 3\vec{c}$.
   (a) Determine $\vec{d} + \vec{e}$.
   (b) Determine $\vec{d} \cdot \vec{e}$.

1.5 The basis $\{\vec{e}_x, \vec{e}_y, \vec{e}_z\}$ is right-handed and orthonormal. The vectors $\vec{a}_x, \vec{a}_y$ and $\vec{a}_z$ are given by: $\vec{a}_x = 4\vec{e}_x + 3\vec{e}_y$; $\vec{a}_y = 3\vec{e}_x - 4\vec{e}_y$ and $\vec{a}_z = \vec{a}_x \times \vec{a}_y$.
   (a) Determine $\vec{a}_z$ expressed in $\vec{e}_x, \vec{e}_y$ and $\vec{e}_z$.
   (b) Determine $|\vec{a}_i|$ for $i = x, y, z$.
   (c) Determine the volume of the parallelepiped defined by $\vec{a}_x, \vec{a}_y$ and $\vec{a}_z$.
   (d) Determine the angle between the lines of action of $\vec{a}_x$ and $\vec{a}_y$.
   (e) Determine the vector $\vec{\alpha}_x$ from $\vec{a}_i = |\vec{a}_i|\vec{\alpha}_i$ for $i = x, y, z$. Is $\{\vec{\alpha}_x, \vec{\alpha}_y, \vec{\alpha}_z\}$ a right-handed, orthonormal vector basis?

(f)   Consider the vector $\vec{b} = 2\vec{e}_x + 3\vec{e}_y + \vec{e}_z$. Determine the column representation of $\vec{b}$ according to the bases $\{\vec{e}_x, \vec{e}_y, \vec{e}_z\}$, $\{\vec{a}_x, \vec{a}_y, \vec{a}_z\}$ and $\{\vec{\alpha}_x, \vec{\alpha}_y, \vec{\alpha}_z\}$.

(g)   Show that: $\vec{a}_x \times \vec{a}_y \cdot \vec{b} = \vec{a}_x \cdot \vec{a}_y \times \vec{b} = \vec{a}_y \cdot \vec{b} \times \vec{a}_x$.

1.6   Assume $\{\vec{e}_x, \vec{e}_y, \vec{e}_z\}$ is an orthonormal vector basis. The following vectors are defined:

$$\vec{a} = 4\vec{e}_x + 3\vec{e}_y - \vec{e}_z$$
$$\vec{b} = 6\vec{e}_y - \vec{e}_z$$
$$\vec{c} = 8\vec{e}_x - \vec{e}_z .$$

Are $\vec{a}, \vec{b}$ and $\vec{c}$ linearly independent? If not, what is the relationship between the vectors?

1.7   The vector bases $\{\vec{e}_x, \vec{e}_y, \vec{e}_z\}$ and $\{\vec{\epsilon}_x, \vec{\epsilon}_y, \vec{\epsilon}_z\}$ are orthonormal and do not coincide:

(a)   What is the effect of $\vec{e}_x\vec{\epsilon}_x + \vec{e}_y\vec{\epsilon}_y + \vec{e}_z\vec{\epsilon}_z$ acting on a vector $\vec{a}$?

(b)   What is the effect of $\vec{\epsilon}_x\vec{e}_x + \vec{\epsilon}_y\vec{e}_y + \vec{\epsilon}_z\vec{e}_z$ acting on a vector $\vec{a}$?

1.8   The vector basis $\{\vec{e}_x, \vec{e}_y, \vec{e}_z\}$ is orthonormal. What is the effect of the following dyadic products if they are applied to a vector $\vec{a}$?

(a) $\vec{e}_x\vec{e}_x$.

(b) $\vec{e}_x\vec{e}_x + \vec{e}_y\vec{e}_y$.

(c) $\vec{e}_x\vec{e}_x + \vec{e}_y\vec{e}_y + \vec{e}_z\vec{e}_z$.

(d) $\vec{e}_x\vec{e}_y - \vec{e}_y\vec{e}_x + \vec{e}_z\vec{e}_z$.

(e) $\vec{e}_x\vec{e}_x - \vec{e}_y\vec{e}_y + \vec{e}_z\vec{e}_z$.

2 The concepts of force and moment

2.1 Introduction

We experience the effects of force in everyday life and have an intuitive notion of force. For example, we exert a force on our body when we lift or push an object while we continuously (fortunately) feel the effect of gravitational forces, for instance while sitting, walking, etc. All parts of the human body in one way or the other are loaded by forces. Our bones provide rigidity to the body and can sustain high loads. The skin is resistant to force, simply pull on the skin to witness this. The cardiovascular system is continuously loaded dynamically due to the pulsating blood pressure. The bladder is loaded and stretched when it fills up. The intervertebral discs serve as flexible force transmitting media that give the spine its flexibility. Beside force we are using levers all the time in our daily life to increase the 'force' that we want to apply to some object, for example by opening doors with the latch, opening a bottle with a bottle-opener. We feel the effect of a lever arm when holding a weight close to our body instead of using a stretched arm. These experiences are the result of the moment that can be exerted by a force. Understanding the impact of force and moment on the human body requires us to formalize the intuitive notion of force and moment. That is the objective of this chapter.

2.2 Definition of a force vector

Imagine pulling on a thin wire that is attached to a wall. The pulling force exerted on the point of application is a vector with a physical meaning, it has

- a **length**: the magnitude of the pulling force
- an **orientation** in space: the direction of the wire
- a **line-of-action**, which is the line through the force vector.

The graphical representation of a force vector, denoted by $\vec{F}$, is given in Fig. 2.1. The 'shaft' of the arrow indicates the orientation in space of the force vector. The point of application of the force vector is denoted by the point P.

Figure 2.1

The force vector $\vec{F}$ and unit vector $\vec{e}$.

Figure 2.2

Force vector $\vec{F}$ written with respect to $\vec{e}_1$ and written with respect to $\vec{e}_2$.

The magnitude of the force vector is denoted by $|\vec{F}|$. If $\vec{e}$ denotes a unit vector the force vector may be written as

$$\vec{F} = F\vec{e}, \tag{2.1}$$

where F may be any rational number (i.e. negative, zero or positive). The absolute value $|F|$ of the number F is equal to the magnitude of force vector:

$$|F| = |\vec{F}|. \tag{2.2}$$

In Fig. 2.2 the force vector $\vec{F}$ is written with respect to either the unit vector $\vec{e}_1$ or with respect to the unit vector $\vec{e}_2$ that has the same working line in space as $\vec{e}_1$ but the opposite direction. Since the unit vector $\vec{e}_1$ has the same direction as the force vector $\vec{F}$:

$$\vec{F} = |\vec{F}|\vec{e}_1 . \tag{2.3}$$

In contrast, the unit vector $\vec{e}_2$ has a direction that is opposed to $\vec{F}$, therefore

$$\vec{F} = -|\vec{F}|\vec{e}_2. \tag{2.4}$$

Example 2.1   Let the force vector $\vec{F}$ be given by

$$\vec{F} = 2\vec{e}_1.$$

If the unit vectors $\vec{e}_1$ and $\vec{e}_2$ have opposite direction:

$$\vec{e}_2 = -\vec{e}_1,$$

then the force vector may also be written as

$$\vec{F} = -2\vec{e}_2.$$

2.3 Newton's Laws

The concepts in this biomechanics textbook are based on the work of Sir Isaac Newton (1643–1727). In his most famous work 'Philosophiae Naturalis Principia Mathematica' he described the law of gravity and what are currently known as the three laws of Newton, forming the basis for classical mechanics. These laws are:

- Every object in a state of uniform motion tends to remain in that state of motion unless an external force is applied to it. This is often termed simply the 'Law of Inertia'.
- In a one-dimensional context the second law states that the force F on an object equals the mass m, with SI unit [kg], of the object multiplied by the acceleration a, with dimension [m s^{-2}], of the object:

$$F = ma. \tag{2.5}$$

Consequently, the force F has the dimension [N] (Newton), with

$$1\,[\mathrm{N}] = 1\,[\mathrm{kg\ m\ s^{-2}}\,].$$

This may be generalized to the three-dimensional space in a straightforward manner. Let the position of a material particle in space be given by the vector $\vec{x}$. If the particle moves in space, this vector will be a function of the time t, i.e.

$$\vec{x} = \vec{x}(t). \tag{2.6}$$

The velocity $\vec{v}$ of the particle is given by

$$\vec{v}(t) = \frac{d\vec{x}}{dt}, \tag{2.7}$$

and the acceleration $\vec{a}$ follows from

$$\vec{a}(t) = \frac{d\vec{v}}{dt} = \frac{d^2\vec{x}}{dt^2}. \tag{2.8}$$

Newton's second law may now be formulated as

$$\vec{F} = m\vec{a}. \tag{2.9}$$

- The third law states that for every action there is an equal and opposite reaction. This law is exemplified by what happens when we step off a boat onto the bank of a lake: if we move in the direction of the shore, the boat tends to move in the opposite direction.

Example 2.2 Let the position of a particle with mass m for $t \geq 0$ be given by

$$\vec{x}(t) = \left(1 + \left(\frac{t}{\tau}\right)^2\right)\vec{x}_0,$$

where $\vec{x}_0$ denotes the position of the particle at $t=0$ and τ is a constant, characteristic time. The velocity of this particle is obtained from

$$\vec{v} = \frac{d\vec{x}}{dt} = \frac{d}{dt}\left((1+(t/\tau)^2)\vec{x}_0\right) = \frac{d(1+(t/\tau)^2)}{dt}\vec{x}_0 = (2t/\tau^2)\vec{x}_0,$$

while the acceleration follows from

$$\vec{a} = \frac{d\vec{v}}{dt} = (2/\tau^2)\vec{x}_0.$$

The force on this particle equals

$$\vec{F} = (2m/\tau^2)\vec{x}_0.$$

2.4 Vector operations on the force vector

Suppose that a force vector is represented by

$$\vec{F}_1 = F_1\vec{e}, \tag{2.10}$$

then another force vector, say $\vec{F}_2$ may be obtained by multiplying the force by a factor α, see Fig. 2.3(a):

$$\vec{F}_2 = \alpha F_1\vec{e} = F_2\vec{e}. \tag{2.11}$$

The force vector $\vec{F}_2$ has the same orientation in space as $\vec{F}_1$, but if $\alpha \neq 1$ it will have a different length, and it may have a direction sense (if $\alpha < 0$).

The net result of two force vectors, say $\vec{F}_1$ and $\vec{F}_2$, acting on the same point P is obtained by the vector sum, graphically represented in Fig. 2.3(b):

$$\vec{F}_3 = \vec{F}_1 + \vec{F}_2. \tag{2.12}$$

The vector $\vec{F}_3$ is placed along the diagonal of the parallelogram formed by the vectors $\vec{F}_1$ and $\vec{F}_2$. This implicitly defines the orientation, sense and magnitude of the resulting force vector $\vec{F}_3$.

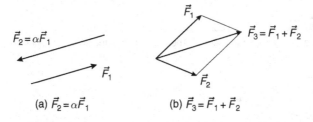

(a) $\vec{F}_2 = \alpha\vec{F}_1$ (b) $\vec{F}_3 = \vec{F}_1 + \vec{F}_2$

Figure 2.3

Graphical representation of the scalar multiplication of a force vector with $\alpha < 0$ (a) and the sum of two force vectors (b).

Clearly, if two force vectors $\vec{F}_1$ and $\vec{F}_2$ are parallel, then the resulting force vector $\vec{F}_3 = \vec{F}_1 + \vec{F}_2$ will be parallel to the vectors $\vec{F}_1$ and $\vec{F}_2$ as well. If $\vec{F}_1 = -\vec{F}_2$, then the addition of these two force vectors yields the so-called zero vector $\vec{0}$, having zero length.

2.5 Force decomposition

Suppose that a bone is loaded with a force $\vec{F}$ as sketched in Fig. 2.4. The principal axis of the bone has a direction indicated by the unit vector $\vec{e}$. The smallest angle between the force vector $\vec{F}$ and the unit vector $\vec{e}$ is denoted by α. It is useful to know, which part of the force $\vec{F}$ acts in the direction of the unit vector $\vec{e}$, indicated by $\vec{F}_t$ and which part of the force acts perpendicular to the bone, indicated by the force vector $\vec{F}_n$. The force vector $\vec{F}$ may, in that case, be written as

$$\vec{F} = \vec{F}_t + \vec{F}_n. \tag{2.13}$$

To determine the vectors $\vec{F}_t$ and $\vec{F}_n$, vector calculus will be used. The inner product of two vectors, say $\vec{a}$ and $\vec{b}$, is defined as

$$\vec{a} \cdot \vec{b} = |\vec{a}|\,|\vec{b}|\,\cos(\alpha), \tag{2.14}$$

where α is the smallest angle between the two vectors $\vec{a}$ and $\vec{b}$, see Fig. 2.5 and Chapter 1 for further details on the properties of the vector inner product. Computation of the inner product requires knowledge of the length of both vectors

Figure 2.4

Bone loaded by the force vector $\vec{F}$. The orientation of the bone is indicated by the unit vector $\vec{e}$.

Figure 2.5

Definition of the angle α.

(i.e. $|\vec{a}|$ and $|\vec{b}|$) and the smallest angle between the two vectors (i.e. α), all physical quantities that can easily be obtained. If the vectors $\vec{a}$ and $\vec{b}$ are perpendicular to each other, hence if $\alpha = \pi/2$, then the inner product equals zero, i.e. $\vec{a} \cdot \vec{b} = 0$. The length of a vector satisfies $|\vec{a}| = \sqrt{\vec{a} \cdot \vec{a}}$.

Now, consider the inner product of an arbitrary vector $\vec{b}$ with a unit vector $\vec{e}$ (i.e. $|\vec{e}| = 1$), then

$$\vec{b} \cdot \vec{e} = |\vec{b}| \cos(\alpha). \tag{2.15}$$

Let the vector $\vec{b}$ be written as the sum of a vector parallel to $\vec{e}$, say $\vec{b}_t$, and a vector normal to $\vec{e}$, say $\vec{b}_n$, such that:

$$\vec{b} = \vec{b}_t + \vec{b}_n, \tag{2.16}$$

as depicted in Fig. 2.6(a).

If the angle α between the unit vector $\vec{e}$ and the vector $\vec{b}$ is acute, hence if $\alpha \leq \pi/2$, it is easy to show that this inner product is equal to the length of the vector $\vec{b}_t$, the component of $\vec{b}$ parallel to the unit vector $\vec{e}$, see Fig. 2.6(a). By definition

$$\cos(\alpha) = \frac{|\vec{b}_t|}{|\vec{b}|}. \tag{2.17}$$

However, from Eq. (2.15) we know that

$$\cos(\alpha) = \frac{\vec{b} \cdot \vec{e}}{|\vec{b}|}, \tag{2.18}$$

hence

$$|\vec{b}_t| = \vec{b} \cdot \vec{e}. \tag{2.19}$$

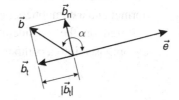

(a) Acute angle α, the length of $\vec{b}_t$ (b) Obtuse angle α, the length of $\vec{b}_t$

Figure 2.6

Vector decomposition in case of (a) an acute and (b) an obtuse angle between the vectors.

Since the angle α is acute, the vector $\vec{b}_t$ has the same sense as the unit vector $\vec{e}$ such that:

$$\vec{b}_t = |\vec{b}_t|\vec{e} = (\vec{b} \cdot \vec{e})\vec{e} . \tag{2.20}$$

If the angle α is obtuse, see Fig. 2.6(b), hence if $\alpha > \pi/2$, we have

$$\cos(\pi - \alpha) = -\cos(\alpha) = \frac{|\vec{b}_t|}{|\vec{b}|}. \tag{2.21}$$

With, according to Eq. (2.15), $\cos(\alpha) = \dfrac{\vec{b} \cdot \vec{e}}{|\vec{b}|}$ this leads to

$$|\vec{b}_t| = -\vec{b} \cdot \vec{e}. \tag{2.22}$$

In this case the sense of the vector $\vec{b}_t$ is opposite to the unit vector $\vec{e}$, such that

$$\vec{b}_t = -|\vec{b}_t|\vec{e} . \tag{2.23}$$

So, clearly, whether the angle α is acute or obtuse, the vector $\vec{b}_t$ parallel to the unit vector $\vec{e}$ is given by

$$\vec{b}_t = (\vec{b} \cdot \vec{e})\vec{e} . \tag{2.24}$$

Recall, that this is only true if $\vec{e}$ has unit length! In conclusion, the inner product of an arbitrary vector $\vec{b}$ with a unit vector $\vec{e}$ defines the magnitude and sense of a vector $\vec{b}_t$ that is parallel to the unit vector $\vec{e}$ such that the original vector $\vec{b}$ may be written as the sum of this parallel vector and a vector normal to the unit vector $\vec{e}$. The vector $\vec{b}_n$ normal to $\vec{e}$ follows automatically from

$$\vec{b}_n = \vec{b} - \vec{b}_t. \tag{2.25}$$

This implicitly defines the unique decomposition of the vector $\vec{b}$ into a component normal and a component parallel to the unit vector $\vec{e}$.

Based on the considerations above, the force vector $\vec{F}$ in Fig. 2.4 can be decomposed into a component parallel to the bone principal axis $\vec{F}_t$ given by

$$\vec{F}_t = (\vec{F} \cdot \vec{e})\vec{e}, \tag{2.26}$$

where $\vec{e}$ denotes a vector of unit length, and a component normal to the principal axis of the bone:

$$\vec{F}_n = \vec{F} - \vec{F}_t. \tag{2.27}$$

2.6 Representation of a vector with respect to a vector basis

Two vectors are called mutually independent if both vectors are non-zero and non-parallel. In a two-dimensional space any vector can be expressed as a linear combination of two mutually independent vectors, say $\vec{a}$ and $\vec{b}$, by using the scalar product and vector sum, for example:

$$\vec{F} = \alpha\vec{a} + \beta\vec{b}, \tag{2.28}$$

see, Fig. 2.7. Clearly, in a three-dimensional space three mutually independent vectors are needed:

$$\vec{F} = \alpha\vec{a} + \beta\vec{b} + \gamma\vec{c} . \tag{2.29}$$

The above mutually independent vectors $\vec{a}$ and $\vec{b}$ are called basis vectors that form a so-called vector basis $\{\vec{a}, \vec{b}\}$ in a two-dimensional space, while in a three-dimensional space three mutually independent vectors, say $\vec{a}$, $\vec{b}$ and $\vec{c}$, are needed to form a basis. It is convenient to have such a vector basis because it facilitates vector manipulation. For example, let

$$\vec{F}_1 = 2\vec{a} + 5\vec{b}, \tag{2.30}$$

and

$$\vec{F}_2 = -\vec{a} + 3\vec{b}, \tag{2.31}$$

then

$$\vec{F}_1 + \vec{F}_2 = \vec{a} + 8\vec{b}, \qquad \vec{F}_1 + 3\vec{F}_2 = -\vec{a} + 14\vec{b}. \tag{2.32}$$

The inner product of $\vec{F}_1$ and $\vec{F}_2$ yields

$$\begin{aligned} \vec{F}_1 \cdot \vec{F}_2 &= (2\vec{a} + 5\vec{b}) \cdot (-\vec{a} + 3\vec{b}) \\ &= -2\vec{a} \cdot \vec{a} + 6\vec{a} \cdot \vec{b} - 5\vec{b} \cdot \vec{a} + 15\vec{b} \cdot \vec{b} \\ &= -2\vec{a} \cdot \vec{a} + \vec{a} \cdot \vec{b} + 15\vec{b} \cdot \vec{b}, \end{aligned} \tag{2.33}$$

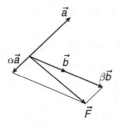

Figure 2.7

An arbitrary vector $\vec{F}$ as a linear combination of two vectors: $\vec{F} = \alpha\vec{a} + \beta\vec{b}$.

since $\vec{a} \cdot \vec{b} = \vec{b} \cdot \vec{a}$. This demonstrates that if vectors are expressed with respect to a vector basis, typical vector operations (such as vector addition and vector inner product) are relatively straightforward to perform. If the vector sum of the basis vectors and the inner products of the basis vectors with respect to each other and themselves are known, vector operations on other vectors are straightforward.

However, expressing an arbitrary vector with respect to the basis vectors may be cumbersome. For a given vector $\vec{F}$ this requires to determine the coefficients α and β in

$$\vec{F} = \alpha\vec{a} + \beta\vec{b}. \tag{2.34}$$

One possibility to realize this, is to take the inner product of $\vec{F}$ with respect to both the basis vectors:

$$\vec{F} \cdot \vec{a} = \alpha\,\vec{a} \cdot \vec{a} + \beta\,\vec{b} \cdot \vec{a}$$
$$\vec{F} \cdot \vec{b} = \alpha\,\vec{a} \cdot \vec{b} + \beta\,\vec{b} \cdot \vec{b}. \tag{2.35}$$

Recall, that each of the above inner products can be computed and simply yield a number. Therefore, the set of Eqs. (2.35) provides two linear equations from which the two unknown coefficients α and β can be obtained. A similar operation (involving three basis vectors) is needed in a three-dimensional space.

Solving for the coefficients α and β would be easy if the basis vectors $\vec{a}$ and $\vec{b}$ have unit length (such that e.g. $\vec{a} \cdot \vec{a} = 1$) and if the basis vectors are mutually perpendicular, hence if $\vec{a} \cdot \vec{b} = 0$. In that case, the set of Eqs. (2.35) would reduce to:

$$\vec{F} \cdot \vec{a} = \alpha$$
$$\vec{F} \cdot \vec{b} = \beta. \tag{2.36}$$

If the vectors of a vector basis are mutually perpendicular but do not have unit length, the vector basis is called orthogonal. If the vectors of an orthogonal vector basis have unit length, then it is called an orthonormal vector basis. If the basis vectors of an orthonormal basis have a so-called right-handed orientation with respect to each other and are independent of the location in three-dimensional space, it is called a Cartesian vector basis.

The **Cartesian** vector basis $\{\vec{e}_x, \vec{e}_y, \vec{e}_z\}$ is used to uniquely specify an arbitrary vector, see Fig. 2.8. An arbitrary force vector in two-dimensional space, say $\vec{F}$, can be expressed with respect to the Cartesian vector basis as

$$\vec{F} = F_x\vec{e}_x + F_y\vec{e}_y. \tag{2.37}$$

The use of a Cartesian vector basis substantially simplifies vector manipulation as is illustrated next.

(a) 2D Cartesian vector basis:
$\vec{F} = F_x \vec{e}_x + F_y \vec{e}_y$

(b) 3D Cartesian vector basis:
$\vec{F} = F_x \vec{e}_x + F_y \vec{e}_y + F_z \vec{e}_z$

Figure 2.8

Decomposition of $\vec{F}$ in a two- or three-dimensional Cartesian basis.

Example 2.3 Clearly, vector addition is straightforward, for example if

$$\vec{F}_1 = 2\vec{e}_x + 5\vec{e}_y, \qquad \vec{F}_2 = -\vec{e}_x + 3\vec{e}_y,$$

then

$$\vec{F}_1 + \vec{F}_2 = \vec{e}_x + 8\vec{e}_y, \qquad \vec{F}_1 + 3\vec{F}_2 = -\vec{e}_x + 14\vec{e}_y.$$

This is similar to using an arbitrary, non-orthonormal, vector basis, see Eq. (2.32).
Taking the inner product of the two vectors is substantially simplified:

$$\vec{F}_1 \cdot \vec{F}_2 = (2\vec{e}_x + 5\vec{e}_y) \cdot (-\vec{e}_x + 3\vec{e}_y)$$
$$= -2\underbrace{\vec{e}_x \cdot \vec{e}_x}_{1} + \underbrace{6\vec{e}_x \cdot \vec{e}_y - 5\vec{e}_x \cdot \vec{e}_y}_{0} + 15\underbrace{\vec{e}_y \cdot \vec{e}_y}_{1}$$
$$= 13.$$

Example 2.4 In the foot, the tendons of the tibialis anterior and the tibialis posterior may be
identified, see Fig. 2.9. Let the magnitude of the force vectors be given by:

$$F_a = |\vec{F}_a| = 50\,[\text{N}], \qquad F_p = |\vec{F}_p| = 60[\text{N}],$$

while the angles α and β are specified by:

$$\alpha = \frac{5\pi}{11}, \qquad \beta = \frac{\pi}{6}.$$

What is the net force acting on the attachment point Q of the two muscles on the
foot?

First, the force vectors $\vec{F}_a$ and $\vec{F}_p$ are written with respect to the Cartesian
coordinate system. Clearly:

Figure 2.9

Forces of the tendons of the tibialis anterior $\vec{F}_a$ and posterior $\vec{F}_p$, respectively.

$$\vec{F}_a = F_a \left[\cos(\alpha + \beta)\vec{e}_x + \sin(\alpha + \beta)\vec{e}_y \right]$$
$$\approx -18.6\vec{e}_x + 46.4\vec{e}_y \ [N],$$

and

$$\vec{F}_p = F_p \left[\cos(\alpha)\vec{e}_x + \sin(\alpha)\vec{e}_y \right]$$
$$\approx 8.5\vec{e}_x + 59.4\vec{e}_y \ [N].$$

Therefore, the net force due to $\vec{F}_a$ and $\vec{F}_p$ acting on point Q equals

$$\vec{F} = \vec{F}_a + \vec{F}_p = -10.1\vec{e}_x + 105.8\vec{e}_y \ [N].$$

Example 2.5 The decomposition of a force vector $\vec{F}$ into a component parallel to a unit vector $\vec{e}$ and a component normal to this vector is also straightforward. For example, let

$$\vec{F} = 2\vec{e}_x + 6\vec{e}_y,$$

and

$$\vec{e} = \frac{1}{\sqrt{13}}(2\vec{e}_x - 3\vec{e}_y).$$

Notice that $|\vec{e}| = 1$. Then, the component of $\vec{F}$ parallel to $\vec{e}$ is obtained from

$$\vec{F}_t = (\vec{F} \cdot \vec{e})\, \vec{e}$$

$$= \underbrace{\left(-\frac{14}{\sqrt{13}}\right)}_{\vec{F} \cdot \vec{e}} \underbrace{\frac{1}{\sqrt{13}}(2\vec{e}_x - 3\vec{e}_y)}_{\vec{e}}$$

$$= -\frac{14}{13}(2\vec{e}_x - 3\vec{e}_y).$$

2.7 Column notation

An arbitrary two-dimensional vector is written as

$$\vec{F} = F_x \vec{e}_x + F_y \vec{e}_y, \tag{2.38}$$

and in a three-dimensional space:

$$\vec{F} = F_x \vec{e}_x + F_y \vec{e}_y + F_z \vec{e}_z. \tag{2.39}$$

The numbers F_x, F_y and, in three dimensions F_x, F_y and F_z may be collected in a column $\underset{\sim}{F}$:

$$\underset{\sim}{F} = \begin{bmatrix} F_x \\ F_y \end{bmatrix}, \qquad \underset{\sim}{F} = \begin{bmatrix} F_x \\ F_y \\ F_z \end{bmatrix}. \tag{2.40}$$

These numbers only have a meaning when associated with a vector basis, in this case the Cartesian vector basis. There is a distinct difference between the vector $\vec{F}$ and the column $\underset{\sim}{F}$. The vector $\vec{F}$ is independent of the choice of the vector basis, while the numbers that are stored in the column $\underset{\sim}{F}$ depend on the vector basis that has been chosen. An example is given in Fig. 2.10. Both the vector basis $\{\vec{e}_x, \vec{e}_y\}$ and the vector basis $\{\vec{e}_x^*, \vec{e}_y^*\}$ are Cartesian, but they have a different orientation in space. Hence, the column associated with each of these bases is different. With respect to the $\{\vec{e}_x, \vec{e}_y\}$ basis it holds that

$$\underset{\sim}{F} = \begin{bmatrix} F_x \\ F_y \end{bmatrix}, \tag{2.41}$$

while with respect to $\{\vec{e}_x^*, \vec{e}_y^*\}$ this column is given by

$$\underset{\sim}{F}^* = \begin{bmatrix} F_x^* \\ F_y^* \end{bmatrix}. \tag{2.42}$$

According to Fig. 2.10 the $\{\vec{e}_x^*, \vec{e}_y^*\}$ basis is rotated by an angle α with respect to the $\{\vec{e}_x, \vec{e}_y\}$ basis. In that case:

Figure 2.10

(a) Basis rotation by angle α (b) $\vec{F}$ with respect to basis $\{\vec{e}_x, \vec{e}_y\}$ (c) $\vec{F}$ with respect to basis $\{\vec{e}_x^*, \vec{e}_y^*\}$.

$$\vec{e}_x = \cos(\alpha)\vec{e}_x^* - \sin(\alpha)\vec{e}_y^*$$
$$\vec{e}_y = \sin(\alpha)\vec{e}_x^* + \cos(\alpha)\vec{e}_y^*, \tag{2.43}$$

or, alternatively:

$$\vec{e}_x^* = +\cos(\alpha)\vec{e}_x + \sin(\alpha)\vec{e}_y$$
$$\vec{e}_y^* = -\sin(\alpha)\vec{e}_x + \cos(\alpha)\vec{e}_y. \tag{2.44}$$

Therefore, if $\vec{F}$ is known with respect to the $\{\vec{e}_x, \vec{e}_y\}$ basis, i.e.

$$\vec{F} = F_x\vec{e}_x + F_y\vec{e}_y, \tag{2.45}$$

this vector can also be expressed with respect to the $\{\vec{e}_x^*, \vec{e}_y^*\}$ basis:

$$\vec{F} = F_x\left(\cos(\alpha)\vec{e}_x^* - \sin(\alpha)\vec{e}_y^*\right) + F_y\left(\sin(\alpha)\vec{e}_x^* + \cos(\alpha)\vec{e}_y^*\right)$$
$$= \left(F_x\cos(\alpha) + F_y\sin(\alpha)\right)\vec{e}_x^* + \left(-F_x\sin(\alpha) + F_y\cos(\alpha)\right)\vec{e}_y^*. \tag{2.46}$$

Therefore, in terms of F_x and F_y:

$$\underset{\sim}{F}^* = \left[\begin{array}{c} F_x\cos(\alpha) + F_y\sin(\alpha) \\ -F_x\sin(\alpha) + F_y\cos(\alpha) \end{array} \right]. \tag{2.47}$$

The magnitude $|\vec{F}|$ of the force vector is obtained from

$$|\vec{F}| = \sqrt{\vec{F} \cdot \vec{F}}. \tag{2.48}$$

In a three-dimensional context, it follows immediately that with respect to a Cartesian vector basis:

$$|\vec{F}| = \sqrt{F_x^2 + F_y^2 + F_z^2}, \tag{2.49}$$

or, equivalently

$$|\vec{F}| = \sqrt{\underset{\sim}{F}^T \underset{\sim}{F}}. \tag{2.50}$$

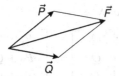

Figure 2.11

The force vector $\vec{F}$ resulting from the two force vectors $\vec{P}$ and $\vec{Q}$.

Suppose that two thin wires are connected to a point. The first wire is loaded with force $\vec{P}$ and on the second wire a force $\vec{Q}$ is applied. The total force vector $\vec{F}$ exerted on the point is calculated from

$$\vec{F} = \vec{P} + \vec{Q}.\tag{2.51}$$

See Fig. 2.11 for a visualization.

The force vectors may be written as

$$\vec{P} = P_x \vec{e}_x + P_y \vec{e}_y + P_z \vec{e}_z\tag{2.52}$$

$$\vec{Q} = Q_x \vec{e}_x + Q_y \vec{e}_y + Q_z \vec{e}_z,\tag{2.53}$$

such that

$$\vec{F} = (P_x + Q_x)\vec{e}_x + (P_y + Q_y)\vec{e}_y + (P_z + Q_z)\vec{e}_z.\tag{2.54}$$

The magnitude of the resulting force vector is given by

$$|\vec{F}| = \sqrt{(P_x + Q_x)^2 + (P_y + Q_y)^2 + (P_z + Q_z)^2}.\tag{2.55}$$

Using the column representation the components defining the force vectors $\vec{P}$ and $\vec{Q}$ with respect to the Cartesian vector basis may be collected into, respectively:

$$\underset{\sim}{P} = \begin{bmatrix} P_x \\ P_y \\ P_z \end{bmatrix}, \quad \underset{\sim}{Q} = \begin{bmatrix} Q_x \\ Q_y \\ Q_z \end{bmatrix},\tag{2.56}$$

such that the force vector $\vec{F}$ is represented by

$$
\vec{F} = \begin{bmatrix} F_x \\ F_y \\ F_z \end{bmatrix} = \begin{bmatrix} P_x \\ P_y \\ P_z \end{bmatrix} + \begin{bmatrix} Q_x \\ Q_y \\ Q_z \end{bmatrix} = \begin{bmatrix} P_x + Q_x \\ P_y + Q_y \\ P_z + Q_z \end{bmatrix}. \tag{2.57}
$$

2.8 Drawing convention

Consider two force vectors, $\vec{F}_1$ and $\vec{F}_2$, both parallel to the unit vector $\vec{e}$ as sketched in Fig. 2.12. In this case the two vectors are identified by numbers F_1 and F_2, rather than by the vector symbols $\vec{F}_1$ and $\vec{F}_2$. These numbers denote the magnitude of the force vector, while the orientation of the arrow denotes the direction of the vector. Consequently this way of drawing and identifying the vectors implicitly assumes

$$
\vec{F}_1 = F_1 \vec{e}, \tag{2.58}
$$

while

$$
\vec{F}_2 = -F_2 \vec{e}. \tag{2.59}
$$

This drawing convention is generally used in combination with a certain vector basis. In this course the Cartesian vector basis is used only. In that case, forces acting in the horizontal plane, hence in the $\vec{e}_x$ direction, are frequently identified by H_i (from Horizontal), while forces acting in vertical direction, hence in the $\vec{e}_y$

Figure 2.12

Force vectors identified by their magnitude (F_1 and F_2).

Figure 2.13

Force vectors.

direction are identified by V_i (from Vertical). For example the vectors drawn in Fig. 2.13 indicate that

$$\vec{H}_1 = H_1\vec{e}_x, \quad \vec{H}_2 = -H_2\vec{e}_x, \quad \vec{V}_3 = V_3\vec{e}_y. \tag{2.60}$$

2.9 The concept of moment

A simple example of the effect of a moment is experienced when holding a tray with a mass on it that exerts a (gravity) force on the tray, see the schematic drawing in Fig. 2.14(a). This force, which acts at a certain distance d, causes a moment at the position of our hand as is shown in Fig. 2.14(b), where the tray has been removed from the drawing and the resulting load on the hand is indicated by the arrow F, representing the force, and additionally the curved arrow M, representing the moment.

Increasing the distance of the mass with respect to our hand or increasing the mass, will increase the moment that we experience. In fact, the moment (or torque if you like) that is felt on our hand equals the distance d multiplied by the force due to the mass F:

$$M = dF. \tag{2.61}$$

The moment has a certain orientation in space. Changing the direction of the force F, as visualized in Fig. 2.15(a), will change the orientation of the moment. If the force acts at a certain angle on the tray, as indicated in Fig. 2.15(b), only the force normal to the tray will generate a moment with respect to the hand:

$$M = dF_n. \tag{2.62}$$

In this chapter this intuitive notion of moment is formalized.

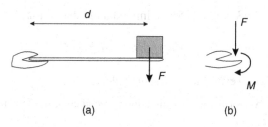

(a)   (b)

Figure 2.14

(a) Weight of an object on a tray (b) Loading on the hand.

(a) Moment due to reversed force F (b) Moment due to oriented force F

Figure 2.15

Moment due to various forces F.

2.10 Definition of the moment vector

A point in space may be identified by its position vector $\vec{x}$, see for instance the three-dimensional example in Fig. 2.16, where O denotes the location of the origin of the Cartesian vector basis $\{\vec{e}_x, \vec{e}_y, \vec{e}_z\}$.

Assume that a force $\vec{F}$ is applied to a point Q with location $\vec{x}_Q$. The moment vector is defined with respect to a point in space, say P having location $\vec{x}_P$. The moment exerted by the force $\vec{F}$ with respect to point P is defined as

$$\vec{M} = (\vec{x}_Q - \vec{x}_P) \times \vec{F} = \vec{d} \times \vec{F}. \tag{2.63}$$

For an interpretation of Eq. (2.63) it is useful to first focus on a two-dimensional configuration. Consider the situation as depicted in Fig. 2.17, where we focus our attention on the plane that is formed by the vector $\vec{d}$ and the force vector $\vec{F}$. Define a Cartesian vector basis $\{\vec{e}_x, \vec{e}_y, \vec{e}_z\}$ with the basis vectors $\vec{e}_x$ and $\vec{e}_y$ in the plane and $\vec{e}_z$ perpendicular to the plane. In this case, vector $\vec{e}_z$ is pointing towards the reader. With respect to the basis the column representations of the vectors $\vec{d}$ and $\vec{F}$ can be given by

$$\underline{d} = \begin{bmatrix} d_x \\ d_y \\ 0 \end{bmatrix}, \quad \underline{F} = \begin{bmatrix} F_x \\ F_y \\ 0 \end{bmatrix}. \tag{2.64}$$

By using Eqs. (2.63) and (1.35) we immediately derive that

$$\vec{M} = (d_x F_y - d_y F_x)\vec{e}_z. \tag{2.65}$$

From this analysis several items become clear:

- The moment vector points in a direction perpendicular to the plane that is formed by the vectors $\vec{d}$ and $\vec{F}$.
- The total moment vector can be written as an addition of the moments $\vec{M}_1 = d_x F_y \vec{e}_z$ and $\vec{M}_2 = -d_y F_x \vec{e}_z$. For both composing moments the force is perpendicular to the working

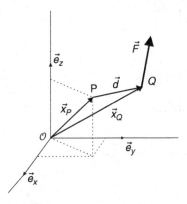

Figure 2.16

A point in space identified by its position vector $\vec{x}$.

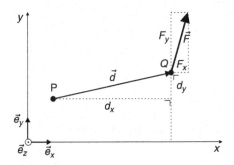

Figure 2.17

The moment of a force acting at point Q with respect to point P.

distance of the forces, i.e. $d_x\vec{e}_x$ is perpendicular to $F_y\vec{e}_y$ and $d_y\vec{e}_y$ is perpendicular to $F_x\vec{e}_x$.

- The directions of the composing moments $\vec{M}_1 = d_xF_y\vec{e}_z$ and $\vec{M}_2 = -d_yF_x\vec{e}_z$ follow from the corkscrew rule. To apply this corkscrew rule correctly, place the tails of the two vectors (e.g. $d_x\vec{e}_x$ and $F_y\vec{e}_y$) at the same location in space, see Fig. 2.18. In the case of the combination $d_x\vec{e}_x$ and $F_y\vec{e}_y$, the rotation of the arm to the force is a counterclockwise movement, leading to a vector that points out of the plane, i.e. in positive $\vec{e}_z$-direction. In the case of the combination $d_y\vec{e}_y$ and $F_x\vec{e}_x$, rotating the arm to the force is a clockwise movement resulting in a moment vector that points into the plane, i.e. in negative $\vec{e}_z$-direction.

In the definition of the moment vector the location of the force vector along the line-of-action is not relevant since only the magnitude of the force, the direction and the distance of the point P to the line-of-action are of interest. This is illustrated in Fig. 2.19. We can decompose the vector $\vec{d}$ pointing from the point P to Q in a vector $\vec{d}_n$, perpendicular to the line of action of the force $\vec{F}$, and a vector

Application of the corkscrew rule.

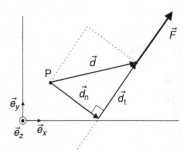

Figure 2.19

The moment of a force acting at point Q with respect to point P.

$\vec{d}_t$, parallel to the line of action. Then, we can write for the moment $\vec{M}$ of vector $\vec{F}$ with respect to point P:

$$\vec{M} = \vec{d} \times \vec{F} = (\vec{d}_n + \vec{d}_t) \times \vec{F} = \vec{d}_n \times \vec{F}. \tag{2.66}$$

The definition in Eq. (2.63) also assures that the resulting moment is the zero vector if the point P is located on the line-of-action of the force vector (in that case $\vec{d}_n = \vec{0}$).

In the general three-dimensional case, see Fig. 2.16, the procedure to determine the moment of the force $\vec{F}$ with respect to the point P is comparable. The column representations of the vectors $\vec{d}$ and $\vec{F}$ in this case are given by

$$\underset{\sim}{d} = \begin{bmatrix} d_x \\ d_y \\ d_z \end{bmatrix}, \quad \underset{\sim}{F} = \begin{bmatrix} F_x \\ F_y \\ F_z \end{bmatrix}, \tag{2.67}$$

and the resulting column representation of the moment follows from Eq. (1.35):

$$\underset{\sim}{M} = \begin{bmatrix} d_y F_z - d_z F_y \\ d_z F_x - d_x F_z \\ d_x F_y - d_y F_x \end{bmatrix}. \tag{2.68}$$

Example 2.6   Let the origin of the Cartesian coordinate system be the point with respect to which the moment vector is computed, i.e.

$$\vec{x}_P = \vec{0}.$$

The point of application of the force vector $\vec{F}$, is denoted by:

$$\vec{x}_Q = 2\vec{e}_x + \vec{e}_y,$$

which means that this point is located in the xy-plane. The force vector is also located in this plane:

$$\vec{F} = 5\vec{e}_y.$$

The moment of the force $\vec{F}$ with respect to the point P follows from

$$
\begin{aligned}
\vec{M} &= (\vec{x}_Q - \vec{x}_P) \times \vec{F} \\
&= (2\vec{e}_x + \vec{e}_y) \times 5\vec{e}_y \\
&= 10 \underbrace{\vec{e}_x \times \vec{e}_y}_{\vec{e}_z} + 5 \underbrace{\vec{e}_y \times \vec{e}_y}_{\vec{0}} \\
&= 10\vec{e}_z.
\end{aligned}
$$

Example 2.7   Let, as before:

$$\vec{x}_Q = 2\vec{e}_x + \vec{e}_y,$$

and

$$\vec{F} = 5\vec{e}_y,$$

but

$$\vec{x}_P = 3\vec{e}_z.$$

Then

$$
\begin{aligned}
\vec{M} &= (\vec{x}_Q - \vec{x}_P) \times \vec{F} \\
&= (2\vec{e}_x + \vec{e}_y - 3\vec{e}_z) \times 5\vec{e}_y \\
&= 10\vec{e}_z + 15\vec{e}_x.
\end{aligned}
$$

2.11 The two-dimensional case

If all forces act in the same plane, the resulting moment vector with respect to any point in that plane is, by definition, perpendicular to this plane. However, it is common practice in this case to indicate a moment as a curved arrow that is showing a clockwise or counterclockwise direction, see Fig. 2.20. Using the notation:

$$\vec{M} = M\vec{e}_z, \tag{2.69}$$

and defining the orientation vector $\vec{e}_z = \vec{e}_x \times \vec{e}_y$ to be pointing out of the plane into the direction of the viewer, a counterclockwise moment corresponds to $M > 0$, while a clockwise moment corresponds to $M < 0$. Fig. 2.20(a) shows a two-dimensional body with a force $\vec{F}$ acting on it at point Q. We can define an arbitrary point P in the body. The moment $\vec{M}$ with respect to P as a result of the force $\vec{F}$ will be a vector perpendicular to the plane of drawing. Using the drawing convention as proposed above Fig. 2.20(a) can be replaced by Fig. 2.20(b). In this case the line-of-action of force $\vec{F}$ is drawn through point P and the resulting moment is given by a curved arrow in counterclockwise direction. The loading of the body according to Figures 2.20(a) and 2.20(b) is statically, completely equivalent. The same is true for Figures 2.20(c) and 2.20(d) for a clockwise direction of the moment.

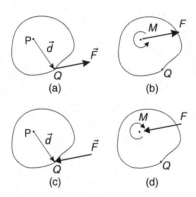

Figure 2.20

Drawing convention of the moment vector for different force vector orientations. Figures (a) and (b) indicate a statically equivalent load for a counterclockwise orientation of the moment vector. Figures (c) and (d) are equivalent for a clockwise orientation of the moment vector.

Figure 2.21

Resulting moment in two dimensions.

Example 2.8   Resulting moment using scalar notation. Following the drawing convention of Section 2.2, the force vectors in Fig. 2.21 are given by

$$\vec{F}_1 = F_1\vec{e}_x$$
$$\vec{F}_2 = -F_2\vec{e}_x$$
$$\vec{F}_3 = F_3\vec{e}_x.$$

Similarly,

$$\vec{d}_1 = 3\vec{e}_x + 4\vec{e}_y$$
$$\vec{d}_2 = -2\vec{e}_x - 2\vec{e}_y$$
$$\vec{d}_3 = 2\vec{e}_x - 5\vec{e}_y.$$

Each of the force vectors $\vec{F}_i$ generates a moment vector with respect to the point P:

$$\vec{M}_i = \vec{d}_i \times \vec{F}_i.$$

Clearly, given the fact that all force vectors are in the plane spanned by the $\vec{e}_x$ and $\vec{e}_y$ vectors, the moment vectors are all in the $\vec{e}_z$ direction:

$$\vec{M}_i = M_i\vec{e}_z.$$

Either using the formal definition of the moment vector or the drawing convention for two-dimensional problems as given above, it follows that

$$M_1 = -4F_1$$
$$M_2 = -2F_2$$
$$M_3 = 5F_3.$$

The force vectors $\vec{F}_1$ and $\vec{F}_2$ produce a clockwise, hence negative, moment, while $\vec{F}_3$ produces a counterclockwise, hence positive, moment. The resulting moment with respect to point P equals

$$M = M_1 + M_2 + M_3 = -4F_1 - 2F_2 + 5F_3.$$

Example 2.9   Resulting moment using vector notation. Consider the forces as depicted in Fig. 2.22. The forces are given by

$$\vec{F}_1 = 3\vec{e}_x + \vec{e}_y$$
$$\vec{F}_2 = 4\vec{e}_x - \vec{e}_y$$
$$\vec{F}_3 = -2\vec{e}_x - 3\vec{e}_y,$$

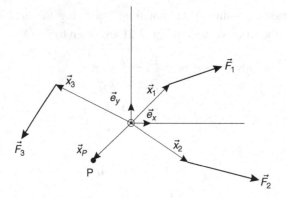

Figure 2.22

Forces and moment.

while the points of application are, respectively:

$$\vec{x}_1 = 2\vec{e}_x + 2\vec{e}_y$$
$$\vec{x}_2 = 3\vec{e}_x - 2\vec{e}_y$$
$$\vec{x}_3 = -4\vec{e}_x + 2\vec{e}_y.$$

The point P has location:

$$\vec{x}_P = -2\vec{e}_x - 2\vec{e}_y .$$

The resulting moment of the forces with respect to the point P follows from

$$\vec{M} = (\vec{x}_1 - \vec{x}_P) \times \vec{F}_1 + (\vec{x}_2 - \vec{x}_P) \times \vec{F}_2 + (\vec{x}_3 - \vec{x}_P) \times \vec{F}_3,$$

hence

$$\vec{M} = (4\vec{e}_x + 4\vec{e}_y) \times (3\vec{e}_x + \vec{e}_y) + 5\vec{e}_x \times (4\vec{e}_x - \vec{e}_y)$$
$$+ (-2\vec{e}_x + 4\vec{e}_y) \times (-2\vec{e}_x - 3\vec{e}_y)$$
$$= -8\vec{e}_z - 5\vec{e}_z + 14\vec{e}_z$$
$$= \vec{e}_z.$$

2.12 Drawing convention of moments in three dimensions

An arrow drawn with two arrowheads, and identified by a scalar, rather than a vector symbol, denotes a moment vector following the right-handed or corkscrew rule. For example the moment vectors drawn in Fig. 2.23 and identified by the scalars M_1, M_2 and M_3, respectively, correspond to the moment vectors:

Figure 2.23

Moment vectors identified by means of scalars.

Figure 2.24

Moment vectors identified by means of vectors and having a single arrowhead.

$$\vec{M}_1 = M_1\vec{e}_x, \quad \vec{M}_2 = -M_2\vec{e}_y, \quad \vec{M}_3 = M_3\vec{e}_z. \tag{2.70}$$

Notice that the moment vector $\vec{M}_2$ is pointing into the negative y-direction. Alternatively, if in the figure the moment vectors are drawn with a single arrowhead, as in Fig. 2.24, they denote actual vectors and are identified with vector symbols.

Exercises

2.1   The vector bases $\{\vec{e}_1, \vec{e}_2, \vec{e}_3\}$ and $\{\vec{\epsilon}_1, \vec{\epsilon}_2, \vec{\epsilon}_3\}$ are orthonormal. The following relations exist:

$$\vec{\epsilon}_1 = \frac{1}{2}\sqrt{2}\vec{e}_1 + \frac{1}{2}\sqrt{2}\vec{e}_2$$

$$\vec{\epsilon}_2 = -\frac{1}{2}\sqrt{2}\vec{e}_1 + \frac{1}{2}\sqrt{2}\vec{e}_2$$

$$\vec{\epsilon}_3 = \vec{e}_3 \ .$$

The force vector $\vec{F}$ is defined with respect to basis $\{\vec{e}_1, \vec{e}_2, \vec{e}_3\}$ according to:
$\vec{F} = 2\vec{e}_1 + 3\vec{e}_2 - 4\vec{e}_3$.

(a)   Determine the decomposition of $\vec{F}$ with respect to basis $\{\vec{\epsilon}_1, \vec{\epsilon}_2, \vec{\epsilon}_3\}$.

(b) Determine the length of $\vec{F}$, using the specifications of $\vec{F}$ expressed in $\{\vec{e}_1, \vec{e}_2, \vec{e}_3\}$ and in $\{\vec{\epsilon}_1, \vec{\epsilon}_2, \vec{\epsilon}_3\}$.

2.2 For the points P, Q and R the following location vectors are given, respectively:

$$\vec{x}_P = \vec{e}_x + 2\vec{e}_y$$
$$\vec{x}_Q = 4\vec{e}_x + 2\vec{e}_y$$
$$\vec{x}_R = 3\vec{e}_x + e_y.$$

The force vector $\vec{F} = 2\vec{e}_x$ acts on point Q.

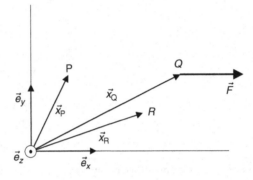

(a) Calculate the moment of the force $\vec{F}$ with respect to point P and with respect to point R.

(b) Write the vectors, mentioned above, in column notation according to the right-handed orthonormal basis $\{\vec{e}_x, \vec{e}_y, \vec{e}_z\}$ and calculate the moment of the force with respect to the points P and R by using Eq. (1.35).

2.3 Calculate for each of the situations given below the resulting moment with respect to point P.

2.4 On an axis a wheel with radius R is fixed to a smaller wheel with radius r. The forces F and f are tangentially applied to the contours of both wheels (as shown in the figure). Calculate the ratio between the forces F and f in the case where the total moment with respect to the centroid P is zero.

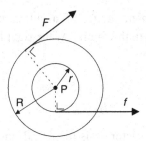

2.5   On the body in the drawing the moments $\vec{M}_1 = -3\vec{e}_x$ and $\vec{M}_2 = 4\vec{e}_x$ and the forces $\vec{F}_1 = -2\vec{e}_z$ and $\vec{F}_2 = \vec{e}_z$ are exerted. The forces are acting at the points $\vec{x}_1 = -2\vec{e}_x + 3\vec{e}_z$ and $\vec{x}_2 = 3\vec{e}_x + \vec{e}_y + 3\vec{e}_z$. Calculate the resulting moment vector with respect to point P, given by the position vector $\vec{x}_P = 2\vec{e}_x + 4\vec{e}_y + 2\vec{e}_z$.

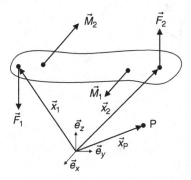

2.6   A person is pulling a rope. This results in a force $\vec{F}$ acting on the hand at point H as depicted in the figure. Calculate the moment in the shoulder (point S) and the elbow (point E) as a result of the force $\vec{F}$. The following vectors are given:

$$\vec{x}_1 = 5\vec{e}_x + \vec{e}_y$$
$$\vec{x}_2 = 4\vec{e}_x + 5\vec{e}_y$$
$$\vec{F} = \vec{e}_x + 2\vec{e}_y.$$

2.7   A very simple segmented model of a sitting person is shown in the schematic visualization in the figure. At the right hand H a force $\vec{F}_H$ and a moment $\vec{M}_H$ are given by:

$$\vec{F}_H = 5\vec{e}_x + \vec{e}_z$$
$$\vec{M}_H = \vec{e}_y + 3\vec{e}_z,$$

related to the Cartesian vector basis $\{\vec{e}_x, \vec{e}_y, \vec{e}_z\}$. In addition the positions of the hand H and shoulder S with respect to the origin at the lower end of the spine are given:

$$\vec{x}_H = 4\vec{e}_x - 2\vec{e}_y + 5\vec{e}_z$$
$$\vec{x}_S = \vec{e}_x - 2\vec{e}_y + 4\vec{e}_z.$$

(a)   Calculate the moment at the shoulder (point S) as a result of the load at the hand H.

(b)   Calculate the moment at the lower end of the spine O as a result of the load at the hand H.

3 Static equilibrium

3.1 Introduction

According to Newton's law the acceleration of the centroid of a body multiplied by its mass equals the total force applied to the body and there will be a spin around the centroid, when there is a resulting moment with respect to the centroid. But in many cases bodies do not move at all when forces are applied to them. In that case the bodies are in static equilibrium. A simple example is given in Fig. 3.1. In Fig. 3.1(a) a body is loaded by two forces of equal size but with an opposite direction. The lines-of-action of the two forces coincide and clearly the body is in equilibrium. If the lines-of-action do not coincide, as in Fig. 3.1(b), the forces have a resulting moment and the body will rotate. To enforce static equilibrium a counteracting moment should be applied to prevent the body rotating, as indicated in Fig. 3.1(c).

3.2 Static equilibrium conditions

If a body moves monotonously (no acceleration of the centroid, no rate of rotation around the centroid), the body is in static equilibrium. If the velocities are zero as well, the body is at rest. In both cases the sum of all forces **and** the sum of all moments (with respect to any point) acting on the body are zero. Suppose that n forces $\vec{F}_i$ ($i = 1, 2, ..., n$) are applied to the body. Each of these forces will have a moment M_i with respect to an arbitrary point P. There may be a number of additional moments $\vec{M}_j$ ($j = 1, 2, ..., m$) applied to the body. Static equilibrium then requires that

$$\sum_{i=1}^{n} \vec{F}_i = \vec{0}$$

$$\sum_{i=1}^{n} \vec{M}_i + \sum_{j=1}^{m} \vec{M}_j = \vec{0}. \tag{3.1}$$

(a) Static equilibrium   (b) **No** static equilibrium   (c) Static equilibrium

Figure 3.1

Examples of satisfaction and violation of static equilibrium.

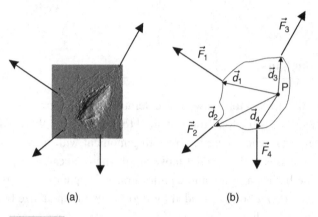

(a)   (b)

Figure 3.2

An image and a model of a cell.

Simply demanding that the sum of all the forces is equal to zero, to assure equilibrium, is insufficient, since a resulting moment may induce a (rate of) rotation of the body. Therefore the sum of the moments must vanish as well. The moment vectors associated with each force vector are computed with respect to some point P. However, if the sum of all forces is zero, the sum of the moment vectors should be zero with respect to **any** point P.

Example 3.1   Fig. 3.2(a) shows an image of a single cell, that was captured by means of an atomic force microscope. Cells attach themselves to the supporting surface in a discrete number of points. The forces acting on these points of the cell are shown as arrows. In Fig. 3.2(b) a model of the cell is given that can be used to examine the equilibrium of forces and moments. In this case, the sum of all the forces should be equal to zero:

$$\sum_{i=1}^{4} \vec{F}_i = \vec{0},$$

while the sum of the moments with respect to the point P (with an arbitrary position $\vec{x}_P$) due to these forces must be zero as well:

$$\sum_{i=1}^{4} (\vec{d}_i \times \vec{F}_i) = \vec{0}.$$

Notice that in this particular case there are no externally applied additional moments.

In the example of Fig. 3.2(b) the moments were determined with respect to point P. If the moment is computed with respect to another point in space, say R, having coordinates $\vec{x}_R = \vec{x}_P + \vec{a}$, then the moment vector with respect to this point R is defined by

$$\vec{M} = \sum_{i=1}^{4}(\vec{d}_i - \vec{a}) \times \vec{F}_i$$

$$= \sum_{i=1}^{4}(\vec{d}_i \times \vec{F}_i) - \sum_{i=1}^{4}(\vec{a} \times \vec{F}_i)$$

$$= \sum_{i=1}^{4}(\vec{d}_i \times \vec{F}_i) - \vec{a} \times \sum_{i=1}^{4}\vec{F}_i,$$

which vanishes if the forces are in equilibrium and the sum of the moments with respect to point P equals $\vec{0}$. This implies that any point can be taken to enforce equilibrium of moments. This argument can be generalized in a straightforward manner to any number of forces.

Example 3.2   An example of pure force equilibrium is given in Fig. 3.3. This figure shows an electron micrograph of an actin network supporting the cell membrane. At the

Figure 3.3

TEM image of the actin network supporting the cell membrane, with forces acting on an interconnection point.

intersection point of the network the molecules are (weakly) cross-linked. Within each of the molecules a (tensile) force is present and at the interconnection point force equilibrium must apply. In Fig. 3.3 the forces acting on one of the interconnection points have been sketched, the sum of these force vectors has to be equal to zero.

With respect to a Cartesian coordinate system equilibrium requires that the sum of all forces in the x-, y- and z-direction is zero. With the decomposition of a force $\vec{F}$, according to $\vec{F} = F_x\vec{e}_x + F_y\vec{e}_y + F_z\vec{e}_z$, that is the case if:

$$\sum_{i=1}^{n} F_{x,i} = 0$$

$$\sum_{i=1}^{n} F_{y,i} = 0$$

$$\sum_{i=1}^{n} F_{z,i} = 0,$$

and that the sum of all moments in the x-, y- and z-direction with respect to an arbitrarily selected point P is zero. Choosing point P to coincide with the point of application of the forces, immediately reveals that the equilibrium of moments is trivially satisfied.

Equilibrium of forces may also be expressed in column notation, according to:

$$\sum_{i=1}^{n} \underset{\sim}{F}_i = \underset{\sim}{0}.$$

3.3 Free body diagram

A free body diagram serves to specify and visualize the complete loading of a body, including the reaction forces and moments acting on the body that is supported in one way or the other. The body may be part of a system of bodies and, using the free body diagram, the reaction forces on the body under consideration imposed by the other bodies may be identified. For this purpose the body is isolated from its surroundings and the proper reaction forces and moments are introduced such as to ensure equilibrium of the body. Clearly, these reaction forces and moments are not known a priori, but the equilibrium conditions may be used to try to compute these unknowns. A distinction must be made between the **statically determinate** and the **statically indeterminate** case.

Requiring force and moment equilibrium provides for a limited number of equations only, and therefore only a limited number of unknowns can be determined. For two-dimensional problems force equilibrium results in two equations, while the requirement of moment equilibrium supplies only one equation, hence three independent equations can be formulated. Only if the number of unknown reaction loads equals three is the solution of the unknowns possible. Likewise, in the three-dimensional case, imposing force and moment equilibrium generates six independent equations, such that six unknown reactions can be computed. If a free body diagram is drawn and all the reactions can be directly identified from enforcing the equilibrium conditions, this is referred to as the statically determinate case.

If the reactions defined on a free body diagram cannot be calculated by imposing the equilibrium conditions, then this is referred to as the statically **in**determinate case. This is dealt with, if more than three forces or moments for two-dimensional problems or more than six forces or moments for three-dimensional problems need to be identified. It should be noted that the equilibrium equations do not suffice if in the two-dimensional case more than one moment is unknown and in the three-dimensional case more than three moments are unknown.

Example 3.3   As a two-dimensional example, consider the body of a single cell as sketched in Fig. 3.4, that is loaded by a known force F_P, while the body is supported at two points, say A and B. The support is such that at point A only a force in the horizontal direction can be transmitted. This is represented by the rollers, that allow point A to freely move in the vertical direction. At point B, however, forces in both the vertical and horizontal direction can be transmitted from the surrounding to the body, in the figure indicated by the hinge. A free body diagram is sketched in Fig. 3.5. The supports are separated from the body. It is assumed that the supports cannot exert a moment on the body, therefore only reaction forces in the horizontal and vertical direction have been introduced. As a naming convention all forces in the horizontal direction have been labelled H_α (the subscript α referring to the point of application), while all vertical forces have been labelled V_α. At each of the attachment points, A and B, reaction forces have been introduced on both the body and the support. According to the third law of Newton (see Section 2.3): action $= -$ reaction, forces are defined in the opposite direction with respect to each other, but have equal magnitude. The (three) reaction forces at point A and point B are, for the time being, unknown. They can be calculated by enforcing force and moment equilibrium of the body. Hence, both the sum of all forces in the horizontal direction as well as the sum of all the forces in the vertical direction acting on the body have to be equal to zero. For this purpose the load F_P has been decomposed into a horizontal force H_P and a vertical force V_P:

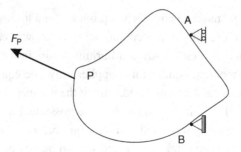

Figure 3.4

A loaded body.

Figure 3.5

Free body diagram of the simple loaded body.

$$-H_P - H_A - H_B = 0$$
$$V_P + V_B = 0.$$

The above expressions represent two equations with three unknowns (H_A, H_B and V_B), being insufficient to determine these. However, the sum of the moments with respect to an arbitrary point has be zero as well. Computing the resulting moment with respect to point A gives

$$-2l_y H_B - l_y H_P - l_x V_P = 0.$$

This yields **one** additional equation such that the unknown reaction forces may be determined.

The above procedure is not unique in the sense that different points can be used with respect to which the sum of moments should be required to be zero. For instance, rather than using the sum of moments with respect to point A, the sum of moments with respect to point P could have been used.

Suppose that

$$l_x = 2 \, [\text{m}], \quad l_y = 1 \, [\text{m}], \quad H_P = 20 \, [\text{N}], \quad V_P = 10 \, [\text{N}].$$

Substitution of these values into the equation above, renders:

$$-20 - H_A - H_B = 0$$
$$10 + V_B = 0$$
$$-2H_B - 20 - 20 = 0.$$

Clearly, from the second equation it follows immediately that $V_B = -10$ [N], while from the last equation it is clear that $H_B = -20$ [N], which leaves the first equation to calculate H_A as $H_A = 0$ [N]. Hence we have the solution:

$$H_A = 0 \, [\text{N}], \quad H_B = -20 \, [\text{N}], \quad V_B = -10 \, [\text{N}].$$

Example 3.4 An example of a statically indeterminate case appears if the rolling support at point A of Fig. 3.4 is replaced by a hinge as in point B. In that case an additional reaction force in the vertical direction must be introduced at point A, see Fig. 3.6. With the same values of the parameters as before, force and moment equilibrium yields

$$-20 - H_A - H_B = 0$$
$$10 + V_A + V_B = 0$$
$$-2H_B - 20 - 20 = 0.$$

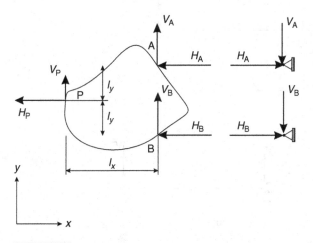

Figure 3.6
Free body diagram of the simple loaded body, statically indeterminate case.

The horizontal reaction forces H_A and H_B can, incidentally, still be calculated, giving, as before:

$$H_A = 0 \text{ [N]}, \quad H_B = -20 \text{ [N]}.$$

But, there is insufficient information to compute V_A and V_B. In fact, there are only three equations to determine the four unknowns (H_A, H_B, V_A and V_B).

The nature of the support defines the possible set of reaction forces that have to be introduced. In the above given two-dimensional statically indeterminate example the supports are assumed to be hinges or pin-connections. Effectively this means that no moments can be exerted on the support and only forces in the x- and y-direction have to be introduced.

Example 3.5   An example of a fixed support is given in Fig. 3.7(a), showing a beam that is clamped at one end and loaded at a distance L from this fixation (cantilever beam). The reaction forces and moment can be computed by enforcing force and moment equilibrium. The sum of the forces in the x- and y-direction has to be equal to zero:

$$H_A = 0$$
$$-F + V_A = 0,$$

while the sum of the moments has to be zero, for instance, with respect to point A:

$$M_A - LF = 0.$$

(a)

(b)

Figure 3.7
Free body diagram of bar fixed at one end and loaded at the other.

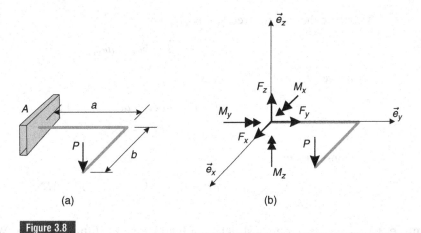

Figure 3.8

A beam construction loaded by a force P and the free body diagram.

Example 3.6   Consider the beam construction, sketched in Fig. 3.8(a), loaded by a force P. The beam is clamped at point A and we want to determine the reaction loads at point A. First of all a coordinate system is introduced and a free body diagram of the loaded beam construction is drawn, as in Fig. 3.8(b). The applied load is represented by the vector:

$$\vec{P} = -P\vec{e}_z.$$

The reaction force vector on the beam construction at point A is denoted by $\vec{F}$ and is decomposed according to:

$$\vec{F} = F_x\vec{e}_x + F_y\vec{e}_y + F_z\vec{e}_z,$$

while the reaction moment vector at point A is written as

$$\vec{M} = M_x\vec{e}_x + M_y\vec{e}_y + M_z\vec{e}_z.$$

The requirement that the sum of all forces is equal to zero implies that

$$\vec{F} + \vec{P} = \vec{0},$$

and consequently

$$F_x = 0, \quad F_y = 0, \quad F_z - P = 0.$$

The requirement that the sum of all moments with respect to A equals zero leads to:

$$\vec{M} + \vec{d} \times \vec{P} = \vec{0},$$

where the distance vector $\vec{d}$ is given by

$$\vec{d} = b\vec{e}_x + a\vec{e}_y,$$

hence

$$\vec{d} \times \vec{P} = (b\vec{e}_x + a\vec{e}_y) \times (-P\vec{e}_z)$$
$$= bP\vec{e}_y - aP\vec{e}_x.$$

Consequently

$$M_x - aP = 0, \qquad M_y + bP = 0, \qquad M_z = 0.$$

Example 3.7 Consider the man sketched in Fig. 3.9 who is lifting a weight. We would like to compute the force $\vec{F}_M$ in the muscle connecting the upper arm to the shoulder. A basis $\{\vec{e}_x, \vec{e}_y\}$ is introduced with the origin in the joint, point J. The basis vector $\vec{e}_x$ has the direction of the arm, while basis vector $\vec{e}_y$ is perpendicular to the arm (see figure). The forces $\vec{W} = -W\vec{e}_y$, due to the weight of the arm, and $\vec{W}_0 = -W_0\vec{e}_y$ due to the lifted weight are both supposed to be known. The reaction force in the joint $\vec{F}_J$ and the force in the muscle $\vec{F}_M$ are both unknown. However, the direction of the force in the muscle is known since this force is oriented with respect to the arm at an angle θ. Consequently

$$\vec{F}_J = F_{Jx}\vec{e}_x + F_{Jy}\vec{e}_y,$$

Figure 3.9

Lifting a weight.

while the force in the muscle is given by

$$\vec{F}_M = -F_M \cos(\theta)\vec{e}_x + F_M \sin(\theta)\vec{e}_y.$$

Notice that both the x- and y-component of the joint reaction force, F_{Jx} and F_{Jy}, respectively, are unknown (the joint is modelled by a hinge), while for the muscle only the magnitude of the muscle force F_M is unknown.

Application of the force balance in the x- and y-direction yields

$$F_{Jx} - F_M \cos(\theta) = 0$$

and

$$F_{Jy} + F_M \sin(\theta) - W - W_0 = 0.$$

Force equilibrium supplies two equations for three unknowns, hence moment equilibrium needs to be enforced as well. With the points A, B and C located at $\vec{x}_A = a\vec{e}_x$, $\vec{x}_B = b\vec{e}_x$ and $\vec{x}_C = c\vec{e}_x$, respectively, moment equilibrium with respect to point J requires:

$$a\,F_M \sin(\theta) - b\,W - c\,W_0 = 0.$$

From this equation it follows that

$$F_M = \frac{b\,W + c\,W_0}{a\,\sin(\theta)}.$$

Hence, from the force balance in the x- and y-direction the joint reaction forces can be computed immediately.

Suppose that

$$\theta = \frac{\pi}{10}, \quad a = 0.1 \text{ [m]}, \quad b = 0.25 \text{ [m]}, \quad c = 0.6 \text{ [m]},$$

and

$$W = 5 \text{ [N]}, \quad W_0 = 10 \text{ [N]},$$

then

$$F_M = 235 \text{ [N]}, \quad F_{Jx} = 223 \text{ [N]}, \quad F_{Jy} = -58 \text{ [N]}.$$

Exercises

3.1   The position vectors of the points P, Q, R and S are given, respectively:

$$\vec{x}_P = \vec{e}_x + 3\vec{e}_y$$
$$\vec{x}_Q = 4\vec{e}_x + 2\vec{e}_y$$

$$\vec{x}_R = 3\vec{e}_x + e_y$$
$$\vec{x}_S = -\vec{e}_x - \vec{e}_y.$$

The force vector $\vec{F}_1 = 2\vec{e}_x$ acts on point Q. The force vector $\vec{F}_2 = -2\vec{e}_x$ acts on point S.

Calculate the resulting moment of the forces $\vec{F}_1$ and $\vec{F}_2$ with respect to the points P and R.

3.2 Determine the reaction forces and moments on the beam construction, experienced at point A due to the fully clamped fixation in point A for both configurations in the figure.

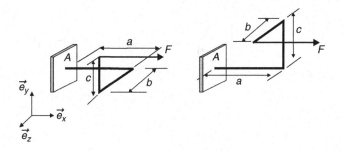

3.3 A person is lying on a board, which is supported at both ends. A vertical reaction force V_A is acting on the board at point A and passes through the origin of the coordinate system. A vertical reaction force V_B is acting at the point located in $\vec{x}_B = \ell\vec{e}_x$. It is known that the weight of the person is F and the weight of the board is P. The vector that determines the centroid C of the person is given by: $\vec{x}_C = \alpha\vec{e}_x + \beta\vec{e}_y$. Determine the reaction forces R_A and R_B as a function of ℓ, α, β, F and P.

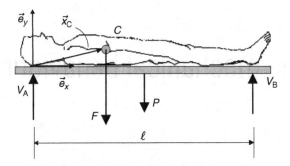

3.4 A swimmer with a weight G is standing at the edge of a diving board. The centre of gravity of the swimmer is just above the edge. The distances a and b are known. Determine the reaction forces on the board at B and C, where the board is supported by rollers and a hinge, respectively.

3.5 Wall bars with length ℓ and weight G are placed under an angle α against a wall. Two ways to model the supports are given in the figure. Calculate the reaction forces on the bar in the supporting points as a function of G and α for both cases.

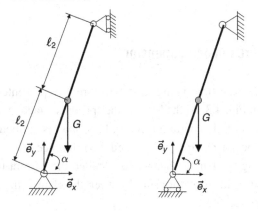

4 The mechanical behaviour of fibres

4.1 Introduction

Fibres and fibre-like structures play an important role in the mechanical properties of biological tissues. Fibre-like structures may be found in almost all human tissues. A typical example is the fibre reinforcement in a heart valve, Fig. 4.1(a). Another illustration is found in the intervertebral disc as shown in Fig. 4.1(b).

Fibre reinforcement, largely inspired by nature, is frequently used in prosthesis design to optimize mechanical performance. An example is found in the aortic valve prosthesis, see Fig. 4.2.

Fibres are long slender bodies and, essentially, have a tensile load bearing capacity along the fibre direction only. The most simple approximation of the, often complicated, mechanical behaviour of fibres is to assume that they behave elastically. In that case fibres have much in common with springs. The objective of this chapter is to formulate a relation between the force in the fibre and the change in length of a fibre. Such a relation is called a **constitutive model**.

4.2 Elastic fibres in one dimension

Assume, for the time being, that the fibre is represented by a simple spring as sketched in Fig. 4.3. At the left end the spring is attached to the wall while the right end is loaded with a certain force F. If no load is applied to the spring (fibre) the length of the spring equals ℓ_0, called the reference or initial length. After loading of the spring the length changes to ℓ, called the current length. It is assumed that there exists a linear relationship between the change in length of the fibre $\ell - \ell_0$ and the applied force:

$$F = a(\ell - \ell_0). \tag{4.1}$$

The constant a reflects the stiffness properties of the spring and can be identified by, for instance, attaching a known weight, i.e. a known force, to the spring in

(a) Fibres in a heart valve. Courtesy Mrs A. Balguid

(b) Fibres in the intervertebral disc

Figure 4.1

Examples of fibre structures.

Figure 4.2

Fibre reinforcement in a stented valve prosthesis [5].

Figure 4.3

Unloaded and loaded spring.

vertical position and measuring the extension of the spring. However, formulating the force-extension relation as in Eq. (4.1) is, although formally correct, not very convenient. If another spring was considered with the same intrinsic properties, while the initial length was different, the coefficient a would change as well. Therefore the relation represented by Eq. (4.1) is scaled with the initial, unloaded, length:

$$F = c\frac{\ell - \ell_0}{\ell_0} = c\left(\frac{\ell}{\ell_0} - 1\right). \tag{4.2}$$

The coefficient c is an intrinsic property of the spring that is independent of the unloaded length of the spring. It is common practice to introduce the so-called **stretch** parameter λ, defined as

$$\lambda = \frac{\ell}{\ell_0},\qquad(4.3)$$

such that

$$F = c(\lambda - 1).\qquad(4.4)$$

The quantity $\lambda - 1$ is usually referred to as **strain**, and it measures the amount of deformation of the spring. Without any elongation of the spring the stretch satisfies $\lambda = 1$, while the strain equals zero. The above force-strain relation represents linear **elastic** behaviour, as depicted in Fig. 4.4(a). If after stretching the fibre returns to its original length the force equals zero and no energy has been dissipated. Since the stretch λ is a dimensionless quantity and the unit of force is [N] (Newton), the constant c also has unit [N].

For relatively small stretches λ the actual behaviour of many biological fibres may indeed be approximated by a linear relation between force and stretch. However, if the stretch exceeds a certain value the force-extension behaviour usually becomes **non-linear**. In fact, in many cases fibres have a **finite extensibility**. If the stretch λ approaches a critical value, say λ_c, the force in the fibre increases sharply. A typical example of such a behaviour is modelled using the following expression for the force stretch relation:

$$F = \frac{c}{1 - \frac{\lambda-1}{\lambda_c-1}}(\lambda - 1).\qquad(4.5)$$

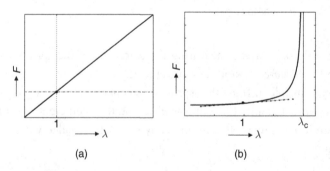

(a)   (b)

Force stretch relation for linear and non-linear spring.

For small extensions ($\lambda \approx 1$) the nominator in this expression satisfies

$$1 - \frac{\lambda - 1}{\lambda_c - 1} \approx 1, \tag{4.6}$$

such that the behaviour is identical to the linear spring. If λ approaches the critical stretch λ_c the force indeed increases rapidly with increasing stretch. This is reflected in Fig. 4.4(b). The solid line represents the non-linear, finite extensibility curve according to Eq. (4.5), while the dashed line represents the linear behaviour according to Eq. (4.4).

4.3 A simple one-dimensional model of a skeletal muscle

The fibres in the skeletal muscle have the unique capability to contract. On a microscopic scale a muscle is composed of muscle fibres and myofibrils. Myofibrils in turn are composed of actin and myosin proteins. The interaction of filaments of these proteins through cross-bridges leads to the contractile properties of the muscle.

The arrangement of these filaments into a sarcomere unit is sketched in Fig. 4.5(a). Upon activation of the muscle the actin and myosin filaments move with respect to each other causing the sarcomere to shorten. Upon de-activation of the muscle the actin and myosin filaments return to their original positions due to the elasticity of the surrounding tissue. In terms of modelling, the change of the sarcomere length implies that the initial, unloaded length of the muscle changes. Let ℓ_0 denote the length of the muscle in the **non-activated** state, while ℓ_c denotes the length of the muscle in the **activated** or **contracted** but **unloaded** state, see Fig. 4.6. Now, in contrast with a simple elastic spring, the contracted length ℓ_c serves as the reference length, such that the force in the muscle may be expressed as:

(a)   (b)

Figure 4.5

(a) Basic structure of a contractile element (sarcomere) of a muscle (b) Cross section of a muscle, vertical stripes correspond to Z-discs.

unloaded, non-activated,

unloaded, activated

loaded, activated

Figure 4.6

Different reference and current lengths of a muscle.

$$F = c \left(\frac{\ell}{\ell_c} - 1 \right). \tag{4.7}$$

For this it is assumed that, despite the contraction, c does not change. The activated, but unloaded, length ℓ_c of the muscle may be expressed in terms of the non-activated length ℓ_0 using a so-called activation or contraction stretch λ_c defined as:

$$\lambda_c = \frac{\ell_c}{\ell_0}. \tag{4.8}$$

Typically $\lambda_c < 1$ since it represents a contractile action. For simplicity it is assumed that λ_c is known for different degrees of activation of the muscle. Using the activation stretch λ_c, the force-stretch relation for a muscle may be rewritten as

$$F = c \left(\frac{\lambda}{\lambda_c} - 1 \right) \quad \text{with} \quad \lambda = \frac{\ell}{\ell_0}. \tag{4.9}$$

Effectively this expression implies that if the muscle is activated, represented by a certain λ_c, and the muscle is not loaded, hence $F = 0$, the muscle will contract such that

$$\lambda = \lambda_c. \tag{4.10}$$

If, on the other hand, the muscle is activated and forced to have constant length ℓ_0, hence $\lambda = 1$, the force in the muscle equals:

$$F = c \left(\frac{1}{\lambda_c} - 1 \right). \tag{4.11}$$

Rather complicated models have been developed to describe the activation of the muscle. A large group of models is based on experimental work by Hill [9] and supply a phenomenological description of the non-linear activated muscle. These models account for the effect of contraction velocity and for the difference in activated and passive state of the muscle. Later, microstructural models were developed, based on the sliding filament theories of Huxley [12]. These models can even account for the calcium activation of the muscle. However, a discussion of these models is beyond the scope of this book.

4.4 Elastic fibres in three dimensions

The above one-dimensional force extension relation can be generalized to a fibre/spring having an arbitrary position in three-dimensional space. The locations of the end points of the spring, say A and B, in the unstretched, initial configuration are denoted by $\vec{x}_{0,A}$ and $\vec{x}_{0,B}$, respectively, see Fig. 4.7.

The initial length of the spring ℓ_0 follows from

$$\ell_0 = |\vec{x}_{0,B} - \vec{x}_{0,A}|. \tag{4.12}$$

The initial orientation of the spring in space is denoted by the vector $\vec{a}_0$ having unit length that follows from

$$\vec{a}_0 = \frac{\vec{x}_{0,B} - \vec{x}_{0,A}}{|\vec{x}_{0,B} - \vec{x}_{0,A}|}. \tag{4.13}$$

In the stretched, current configuration, the positions of the end points of the spring are denoted by $\vec{x}_A$ and $\vec{x}_B$. Therefore the current length ℓ of the spring can be computed from

$$\ell = |\vec{x}_B - \vec{x}_A|, \tag{4.14}$$

while the current orientation in space of the spring may be characterized by the vector $\vec{a}$ of unit length:

$$\vec{a} = \frac{\vec{x}_B - \vec{x}_A}{|\vec{x}_B - \vec{x}_A|}. \tag{4.15}$$

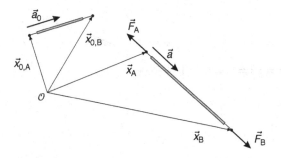

Figure 4.7

Spring in three-dimensional space.

Clearly, in analogy with the scalar one-dimensional case, the stretch of the spring λ is defined as

$$\lambda = \frac{\ell}{\ell_0}. \tag{4.16}$$

To cause a stretch of the spring, a force must be applied to the end points A and B. The forces applied on the end points are vectors represented by $\vec{F}_A$ and $\vec{F}_B$. They have equal magnitude but opposite direction:

$$\vec{F}_B = -\vec{F}_A$$

and are parallel to the orientation vector $\vec{a}$:

$$\vec{F}_B = F\vec{a}. \tag{4.17}$$

The scalar F represents the magnitude of the force vector $\vec{F}_B$ and for linearly elastic springs this magnitude follows from the one-dimensional relation Eq. (4.4):

$$F = c(\lambda - 1). \tag{4.18}$$

Therefore, the force vector acting on point B is given by

$$\vec{F}_B = c(\lambda - 1)\vec{a}, \tag{4.19}$$

while the force vector acting on point A is given by

$$\vec{F}_A = -c(\lambda - 1)\vec{a}. \tag{4.20}$$

Example 4.1 Suppose a spring is mounted as depicted in Fig. 4.8. The spring is fixed in space at point A while it is free to translate in the vertical direction at point B. A Cartesian coordinate system is attached to point A, as depicted in Fig. 4.8. If point B is moved in the vertical direction, the force on the spring at point B is computed assuming linear elasticity according to Eq. (4.19). The length of the spring in the

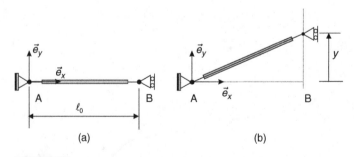

(a)   (b)

Figure 4.8

Linear elastic spring, fixed at A and free to translate in the $\vec{e}_y$ direction at point B. (a) undeformed configuration (b) deformed configuration.

undeformed configuration is denoted by ℓ_0. In the undeformed configuration the force in the spring equals zero. In the current, deformed configuration the position of point B follows from

$$\vec{x}_B = \ell_0 \vec{e}_x + y \vec{e}_y.$$

Point A is positioned at the origin:

$$\vec{x}_A = \vec{0}.$$

The force vector acting on the spring at point B is written as

$$\vec{F}_B = c(\lambda - 1)\vec{a},$$

while the current length ℓ is written as

$$\ell = |\vec{x}_B - \vec{x}_A| = \sqrt{\ell_0^2 + y^2}.$$

The stretch λ of the spring follows from

$$\lambda = \frac{\ell}{\ell_0} = \frac{\sqrt{\ell_0^2 + y^2}}{\ell_0}.$$

The orientation of the spring as represented by the unit vector $\vec{a}$ is given by

$$\vec{a} = \frac{\vec{x}_B - \vec{x}_A}{|\vec{x}_B - \vec{x}_A|} = \frac{\ell_0 \vec{e}_x + y \vec{e}_y}{\sqrt{\ell_0^2 + y^2}}.$$

So, in conclusion the force vector $\vec{F}_B$ applied to the spring at point B equals

$$\vec{F}_B = c\left(\frac{\sqrt{\ell_0^2 + y^2}}{\ell_0} - 1\right) \frac{\ell_0 \vec{e}_x + y \vec{e}_y}{\sqrt{\ell_0^2 + y^2}}.$$

Given an initial length $\ell_0 = 10$ [cm] and a spring constant $c = 0.5$ [N], the force components of $\vec{F}_B$ in the x- and y-direction (F_x and F_y, respectively) are represented in Fig. 4.9 as a function of the y-location of point B. Notice that both F_x and F_y are non-linear functions of y even though the spring is linearly elastic. The non-linearity stems from the fact that the stretch λ, is a non-linear function of y. A non-linear response of this type is called **geometrically non-linear** since it originates from geometrical effects rather than an intrinsic non-linear physical response of the spring.

Example 4.2   Consider a spring, in the undeformed configuration, mounted at an angle α_0 with respect to the x-axis, as depicted in Fig. 4.10. The spring is fixed in space at point A while it is free to translate in the vertical direction at point B. A Cartesian

Figure 4.9

Forces in the horizontal and vertical direction exerted on the spring in point B to displace point B in the y-direction.

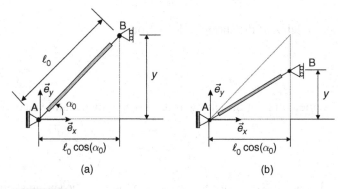

Figure 4.10

(a) Linearly elastic spring in the undeformed configuration oriented at an angle α_0 with respect to $\vec{e}_x$, fixed at A and free to translate in the $\vec{e}_y$-direction at point B (b) Spring in the deformed configuration.

coordinate system is located at point A. If point B is moved in the vertical direction, the force in the spring is computed assuming linear elasticity according to Eq. (4.19). The unstretched length of the spring is denoted by ℓ_0 such that the current position of point B follows from

$$\vec{x}_B = \ell_0 \cos(\alpha_0)\,\vec{e}_x + y\vec{e}_y.$$

Point A is positioned at the origin:

$$\vec{x}_A = \vec{0}.$$

The current length of the spring satisfies

$$\ell = |\vec{x}_B - \vec{x}_A| = \sqrt{(\ell_0 \cos(\alpha_0))^2 + y^2}.$$

The force vector acting on the spring at point B is written as

$$\vec{F}_B = c(\lambda - 1)\vec{a}.$$

The stretch λ of the spring follows from

$$\lambda = \frac{\ell}{\ell_0} = \frac{\sqrt{(\ell_0 \cos(\alpha_0))^2 + y^2}}{\ell_0}.$$

The orientation of the spring as represented by the unit vector $\vec{a}$ is given by

$$\vec{a} = \frac{\vec{x}_B - \vec{x}_A}{|\vec{x}_B - \vec{x}_A|} = \frac{\ell_0 \cos(\alpha_0)\vec{e}_x + y\vec{e}_y}{\sqrt{(\ell_0 \cos(\alpha_0))^2 + y^2}}.$$

So, in conclusion the force vector applied to the spring at point B, $\vec{F}_B$, equals

$$\vec{F}_B = c\left(\frac{\sqrt{(\ell_0 \cos(\alpha_0))^2 + y^2}}{\ell_0} - 1\right)\frac{\ell_0 \cos(\alpha_0)\vec{e}_x + y\vec{e}_y}{\sqrt{(\ell_0 \cos(\alpha_0))^2 + y^2}}.$$

Given an initial length $\ell_0 = 1$ [mm], a spring constant $c = 0.5$ [N] and an initial orientation $\alpha_0 = \pi/4$, the force components of $\vec{F}_B$ in the x- and y-direction (F_x and F_y, respectively) are represented in Fig. 4.11 as a function of the y-location of point B. Notice that, as in the previous example, both F_x and F_y are non-linear functions of y even though the spring is linearly elastic. It is remarkable to see that with decreasing y, starting at the initial position $y_0 = \ell_0 \sin(\alpha_0)$, the magnitude of the force in the y-direction $|F_y|$ first increases and thereafter decreases. This demonstrates a so-called snap-through behaviour. If the translation of point B is

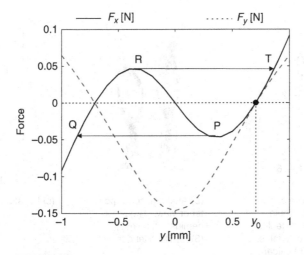

Figure 4.11

Forces in the horizontal and vertical direction exerted on the spring to displace point B in the y-direction. Snap-through behaviour.

driven by an externally applied force, and point P is reached in the force versus y-position curve, the y-coordinate of point B will suddenly move to point Q with equal force magnitude. During the reverse path a snap through will occur from point R to point T.

Example 4.3   The Achilles tendon is attached to the rear of the ankle (the calcaneus) and is connected to two muscle groups: the gastrocnemius and the soleus, which, in turn, are connected to the tibia, see Fig. 4.12(a). A schematic drawing of this, using a lateral view, is given in Fig. 4.12(c). If the ankle is rotated with respect to the pivot point O, i.e. the origin of the coordinate system, the attachment point A is displaced causing a length change of the muscle system. The position of the attachment point A is given by

$$\vec{x}_A = -R\sin(\alpha)\vec{e}_x - R\cos(\alpha)\vec{e}_y,$$

where R is the constant distance of the attachment point A to the pivot point. The angle α is defined in clockwise direction. The muscles are connected to the tibia at point B, hence:

$$\vec{x}_B = H\vec{e}_y,$$

with H the distance of point B to the pivot point. The positions in the undeformed, unstretched configuration of these points are

$$\vec{x}_{0,A} = -R\sin(\alpha_0)\vec{e}_x - R\cos(\alpha_0)\vec{e}_y$$

(a) The ankle and foot.
(1) tibia, (2) fibula,
(3) medial malleolus,
(4) lateral malleolus,
(5) talus, (6) calcaneus

(b) Ankle muscles, pos-
terior view. (1) gastroc-
nemius, (2) soleus,
(3) Achillles tendon

(c) Location of muscle

Figure 4.12

Muscle attached to tibia and calcaneus.

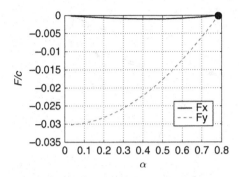

Force in muscle.

and

$$\vec{x}_{0,B} = H\vec{e}_y.$$

Hence, the stretch of the muscle follows from

$$\lambda = \frac{|\vec{x}_A - \vec{x}_B|}{|\vec{x}_{0,A} - \vec{x}_{0,B}|} = \frac{\sqrt{(R\sin(\alpha))^2 + (R\cos(\alpha) + H)^2}}{\sqrt{(R\sin(\alpha_0))^2 + (R\cos(\alpha_0) + H)^2}},$$

while the orientation of the muscle is given by

$$\vec{a} = \frac{R\sin(\alpha)\vec{e}_x + (R\cos(\alpha) + H)\vec{e}_y}{\sqrt{(R\sin(\alpha))^2 + (R\cos(\alpha) + H)^2}}.$$

From these results the force acting on the muscle at point B may be computed:

$$\vec{F}_B = c(\lambda - 1)\vec{a}.$$

The force components in the x- and y-direction, scaled by the constant c, are depicted in Fig. 4.13 in case $R = 5$ [cm], $H = 40$ [cm] and an initial angle $\alpha_0 = \pi/4$.

4.5 Small fibre stretches

As illustrated by the above example the finite displacements of the end points of a spring may cause a complicated non-linear response. In the limit of small displacements of the end points a more manageable relation for the force in the spring results. To arrive at the force versus displacement expression the concept of displacement first needs to be formalized.

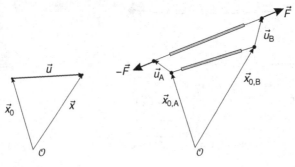

(a) Displacement
vector $\vec{u}$

(b) Small extension of the spring

Figure 4.14

Displacement vector and spring extension.

If, as before, the reference position of a certain point is denoted by $\vec{x}_0$, and the current position by $\vec{x}$, then the displacement vector $\vec{u}$ of this point is defined as the difference between the current and initial position of the point, see Fig. 4.14(a):

$$\vec{u} = \vec{x} - \vec{x}_0. \tag{4.21}$$

If a spring, as in Fig. 4.14(b), is loaded by a force vector $\vec{F}$, the end points A and B will displace by the displacement vectors $\vec{u}_A$ and $\vec{u}_B$, respectively. The current position of the end points can be written as

$$\vec{x}_A = \vec{x}_{0,A} + \vec{u}_A$$
$$\vec{x}_B = \vec{x}_{0,B} + \vec{u}_B . \tag{4.22}$$

The current length of the spring ℓ may also be expressed in terms of the end-point displacements:

$$\ell = |(\vec{x}_{0,B} + \vec{u}_B) - (\vec{x}_{0,A} + \vec{u}_A)|. \tag{4.23}$$

Define $\vec{R}$ as the end-to-end vector from point A to point B in the initial, unloaded, configuration, hence

$$\vec{R} = \vec{x}_{0,B} - \vec{x}_{0,A} \tag{4.24}$$

and let $\vec{\delta}$ denote the difference in displacement of point B and point A:

$$\vec{\delta} = \vec{u}_B - \vec{u}_A. \tag{4.25}$$

The current length ℓ can be expressed in terms of the vectors $\vec{R}$ and $\vec{\delta}$:

$$\ell = |\vec{R} + \vec{\delta}|. \tag{4.26}$$

This implies that (recall that $\vec{a}\cdot\vec{a} = |\vec{a}||\vec{a}| = |\vec{a}|^2$)

$$\ell^2 = (\vec{R}+\vec{\delta})\cdot(\vec{R}+\vec{\delta})$$
$$= \vec{R}\cdot\vec{R} + 2\vec{R}\cdot\vec{\delta} + \vec{\delta}\cdot\vec{\delta}. \tag{4.27}$$

If the end point displacements are sufficiently small, the inner product $\vec{\delta}\cdot\vec{\delta}$ will be small compared to the other inner products in the above expression and may be neglected. Therefore, to a good approximation we have

$$\ell^2 \approx \vec{R}\cdot\vec{R} + 2\vec{R}\cdot\vec{\delta}. \tag{4.28}$$

Consequently, the current length may be written as

$$\ell = \sqrt{\vec{R}\cdot\vec{R} + 2\vec{R}\cdot\vec{\delta}}. \tag{4.29}$$

This may be rewritten in a more convenient form, bearing in mind that each of the inner products yields a scalar:

$$\ell = \sqrt{\vec{R}\cdot\vec{R}\left(1 + 2\frac{\vec{R}\cdot\vec{\delta}}{\vec{R}\cdot\vec{R}}\right)}$$
$$= \sqrt{\vec{R}\cdot\vec{R}}\sqrt{1 + 2\frac{\vec{R}\cdot\vec{\delta}}{\vec{R}\cdot\vec{R}}}.$$

If α is a small number, then $\sqrt{1+2\alpha} \approx 1+\alpha$, hence if $\vec{\delta}$ is sufficiently small the current length ℓ may be approximated by

$$\ell \approx \sqrt{\vec{R}\cdot\vec{R}}\left(1 + \frac{\vec{R}\cdot\vec{\delta}}{\vec{R}\cdot\vec{R}}\right). \tag{4.30}$$

Using

$$\ell_0 = |\vec{R}| = \sqrt{\vec{R}\cdot\vec{R}}, \tag{4.31}$$

this can be rewritten as:

$$\ell = \ell_0 + \frac{\vec{R}\cdot\vec{\delta}}{\ell_0}. \tag{4.32}$$

The stretch λ may now be expressed as:

$$\lambda = \frac{\ell}{\ell_0} = 1 + \frac{\vec{R}\cdot\vec{\delta}}{\ell_0^2}. \tag{4.33}$$

Recall that the force-stretch relation for a spring is given by:

$$\vec{F}_{\mathrm{B}} = c(\lambda - 1)\vec{a}, \tag{4.34}$$

where $\vec{a}$ denotes the vector of unit length pointing from point A to point B and where $\vec{F}_{\mathrm{B}}$ is the force acting on point B.

For sufficiently small displacements of the end points this vector may be approximated by:

$$\vec{a} \approx \frac{\vec{R}}{|\vec{R}|} = \frac{\vec{R}}{\ell_0} = \vec{a}_0. \tag{4.35}$$

Consequently, the following expression of the force vector $\vec{F}_B$ is obtained

$$\vec{F}_B = c \underbrace{\frac{\vec{R} \cdot \vec{\delta}}{\ell_0^2}}_{\lambda-1} \underbrace{\frac{\vec{R}}{\ell_0}}_{\vec{a}_0} = c \left(\vec{a}_0 \cdot \frac{\vec{\delta}}{\ell_0} \right) \vec{a}_0, \tag{4.36}$$

or written in terms of the end point displacements

$$\vec{F}_B = c \, \frac{\vec{a}_0 \cdot (\vec{u}_B - \vec{u}_A)}{\ell_0} \, \vec{a}_0. \tag{4.37}$$

In this expression three parts may be recognized. First the stiffness of the spring, c, second the amount of stretch in the direction of the spring, see Fig. 4.15, also called fibre strain ε:

$$\varepsilon = \lambda - 1 = \frac{\vec{a}_0 \cdot (\vec{u}_B - \vec{u}_A)}{\ell_0} \tag{4.38}$$

and third the orientation of the spring, represented by the unit vector $\vec{a}_0$.

Example 4.4 An immediate consequence of the linearization process is that, if the displacement difference $\vec{u}_B - \vec{u}_A$ is normal to the fibre axis, i.e. $(\vec{u}_B - \vec{u}_A) \cdot \vec{a}_0 = 0$, the force in the fibre equals zero. Two examples of this are given in Fig. 4.16.

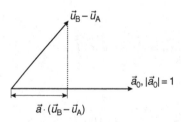

Figure 4.15

Measure of the elongation of the spring.

Figure 4.16

Examples of $(\vec{u}_B - \vec{u}_A) \cdot \vec{a}_0 = 0$ leading to $\vec{F} = \vec{0}$.

(a) Two fibre configuration   (b) Free body diagram

Figure 4.17

Two fibre configuration and free body diagram of point P.

Example 4.5 Suppose that two fibres have been arranged according to Fig. 4.17(a). Both fibres have the same unloaded length ℓ_0 and elastic property c. At point P the two fibres have been connected. A force $\vec{F}$ is applied to this point. The free body diagram with respect to point P is given in Fig. 4.17(b). The orientation of each of the fibres in space is represented by the vector $\vec{a}_i$ ($i = 1, 2$). These vectors have been chosen such that they point from the supports to point P, hence

$$\vec{a}_1 = \vec{e}_x$$
$$\vec{a}_2 = \frac{1}{2}\sqrt{2}(-\vec{e}_x + \vec{e}_y).$$

If $\vec{u}_P = u_x\vec{e}_x + u_y\vec{e}_y$ denotes the displacement of point P, the force vectors $\vec{F}_1$ and $\vec{F}_2$ are described by

$$\vec{F}_1 = \frac{c}{\ell_0}(\vec{a}_1 \cdot \vec{u}_P)\vec{a}_1$$
$$= \frac{c}{\ell_0}u_x\vec{e}_x.$$

and

$$\vec{F}_2 = \frac{c}{\ell_0}(\vec{a}_2 \cdot \vec{u}_P)\vec{a}_2$$
$$= \frac{c}{\ell_0}\frac{1}{2}(-u_x + u_y)(-\vec{e}_x + \vec{e}_y).$$

The requirement of force equilibrium at point P implies

$$\vec{F} - \vec{F}_1 - \vec{F}_2 = \vec{0}.$$

With $\vec{F} = F_x\vec{e}_x + F_y\vec{e}_y$ it follows that in the x-direction:

$$F_x + \frac{c}{2\ell_0}(-3u_x + 2u_y) = 0,$$

while in the y-direction:

$$F_y - \frac{c}{\ell_0}(-u_x + u_y) = 0.$$

This gives two equations from which the two unknowns (u_x and u_y) can be solved.

Exercises

4.1   The length change of a muscle, with respect to the length ℓ_0 in the relaxed state, can be written as $\delta = \ell - \ell_0$.
    (a)   Give an expression for the force F in the activated muscle as a function of c, δ, ℓ_0 and ℓ_c.
    (b)   What is the magnitude of the force F when $\delta = 0$?

4.2   Give a sketch of the force as a function of the active muscle length in the diagram below

4.3   In reality the force length equation for an activated muscle in Eq. (4.9) is only valid in a very limited range of extension ratios. The force that a sarcomere (and thus a skeletal muscle) can exert has a maximum, as depicted in the figure below. Sarcomere lengths at several interesting points in the graph are depicted by L_i, with $i = 1, \ldots, 5$.

(a) Explain why the force decreases when the sarcomere length exceeds L4.

(b) The force versus length relation between L1 and L2 can be described exactly with Eq. (4.9). Determine the value of c in the case where the maximum force as given in the graph is 100 [N].

4.4 A muscle-tendon complex is loaded with a force F. The combination of the muscle-tendon complex can schematically be depicted as given in the figure.

The force versus length relation in the muscle can be written as

$$F^{\mathrm{m}} = c^{\mathrm{m}} \left(\frac{\ell^{\mathrm{m}}}{\ell_c^{\mathrm{m}}} - 1 \right).$$

The force versus length relation in the tendon can be written as

$$F^{\mathrm{t}} = c^{\mathrm{t}} \left(\frac{\ell^{\mathrm{t}}}{\ell_0^{\mathrm{t}}} - 1 \right).$$

Determine the total length change of the muscle-tendon complex as a result of the load F.

4.5 A fibre is marked on two sides with small dots. The position of these dots is measured in an unloaded reference configuration. From these measurements it appears that the positions are given as

$$\vec{x}_{0,\mathrm{A}} = -\vec{e}_x + 3\vec{e}_y$$
$$\vec{x}_{0,\mathrm{B}} = 2\vec{e}_x + 3\vec{e}_y.$$

The fibre moves and in the current (deformed) configuration the positions of points A and B are measured again:

$$\vec{x}_{\mathrm{A}} = 4\vec{e}_x + 2\vec{e}_y$$
$$\vec{x}_{\mathrm{B}} = 8\vec{e}_x - \vec{e}_y.$$

The constant in the force versus extension relation is $c = 300$ [N]. Determine the force vectors $\vec{F}_{\mathrm{A}}$ and $\vec{F}_{\mathrm{B}}$ in the deformed configuration.

4.6 In the situation that is depicted in the drawing below only the gluteus medius is active. During adduction of the bone the femur rotates. The point of application of this muscle on the trochanter B follows a circular path. The point of application on the acetabulum A is given by the vector $\vec{x}_A = L\vec{e}_y$. When the adduction angle $\phi = 0$ the length of the muscle is $\ell = \ell_c$. Calculate the force $\vec{F}$ as a function of the adduction angle ϕ.

5 Fibres: time-dependent behaviour

5.1 Introduction

In the previous chapter on fibres the material behaviour was constantly considered to be **elastic**, meaning that a unique relation exists between the extensional force and the deformation of the fibre. This implies that the force versus stretch curves for the loading and unloading path are indentical. There is no history dependency and all energy that is stored into the fibre during deformation is regained during the unloading phase. This also implies that the rate of loading or unloading does not affect the force versus stretch curves. However, most biological materials do **not** behave elastically!

An example of a loading history and a typical response of a biological material is shown in Figures 5.1(a) and (b). In Fig. 5.1(a) a deformation history is given that might be used in an experiment to mechanically characterize some material specimen. The specimen is stretched fast to a certain value, then the deformation is fixed and after a certain time restored to zero. After a short resting period, the stretch is applied again but to a higher value of the stretch. This deformation cycle is repeated several times. In this case the length change is prescribed and the associated force is measured. Fig. 5.1(b) shows the result of such a measurement. When the length of the fibre is kept constant, the force decreases in time. This phenomenon is called **relaxation**. Reversely, if a constant load is applied, the length of the fibre will increase. This is called **creep**.

When the material is subjected to cyclic loading, the force versus stretch relation in the loading process is usually somewhat different from that in the unloading process. This is called **hysteresis** and is demonstrated in Fig. 5.2. The difference in the response paths during loading and unloading implies that energy is dissipated, usually in the form of heat, during the process. Most biological materials show more or less the above given behaviour, which is called **visco-elastic** behaviour. The present chapter discusses how to describe this behaviour mathematically. Pure viscous behaviour, as can be attributed to an ideal fluid is considered first. The description will be extended to linear visco-elastic behaviour,

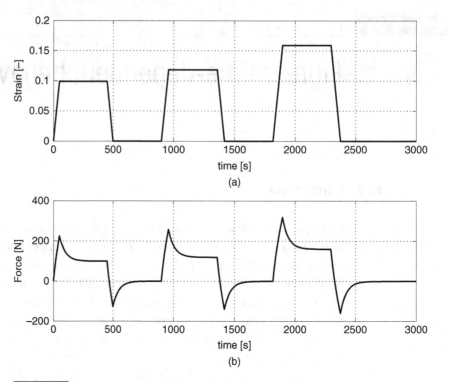

Figure 5.1

Loading history in a relaxation experiment. The deformation of the tissue specimen is prescribed.

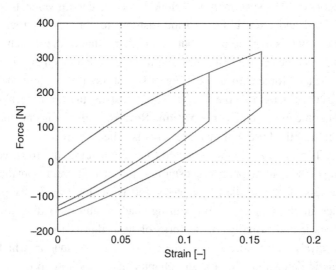

Figure 5.2

Force/strain curve for cyclic loading of a biological material.

followed by a discussion on harmonic excitation, a technique that is often used to determine material properties of visco-elastic materials.

5.2   Viscous behaviour

It is not surprising that biological tissues do not behave purely elastically, since a large percentage of most tissues is water. The behaviour of water can be characterized as 'viscous'. Cast in a one-dimensional format, viscous behaviour during elongation (as in a fibre) may be represented by

$$F = c_\eta \frac{1}{\ell} \frac{d\ell}{dt}, \tag{5.1}$$

where c_η is the damping coefficient in $[\,\text{Ns}\,]$ and $d\ell/dt$ measures the rate of change of the length of the fibre. Mechanically this force-elongational rate relation may be represented by a dashpot (see Fig. 5.3). Generally:

$$D = \frac{1}{\ell} \frac{d\ell}{dt} \tag{5.2}$$

is called the rate of deformation, which is related to the stretch parameter λ. Recall that

$$\lambda = \frac{\ell}{\ell_0}, \tag{5.3}$$

such that

$$D = \frac{1}{\ell} \frac{d\ell}{dt} = \frac{1}{\lambda} \frac{d\lambda}{dt}. \tag{5.4}$$

Ideally, a fluid stretching experiment should create a deformation pattern as visualized in Fig. 5.4(a). In practice this is impossible, because the fluid has to be spatially fixed and loaded, for instance via end plates, as depicted in Fig. 5.4(b). In this experiment a fluid is placed between two parallel plates at an initial distance ℓ_0. Next, the end plates are displaced and the force on the end plates is measured. A typical example of a stretched filament is shown in Fig. 5.5. Although this seems to be a simple experiment, it is rather difficult to perform in practice.

Figure 5.3

Mechanical representation of a viscous fibre by means of a dashpot.

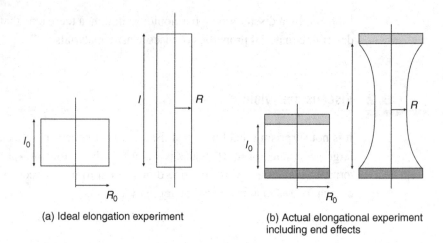

(a) Ideal elongation experiment

(b) Actual elongational experiment including end effects

Figure 5.4

Schematic representation of an elongation experiment for fluids.

time

Figure 5.5

Example of uniaxial testing experiment with a fluid.

Figure 5.6

Point B is moved with a constant velocity v.

Fig. 5.5 shows an experiment, where the fluid is a little extended initially, after which gravitational sag continues the filament stretching process. Ideally, a filament stretching experiment should be performed at a constant elongational rate. This is not trivial to achieve. For instance, let one end of the dashpot, say point A positioned at the origin (i.e. $x_A = 0$), be fixed in space, while the other end, point B is displaced with a constant velocity v, as depicted in Fig. 5.6. In that case the position of point B is given by

$$x_B = \ell_0 + vt, \tag{5.5}$$

with ℓ_0 the initial length of the dashpot. The actual length ℓ at time t is given by:

$$\ell = x_B - x_A = \ell_0 + vt. \tag{5.6}$$

Hence, the elongational rate is given by

$$D = \frac{1}{\ell}\frac{d\ell}{dt} = \frac{v}{\ell_0 + vt}. \tag{5.7}$$

This shows, that if one end is moved with a constant velocity, the elongational rate decays with increasing time t. Maintaining a constant elongational rate is possible if the velocity of point B is adjusted as a function of time. Indeed, a constant elongational rate D implies that the length ℓ must satisfy

$$\frac{1}{\ell}\frac{d\ell}{dt} = D, \tag{5.8}$$

subject to the initial condition $\ell = \ell_0$ at $t = 0$, while D is constant. Since

$$\frac{d\ln(\ell)}{dt} = \frac{1}{\ell}\frac{d\ell}{dt} = D, \tag{5.9}$$

the solution of Eq. (5.8) is given by

$$\ell = \ell_0\, e^{Dt}. \tag{5.10}$$

This means that to maintain a constant elongational rate, point B has to be displaced exponentially in time, which is rather difficult to achieve in practice (moreover, try to imagine how the force can be measured during this type of experiment!).

5.2.1 Small stretches: linearization

If u_B and u_A denote the end point displacements, introduce:

$$\Delta\ell = u_B - u_A. \tag{5.11}$$

The stretch λ may be expressed as

$$\lambda = \frac{\ell_0 + \Delta\ell}{\ell_0}. \tag{5.12}$$

Introducing the strain ε as

$$\varepsilon = \frac{\Delta\ell}{\ell_0}, \tag{5.13}$$

the stretch is written as

$$\lambda = 1 + \varepsilon. \tag{5.14}$$

For sufficiently small strain levels, i.e. $|\varepsilon| \ll 1$, and using the notation $\dot{\varepsilon} = d\varepsilon/dt$ it can be written:

$$D = \frac{1}{\lambda}\frac{d\lambda}{dt} = \frac{1}{1+\varepsilon}\dot{\varepsilon} \approx (1-\varepsilon)\dot{\varepsilon} \approx \dot{\varepsilon}. \tag{5.15}$$

Consequently, if $|\varepsilon| \ll 1$ then

$$D = \frac{1}{\ell}\frac{d\ell}{dt} \approx \dot{\varepsilon} = \frac{1}{\ell_0}\frac{d\ell}{dt}, \tag{5.16}$$

such that Eq. (5.1) reduces to

$$F = c_\eta\dot{\varepsilon}. \tag{5.17}$$

Remark In the literature, the symbol $\dot{\varepsilon}$ is frequently used to denote the elongational rate for large filament stretches instead of D.

5.3 Linear visco-elastic behaviour

5.3.1 Continuous and discrete time models

Biological tissues usually demonstrate a combined viscous-elastic behaviour as described in the introduction. In the present section we assume geometrically and physically linear behaviour of the material. This means that the theory leads to linear relations, expressing the force in the deformation(-rate), and that the constitutive description satisfies two conditions:

- **superposition** The response on combined loading histories can be described as the summation of the responses on the individual loading histories.
- **proportionality** When the strain is multiplied by some factor the force is multiplied by the same factor (in fact proportionality is a consequence of superposition).

To study the effect of these conditions a unit-step function for the force is introduced, defined as H(t) (Heaviside function):

$$\mathrm{H}(t) = \begin{cases} 0 & \text{if } t < 0 \\ 1 & \text{if } t \geq 0 \end{cases}. \tag{5.18}$$

Assume that a unit-step in the force $F(t) = \mathrm{H}(t)$ is applied to a linear visco-elastic material. The response $\varepsilon(t)$ might have an evolution as given in Fig. 5.7. This response denoted by $J(t)$ is called the **creep compliance** or **creep function**.

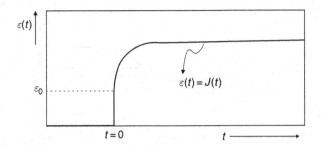

Figure 5.7

Typical example of the strain response after a unit-step in the force.

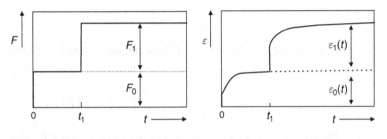

Figure 5.8

Superposition of responses for a linear visco-elastic material.

Proportionality means, that increasing the force with some factor F_0 leads to a proportional increase in the strain:

$$F(t) = \mathrm{H}(t)\, F_0 \quad \rightarrow \quad \varepsilon(t) = J(t)\, F_0. \tag{5.19}$$

Superposition implies, that applying a load step F_0 at $t = 0$, with response:

$$\varepsilon_0(t) = J(t)\, F_0, \tag{5.20}$$

followed by a load step F_1 at $t = t_1$ with individual response:

$$\varepsilon_1(t) = J(t - t_1)\, F_1, \tag{5.21}$$

leads to a total response, which is a summation of the two:

$$\varepsilon(t) = \varepsilon_0(t) + \varepsilon_1(t) = J(t)\, F_0 + J(t - t_1)\, F_1. \tag{5.22}$$

This is graphically shown in Fig. 5.8.

Both principles can be used to derive a more general constitutive equation for linear visco-elastic materials. Assume, we have an arbitrary excitation as sketched in Fig. 5.9. This excitation can be considered to be built up by an infinite number of small steps in the force.

Figure 5.9

An arbitrary force history in a creep test.

The increase ΔF of the force F between time steps $t = \xi$ and $t = \xi + \Delta\xi$ is equal to

$$\Delta F \approx \left(\frac{dF}{d\xi}\right)\Delta\xi = \dot{F}(\xi)\,\Delta\xi. \tag{5.23}$$

The response at time t as a result of this step at time ξ is given by

$$\Delta\varepsilon(t) = \dot{F}(\xi)\,\Delta\xi\,J(t - \xi). \tag{5.24}$$

The time-dependent force $F(t)$ as visualized in Fig. 5.9 can be considered as a composition of sequential small steps. By using the superposition principle we are allowed to add the responses on all these steps in the force (for each ξ). This will lead to the following integral expression, with all intervals $\Delta\xi$ taken as infinitesimally small:

$$\varepsilon(t) = \int_{\xi=-\infty}^{t} J(t - \xi)\dot{F}(\xi)\,d\xi. \tag{5.25}$$

This integral was derived first by Boltzmann in 1876.

In the creep experiment the load is prescribed and the resulting strain is measured. Often, the experimental set-up is designed to prescribe the strain and to measure the associated, required force. If the strain is applied as a step, this is called a relaxation experiment, because after a certain initial increase the force will gradually decrease in time. The same strategy as used to derive Eq. (5.25) can be pursued for an imposed strain history, leading to

$$F(t) = \int_{\xi=-\infty}^{t} G(t - \xi)\dot{\varepsilon}(\xi)\,d\xi, \tag{5.26}$$

with $G(t)$ the relaxation function.

The functions J and G have some important physical properties:

- For ordinary materials J increases, G decreases in time (for $t > 0$):

$$\dot{J}(t) > 0 \quad \dot{G}(t) < 0 \tag{5.27}$$

- The absolute value of the slope of both functions decreases:

$$d^2 J/dt^2 < 0 \quad d^2 G/dt^2 > 0 \tag{5.28}$$

- In the limiting case for $t \to \infty$ the time derivative of $G(t)$ will approach zero:

$$\lim_{t \to \infty} \dot{G}(t) = 0 \tag{5.29}$$

- In the limiting case for $t \to \infty$ the time derivative of $J(t)$ will be greater than or equal zero:

$$\lim_{t \to \infty} \dot{J}(t) \geq 0 \tag{5.30}$$

It will be clear that there must be a relation between $G(t)$ and $J(t)$, because both functions describe the behaviour of the same material. This relationship can be determined by using Laplace transformation (for a summary of definitions and properties of Laplace transformations see Appendix 5.5). The Laplace transformation $\hat{x}(s)$ of the time function $x(t)$ is defined as

$$\hat{x}(s) = \int_0^\infty x(t) e^{-st} dt. \tag{5.31}$$

Assuming, that the creep and relaxation functions are zero for $t < 0$, the Laplace transforms of Eqs. (5.25) and (5.26) are

$$\hat{\varepsilon}(s) = \hat{J}(s) \, s \, \hat{F}(s) \tag{5.32}$$

$$\hat{F}(s) = \hat{G}(s) \, s \, \hat{\varepsilon}(s). \tag{5.33}$$

From these equations it is easy to derive that

$$\hat{G}(s) \hat{J}(s) = \frac{1}{s^2}. \tag{5.34}$$

Back transformation leads to

$$\int_{\xi=0}^t J(t - \xi) G(\xi) \, d\xi = t. \tag{5.35}$$

When $G(t)$ or $J(t)$ is known, the material behaviour, at least for one-dimensional tests, is specified. There are several reasons why a full explicit description of J and G is very difficult. It is not possible to enforce infinitely fast steps in the load, so it it not possible to realize a perfect step. Consequently, it is almost not feasible to determine $G(t)$ and $J(t)$ for very small values of t. At the other side of the time domain the problem is encountered that it is not possible to carry out measurements for an unlimited (infinite) period of time.

Both functions $J(t)$ and $G(t)$ are continuous with respect to time. Often these functions are approximated by discrete spectra. Examples of such spectra are

$$J(t) = J_0 + \sum_{k=1}^{N} f_k[1 - e^{-t/\tau_k}] + t/\eta, \tag{5.36}$$

with J_0, f_k, τ_k ($k = 1, ..., N$) and η material parameters (constants) or

$$G(t) = G_\infty + \sum_{j=1}^{M} g_j\, e^{-t/\tau_j}, \tag{5.37}$$

with G_∞, g_j, τ_j ($j = 1, ..., M$) material constants.

These discrete descriptions can be derived from spring-dashpot models, which will be the subject of the next section.

5.3.2 Visco-elastic models based on springs and dashpots: Maxwell model

An alternative way of describing linear visco-elastic materials is by assembling a model using the elastic and viscous components as discussed before. Two examples are given, while only small stretches are considered. In that case the constitutive models for the elastic spring and viscous dashpot are given by

$$F = c\varepsilon, \quad F = c_\eta\dot{\varepsilon}. \tag{5.38}$$

In the Maxwell model according to the set-up of Fig. 5.10(a) the strain ε is additionally composed of the strain in the spring (ε_s) and the strain in the damper (ε_d):

$$\varepsilon = \varepsilon_s + \varepsilon_d, \tag{5.39}$$

implying that

$$\dot{\varepsilon} = \dot{\varepsilon}_s + \dot{\varepsilon}_d. \tag{5.40}$$

(a) Spring-dashpot in series   (b) Spring-dashpot in parallel

Figure 5.10

A Maxwell (a) and Kelvin-Voigt (b) arrangement of the spring and dashpot.

The forces in both the spring and the damper must be the same, therefore, based on Eq. (5.38), the strain rates in the spring and dashpot satisfy

$$\dot{\varepsilon}_s = \frac{1}{c}\dot{F} \tag{5.41}$$

and

$$\dot{\varepsilon}_d = \frac{F}{c_\eta}. \tag{5.42}$$

Eq. (5.40) reveals:

$$\dot{\varepsilon} = \frac{1}{c}\dot{F} + \frac{F}{c_\eta}. \tag{5.43}$$

This is rewritten (by multiplication with c) as

$$\dot{F} + \frac{c}{c_\eta}F = c\dot{\varepsilon}, \tag{5.44}$$

and with introduction of the so-called **relaxation time** τ:

$$\tau = \frac{c_\eta}{c}, \tag{5.45}$$

the final expression is obtained:

$$\dot{F} + \frac{1}{\tau}F = c\dot{\varepsilon}. \tag{5.46}$$

This differential equation is subject to the condition that for $t < 0$ the force F and the strain rate $\dot{\varepsilon}$ vanish. To find a solution of this differential equation, the force F is split into a solution F_h of the homogeneous equation:

$$\dot{F}_h + \frac{1}{\tau}F_h = 0 \tag{5.47}$$

and one particular solution F_p of the inhomogeneous Eq. (5.46):

$$F = F_h + F_p. \tag{5.48}$$

The homogeneous solution is of the form:

$$F_h = c_1 e^{c_2 t}, \tag{5.49}$$

with c_1 and c_2 constants. Substitution into Eq. (5.47) yields

$$c_1 c_2 e^{c_2 t} + \frac{1}{\tau}c_1 e^{c_2 t} = 0 \quad \longrightarrow \quad c_2 = -\frac{1}{\tau} \quad \longrightarrow \quad F_h = c_1 e^{-t/\tau}. \tag{5.50}$$

The solution F_p is found by selecting

$$F_p = C(t)e^{-t/\tau}, \tag{5.51}$$

with $C(t)$ to be determined. Substitution into Eq. (5.46) yields

$$\underbrace{\frac{dC}{dt}e^{-t/\tau} - \frac{C}{\tau}e^{-t/\tau}}_{\frac{dF_p}{dt}} + \underbrace{\frac{C}{\tau}e^{-t/\tau}}_{\frac{1}{\tau}F_p} = c\dot{\varepsilon} \quad\longrightarrow\quad \frac{dC}{dt} = c\,e^{t/\tau}\dot{\varepsilon}, \qquad (5.52)$$

hence

$$C = \int c\,e^{\xi/\tau}\dot{\varepsilon}(\xi)\,d\xi. \qquad (5.53)$$

Because the strain rate $\dot{\varepsilon} = 0$ for all $t < 0$, it follows that

$$F_p = \left(\int_0^t c\,e^{\xi/\tau}\dot{\varepsilon}(\xi)\,d\xi \right) e^{-t/\tau}. \qquad (5.54)$$

Combining Eqs. (5.49) and (5.54), the solution F is given by

$$F = c_1 e^{-t/\tau} + \left(\int_0^t c\,e^{\xi/\tau}\dot{\varepsilon}(\xi)\,d\xi \right) e^{-t/\tau}. \qquad (5.55)$$

Requiring that for all $t < 0$ the force satisfies $F = 0$ leads to

$$F(t = 0) = c_1, \quad c_1 = 0. \qquad (5.56)$$

Consequently, the solution of the first-order differential equation Eq. (5.46), is given by

$$F(t) = c \int_0^t e^{-(t-\xi)/\tau}\dot{\varepsilon}(\xi)\,d\xi. \qquad (5.57)$$

Apparently, the integral equation as introduced in the previous section, Eq. (5.26), can be considered as a general solution of a differential equation. In the present case the relaxation spectrum, as defined in Eq. (5.37) is built up by just one single Maxwell element and in this case: $G(t) = e^{-t/\tau}$.

To understand the implications of this model, consider a strain history as specified in Fig. 5.11, addressing a spring-dashpot system in which one end point is fixed while the other end point has a prescribed displacement history. The force response is given in Fig. 5.12 in case $t^* = 5\tau$. Notice that in this figure the time has been scaled with the relaxation time τ, while the force has been scaled with $c_\eta r$, with r the strain rate, see Fig. 5.11. Two regimes may be distinguished.

(i) For $t < t^*$ the strain proceeds linearly in time leading to a **constant** strain rate r. In this case the force response is given by (recall that $c_\eta = \tau c$)

$$F = c_\eta r(1 - e^{-\frac{t}{\tau}}). \qquad (5.58)$$

For $t \ll \tau$ it holds that

$$e^{-\frac{t}{\tau}} \approx 1 - \frac{t}{\tau}, \qquad (5.59)$$

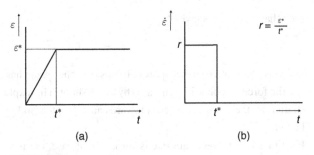

Figure 5.11

Strain(a) and strain rate (b) as a function of time.

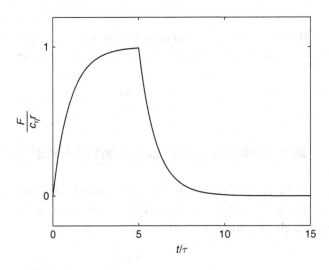

Figure 5.12

Force response of the Maxwell model.

such that the force in that case is given by

$$F = c_\eta \frac{t}{\tau} r = c t r. \tag{5.60}$$

So, for relatively small times t, and constant strain rate, the response is dominantly elastic. This is consistent with the spring-dashpot configuration. For small t only the spring is extended while the dashpot is hardly active. Furthermore, at $t = 0$:

$$\dot{F} = cr, \tag{5.61}$$

which implies that the line tangent to the force versus time curve should have a slope of cr. For larger values of t, but still smaller than t^* we have

$$e^{-\frac{t}{\tau}} \to 0, \tag{5.62}$$

such that the force is given by

$$F = c_\eta r, \tag{5.63}$$

which is a purely viscous response. In this case the spring has a constant extension, and the force response is dominated by the dashpot. This explains why, at a constant strain rate, the force curve tends towards an asymptote in Fig. 5.12 for sufficiently large t.

(ii) For times $t > t^*$ the strain rate is zero. In that case, the force decreases exponentially in time. This is called **relaxation**. If F^* denotes the force $t = t^*$, the force for $t > t^*$ is given by

$$F = F^* e^{-(t-t^*)/\tau}. \tag{5.64}$$

The rate of force relaxation is determined by τ, which explains why τ is called a relaxation time. At $t = t^*$, the slope of the tangent to the force curve equals

$$\dot{F}\big|_{t=t^*} = -\frac{F^*}{\tau}. \tag{5.65}$$

5.3.3 Visco-elastic models based on springs and dashpots: Kelvin–Voigt model

A second example of combined viscous and elastic behaviour is obtained for the set-up of Fig. 5.10(b). In this case the total force F equals the sum of the forces due to the elastic spring and the viscous damper:

$$F = c\varepsilon + c_\eta \dot{\varepsilon}, \tag{5.66}$$

or, alternatively after dividing by c_η:

$$\frac{F}{c_\eta} = \frac{c}{c_\eta}\varepsilon + \dot{\varepsilon}. \tag{5.67}$$

Introducing the **retardation** time:

$$\tau = \frac{c_\eta}{c}, \tag{5.68}$$

Eq. (5.67) may also be written as

$$\frac{F}{c_\eta} = \frac{1}{\tau}\varepsilon + \dot{\varepsilon}. \tag{5.69}$$

The set-up according to Fig. 5.10(b) is known as the Kelvin–Voigt model. In analogy with the Maxwell model, the solution of this differential equation is given by

$$\varepsilon(t) = \frac{1}{c_\eta} \int_0^t e^{-(t-\xi)/\tau} F(\xi) \, d\xi. \tag{5.70}$$

In the case of a **constant** force F, the strain response is given by

$$\varepsilon(t) = \frac{F}{c}(1 - e^{-t/\tau}).$$ (5.71)

This phenomenon of an increasing strain with a constant force (up to a maximum of F/c) is called **creep**. For $t \ll \tau$:

$$\varepsilon \approx \frac{Ft}{c_\eta},$$ (5.72)

corresponding to a viscous response, while for $t \gg \tau$:

$$\varepsilon \approx \frac{F}{c},$$ (5.73)

reflecting a purely elastic response.

5.4 Harmonic excitation of visco-elastic materials

5.4.1 The Storage and the Loss Modulus

In this section some methods will be described that can be used to calculate the response of a visco-elastic material for different excitations. The section is aimed at closed form solutions for the governing equations. Fourier and Laplace transforms and complex function theory are used. First, the methods will be outlined in a general context. After that, an example of a standard linear model will be discussed.

In the previous section first-order differential equations appeared to describe the behaviour of simple visco-elastic models, however, in general, a linear visco-elastic model is characterized by either a higher-order differential equation (or a set of first-order differential equations):

$$p_0 F + p_1 \frac{dF}{dt} + \cdots + p_M \frac{d^M F}{dt^M} = q_0 \varepsilon + q_1 \frac{d\varepsilon}{dt} + \cdots + q_N \frac{d^N \varepsilon}{dt^N},$$ (5.74)

or an integral equation:

$$\varepsilon(t) = \int_0^t J(t - \xi) \dot{F}(\xi) \, d\xi$$ (5.75)

$$F(t) = \int_0^t G(t - \xi) \dot{\varepsilon}(\xi) \, d\xi.$$ (5.76)

Both types of formulation can be derived from each other.

When a model is used consisting of a number of springs and dashpots, the creep and relaxation functions can be expressed by a series of exponential functions. It is said, that the functions form a discrete spectrum. It is possible, and for biological materials sometimes necessary [8], to use continuous functions for $G(t)$ and $J(t)$ thus establishing a more general identification than the differential formulation with limited M and N.

To determine the response to some arbitrary force or strain history several solution methods are available. In the case of a differential model a usual way is to determine the homogeneous solution and after that a particular solution. The general solution is the summation of both. This is a method that is applied in the time domain. Another approach is to use Laplace transforms. In this case the differential equation is replaced by an algebraic equation in the Laplace domain which usually is easy to solve. This solution is then transformed back into the time domain (often the harder part). This approach is usually used for functions that are one-sided, meaning that the functions are zero up to a certain time and finite after that time.

The result of the integral formulation can, for the case of discrete spectra, be considered as the general solution of the associated differential equation and sometimes it is possible to derive closed form expressions for the integrals. This depends strongly on the spectrum and the load history. If no closed form solutions are available numerical methods are necessary to calculate the integrals, which is usually the case for realistic loading histories.

Closed form solutions can only be generated for simple loading histories. Strictly speaking the above methods are only applicable for transient signals (zero for $t < 0$ and finite for $t \geq 0$). For examples see Fig. 5.13.

A frequently used way of excitation in practice is **harmonic excitation**. In that case the applied loading history has the form of a sine or cosine. Let us assume that the prescribed strain is harmonic according to

$$\varepsilon(t) = \varepsilon_0 \cos(\omega t). \tag{5.77}$$

In case of a linear visco-elastic model the output, i.e. the force, will also be a harmonic:

$$F(t) = F_0 \cos(\omega t + \delta), \tag{5.78}$$

or equivalently:

$$F(t) = F_0 \cos(\delta) \cos(\omega t) - F_0 \sin(\delta) \sin(\omega t). \tag{5.79}$$

This equation reveals that the force can be decomposed into two terms: the first with amplitude $F_0 \cos(\delta)$ in phase with the applied strain (called the elastic response), the second with amplitude $F_0 \sin(\delta)$, which is 90° out of phase with the applied strain (called the viscous response). Eq. (5.79) can also be written as

Some 'simple' functions that can be used as loading histories and for which closed form solutions exist for linear visco-elastic models.

$$F(t) = \varepsilon_0 E_1 \cos(\omega t) - \varepsilon_0 E_2 \sin(\omega t), \qquad (5.80)$$

with: $E_1 = (F_0/\varepsilon_0)\cos(\delta)$ the **Storage Modulus** and $E_2 = (F_0/\varepsilon_0)\sin(\delta)$ the **Loss Modulus**. These names will become clear after considering the amount of energy (per unit length of the sample considered) dissipated during one single loading cycle. The necessary amount of work for such a cycle is

$$W = \int_0^{2\pi/\omega} F\dot{\varepsilon}\,dt$$

$$= -\int_0^{2\pi/\omega} [\varepsilon_0 E_1 \cos(\omega t) - \varepsilon_0 E_2 \sin(\omega t)]\,\varepsilon_0 \omega \sin(\omega t)\,dt$$

$$= \pi \varepsilon_0^2 E_2. \qquad (5.81)$$

It is clear that part of the work is dissipated as heat. This part, given by Eq. (5.81) is determined by E_2, the Loss Modulus. During loading the E_1 related part of F also contributes to the stored work, however this energy is released during unloading. That is why E_1 is called the Storage Modulus.

5.4.2 The Complex Modulus

In literature on visco-elasticity the **Complex Modulus** is often used, which is related to the Storage and Loss Modulus. To identify this relation a more formal way to study harmonic excitation is pursued. In the case of a harmonic signal the Boltzmann integral for the relaxation function can be written as

$$F(t) = \int_{-\infty}^{t} G(t - \xi)\dot{\varepsilon}(\xi)\,d\xi. \tag{5.82}$$

The domain of the integral in Eq. (5.82) starts at $-\infty$ because it is assumed that at the current time t the harmonic strain is applied for such a long time, that all effects from switching on the signal have disappeared (also meaning that using a Laplace transform for these type of signals is not recommended). The time t in the upper boundary of the integral can be removed by substitution of $\xi = t - s$ in Eq. (5.82), thus replacing the integration variable ξ by s:

$$F(t) = -\int_{\infty}^{0} G(s)\dot{\varepsilon}(t - s)\,ds = \int_{0}^{\infty} G(s)\dot{\varepsilon}(t - s)\,ds. \tag{5.83}$$

Substitution of (5.77) into this equation leads to

$$F(t) = \varepsilon_0 \cos(\omega t)\left[\omega \int_{0}^{\infty} G(s)\sin(\omega s)\,ds\right]$$
$$- \varepsilon_0 \sin(\omega t)\left[\omega \int_{0}^{\infty} G(s)\cos(\omega s)\,ds\right]. \tag{5.84}$$

The terms between brackets [ ] are only functions of the frequency and not of the time. These terms are solely determined by the type of material that is being considered and can be measured. We can write Eq. (5.84) as

$$F(t) = \varepsilon_0 E_1(\omega)\cos(\omega t) - \varepsilon_0 E_2(\omega)\sin(\omega t), \tag{5.85}$$

where the formal definitions for the Storage and Loss Modulus:

$$E_1(\omega) = \omega \int_{0}^{\infty} G(s)\sin(\omega s)\,ds \tag{5.86}$$

$$E_2(\omega) = \omega \int_{0}^{\infty} G(s)\cos(\omega s)\,ds, \tag{5.87}$$

can be recognized.

In the case of harmonic excitation it is worthwhile to use complex function theory. Instead of (5.77) we write

$$\varepsilon(t) = \text{Re}\{\varepsilon_0 e^{i\omega t}\}. \tag{5.88}$$

The convolution integral for the force, Eq. (5.82) can be written as

$$F(t) = \int_{\xi=-\infty}^{\infty} G(t - \xi)\dot{\varepsilon}(\xi)\,d\xi = \int_{s=-\infty}^{\infty} G(s)\dot{\varepsilon}(t - s)\,ds. \tag{5.89}$$

The upper limit t is replaced by ∞. This is allowed because $G(t)$ is defined such that $G(t - \xi) = 0$ for $\xi > t$. After that we have substituted $\xi = t - s$. Substitution of (5.88) into (5.89) leads to

$$F(t) = i\omega\varepsilon_0 e^{i\omega t} \int_{-\infty}^{\infty} G(s) e^{-i\omega s} ds = i\omega\varepsilon_0 G^*(\omega) e^{i\omega t}. \qquad (5.90)$$

In this equation $G^*(\omega)$ can be recognized as the Fourier transform of $G(t)$ (see Appendix 5.5). Apparently the force has the same form as the strain, only the force has a complex amplitude. If we define the **Complex Modulus** $E^*(\omega)$ as

$$E^*(\omega) = i\omega G^*(\omega) = E_1(\omega) + iE_2(\omega), \qquad (5.91)$$

substitution of (5.91) into (5.90) gives the real part of $F(t)$:

$$F(t) = \varepsilon_0 E_1 \cos(\omega t) - \varepsilon_0 E_2 \sin(\omega t). \qquad (5.92)$$

It is clear again that E_1 and E_2 are the Storage and Loss modulus. The above expression specifies the form of the force output in the time domain. We can also directly derive the relation between $E(\omega)$ and $G(\omega)$ by using a Fourier transform of Eq. (5.89):

$$F^*(\omega) = G^*(\omega) i\omega\varepsilon^*(\omega) = E^*(\omega)\varepsilon^*(\omega). \qquad (5.93)$$

The Complex Modulus is similar to the transfer function in system theory. The Storage Modulus is the real part of the transfer function, the Loss Modulus is the imaginary part.

When experiments are performed to characterize visco-elastic, biological materials, the results are often presented in the form of either the Storage and the Loss Modulus as a function of the excitation frequency, or by using the absolute value of the Complex Modulus, in combination with the phase shift between input (strain) and output (force), as a function of the frequency. In the case of **linear** visco-elastic behaviour these properties give a good representation of the material (this has to be tested first). As a second step often a model is proposed, based on a combination of springs and dashpots, to 'fit' on the given moduli. If this is possible, the material behaviour can be described with a limited number of material parameters (the properties of the springs and dashpots) and all possible selections of properties to describe the material under consideration can be derived from each other. This will be demonstrated in the next subsection for a particular, but frequently used model, the standard linear model.

5.4.3 The standard linear model

The standard linear model can be represented with one dashpot and two springs, as shown in Fig. 5.14. The upper part is composed of a linear spring, the lower part shows a Maxwell element.

Figure 5.14

The 3-parameter standard linear visco-elastic model.

Similar to the procedure as used in Section 5.3.2 the total strain of the Maxwell element is considered as an addition of the strain in the dashpot (ε_d) and the strain in the spring ($\varepsilon_s = \varepsilon - \varepsilon_d$). The following relations can be proposed for variables that determine the standard linear model:

$$F = F_1 + F_2$$
$$F_1 = c_2\varepsilon$$
$$F_2 = c_\eta\dot{\varepsilon}_d$$
$$F_2 = c_1(\varepsilon - \varepsilon_d). \tag{5.94}$$

Elimination of $\dot{\varepsilon}_d$, F_1 and F_2 from this set of equations leads to

$$F + \tau_R\dot{F} = c_2\varepsilon + (c_1 + c_2)\,\tau_R\dot{\varepsilon}, \tag{5.95}$$

with $\tau_R = c_\eta/c_1$ the characteristic relaxation time. The force response to a step $\varepsilon(t) = \varepsilon_0\,H(t)$ in the strain yields

$$F(t) = F(t)_{\text{hom}} + F(t)_{\text{part}} = \alpha e^{-t/\tau_R} + c_2\varepsilon_0, \tag{5.96}$$

with α an integration constant to be determined from the initial conditions. Determining the initial condition at $t = 0$ for this problem is not trivial. It is a jump condition with a discontinuous force F and strain ε at $t = 0$. A way to derive this jump condition is by using the definition of the time derivative:

$$\dot{F} = \lim_{\Delta t \to 0} \frac{F(t + \Delta t) - F(t)}{\Delta t}. \tag{5.97}$$

Let us take two time points a distance Δt apart, one point on the time axis left of $t = 0$, which we call $t = 0^-$ and one point on the right side of $t = 0$ which we call $t = 0^+$. In that case Eq. (5.95) can be written as

$$F(0) + \tau_R\frac{F(0^+) - F(0^-)}{\Delta t} = c_2\varepsilon(0) + (c_1 + c_2)\,\tau_R\frac{\varepsilon(0^+) - \varepsilon(0^-)}{\Delta t}, \tag{5.98}$$

or

$$F(0)\,\Delta t + \tau_R(F(0^+) - F(0^-))$$
$$= c_2\varepsilon(0)\,\Delta t + (c_1 + c_2)\,\tau_R(\varepsilon(0^+) - \varepsilon(0^-)). \tag{5.99}$$

Response of the 3-parameter model to a step in the strain.

Because $F(0^-) = 0$ and $\varepsilon(0^-) = 0$ and the terms with Δt vanish when $\Delta t \to 0$ it is found that

$$F(0^+) = (c_1 + c_2)\,\varepsilon_0, \tag{5.100}$$

so $\alpha = c_1\varepsilon_0$. The solution is shown in Fig. 5.15:

$$F(t) = \varepsilon_0(c_2 + c_1 e^{-t/\tau_R}). \tag{5.101}$$

With Eq. (5.101) the step response $G(t)$ is known. Using the Boltzmann integral this leads to the general solution of Eq. (5.95):

$$F(t) = \int_{-\infty}^{t} \left(c_2 + c_1 e^{-(t-\xi)/\tau_R} \right) \dot{\varepsilon}(\xi)\,d\xi. \tag{5.102}$$

There are several ways to determine the creep function. We can solve Eq. (5.95) for a step in the force. This can be done by determining a homogeneous and a particular solution as was done for the relaxation problem. However, it can also be done by means of Laplace transformation of the differential equation. This leads to an algebraic equation that can be solved. The result can be transformed back from the Laplace domain to the time domain.

Instead of again solving the differential equation we can use the relation that exists between the Creep function and the Relaxation function, Eq. (5.34). A Laplace transformation of G(t) leads to

$$\hat{G}(s) = \frac{c_2}{s} + \frac{c_1}{s + 1/\tau_R} = \frac{c_2(s + 1/\tau_R) + c_1 s}{s(s + 1/\tau_R)}. \tag{5.103}$$

With Eq. (5.34) the Laplace transform of J is found:

$$\hat{J}(s) = \frac{1}{s^2\hat{G}(s)} = \frac{s + 1/\tau_R}{[(c_1 + c_2)s + c_2/\tau_R]s}. \tag{5.104}$$

$$\hat{J}(s) = -\frac{c_1/c_2}{[(c_1 + c_2)s + c_2/\tau_R]} + \frac{1}{c_2 s}. \tag{5.105}$$

Back transformation leads to

$$J(t) = \frac{1}{c_2}\left(1 - \frac{c_1}{c_2 + c_1}e^{-t/\tau_K}\right), \tag{5.106}$$

with $\tau_K = (c_1 + c_2)\tau_R/c_2$ the characteristic creep time. It is striking that the characteristic creep time is different from the characteristic relaxation time. To be complete, the integral equation for force controlled problems is given:

$$\varepsilon(t) = \int_{-\infty}^{t} \frac{1}{c_2}\left(1 - \frac{c_1}{c_1 + c_2}e^{-(t-\xi)/\tau_K}\right)\dot{F}(\xi)\,d\xi. \tag{5.107}$$

At the current point it is opportune to mention some terminology from system dynamics. A linear system can be defined by a **transfer function**. For a harmonic excitation the transfer function is found by a Fourier transform of the original differential Eq. (5.95):

$$F^*(\omega) = E^*(\omega)\varepsilon^*(\omega), \tag{5.108}$$

with

$$E^*(\omega) = \frac{c_2 + (c_2 + c_1)i\omega\tau_R}{1 + i\omega\tau_R}. \tag{5.109}$$

This can be rewritten as

$$E^*(\omega) = c_2 \frac{1 + i\omega\tau_K}{1 + i\omega\tau_R}. \tag{5.110}$$

In system dynamics it is customary to plot a Bode diagram of these functions. For this we need the absolute value of $E^*(\omega)$:

$$|E^*(\omega)| = c_2 \frac{\sqrt{1 + (\omega\tau_K)^2}}{\sqrt{1 + (\omega\tau_R)^2}}. \tag{5.111}$$

The phase shift $\phi(\omega)$ is

$$\phi(\omega) = \arctan(\omega\tau_K) - \arctan(\omega\tau_R). \tag{5.112}$$

In our case $\tau_K > \tau_R$ because c_1 and c_2 are always positive. The result is given in Fig. 5.16.

The Storage and the Loss Modulus are the real and imaginary part of the Complex Modulus $E^*(\omega)$:

$$E_1(\omega) = c_2 \frac{1 + \omega^2\tau_K\tau_R}{1 + \omega^2\tau_R^2} \tag{5.113}$$

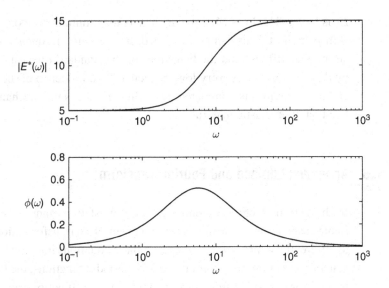

Figure 5.16

The Complex Modulus and phase shift for a standard linear model with two springs ($c_1 = 10$, $c_2 = 5$) and one dashpot ($\tau_R = 0.1$).

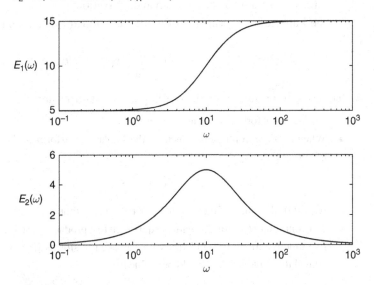

Figure 5.17

The Storage and Loss Modulus for the standard linear model.

$$E_2(\omega) = c_2 \frac{\omega\tau_K - \omega\tau_R}{1 + \omega^2\tau_R^2}. \tag{5.114}$$

The result is given in Fig. 5.17. E_1 has a similar shape as $|E^*(\omega)|$, because the asymptotic values for $\omega \to 0$ and $\omega \to \infty$ are the same. However, the slope of

E_1 is larger. The Loss Modulus E_2 has its maximum at the point where the phase shift is highest. This can be explained. At very high frequencies the dashpot has an infinite stiffness and the behaviour of the standard linear model is dominated by the two springs. At very low frequencies the influence of the dashpot is small and the behaviour is dominated by c_2. In these areas the mechanical behaviour is like that of an elastic material.

5.5 Appendix: Laplace and Fourier transforms

In the current section a summary of the most important issues with regard to Laplace and Fourier transforms will be given. Both transformations can be applied to differential equations and transform these equations into algebraic equations. In general the Fourier transform is used for periodic functions, the Laplace transform is used for one-sided functions, meaning that the functions are zero up to a certain time and finite after that time. In terms of visco-elasticity this means that the Fourier transform is used as a tool to describe harmonic excitation and the Laplace transform is used to describe creep and relaxation.

The Laplace transform $\hat{x}(s)$ of a time function $x(t)$ is defined as

$$\hat{x}(s) = \int_0^\infty x(t)\,e^{-st}dt. \tag{5.115}$$

The most important properties of Laplace transforms are:

- Laplace transform is a linear operation.
- When $x(t)$ is a continuous function, the Laplace transform of the time derivative $\dot{x}(t)$ of $x(t)$ is given by

$$\hat{\dot{x}}(t) = s\,\hat{x}(s) - x(0), \tag{5.116}$$

with $x(0)$ the value of the original function $x(t)$ at time $t = 0$.
- Convolution in the time domain is equivalent to a product in the Laplace domain. Using two time functions $x(t)$ and $y(t)$ with Laplace transforms $\hat{x}(s)$ and $\hat{y}(s)$, the following convolution integral $I(t)$ could be defined:

$$I(t) = \int_{\tau=-\infty}^\infty x(\tau)y(t-\tau)d\tau. \tag{5.117}$$

In that case the Laplace transform of this integral can be written as

$$\hat{I}(s) = \hat{x}(s)\,\hat{y}(s). \tag{5.118}$$

- If a function $x(t)$ has a Laplace transform $\hat{x}(s)$, then the Laplace transform of the function $t^n x(t)$, with $n = 1, 2, 3, \ldots$ can be written as

$$\overline{t^n x(t)} = (-1)^n \frac{d^n \hat{x}(s)}{ds^n}. \tag{5.119}$$

- If a function $x(t)$ has Laplace transform $\hat{x}(s)$, then the Laplace transform of $x(t)/t$, assuming that $\lim_{t \to 0} x(t)/t$ exists, is given by

$$\overline{(\hat{x(t)/t})} = \int_s^\infty x(a)\,da. \tag{5.120}$$

The Fourier transform $x^*(t)$ of time function $x(t)$ is given by

$$x^*(\omega) = \int_{-\infty}^\infty x(t)\,e^{i\omega t}dt, \tag{5.121}$$

with $i = \sqrt{-1}$. The Fourier transform has similar properties to the Laplace transform. The most important properties of Fourier transforms are:

- Fourier transform is a linear operation.
- When $x(t)$ is a continuous function, the Fourier transform of the time derivative $\dot{x}(t)$ of $x(t)$ is given by

$$\dot{x}^*(t) = i\omega\, x^*(\omega). \tag{5.122}$$

- Convolution in the time domain is equivalent to a product in the Fourier domain. Using two time functions $x(t)$ and $y(t)$ with Fourier transforms $x^*(\omega)$ and $y^*(\omega)$, the following convolution integral $I(t)$ could be defined:

$$I(t) = \int_{-\infty}^\infty x(\tau)y(t-\tau)\,d\tau. \tag{5.123}$$

In that case the Fourier transform of this integral can be written as

$$I^*(\omega) = x^*(\omega)\, y^*(\omega). \tag{5.124}$$

- If a function $x(t)$ has a Fourier transform $x^*(\omega)$, then the Fourier transform of the function $t^n x(t)$, with $n = 1, 2, 3, \ldots$ can be written as

$$\left(t^n x(t)\right)^* = (i)^n \frac{d^n x^*(\omega)}{d\omega^n}. \tag{5.125}$$

- If a function $x(t)$ has Fourier transform $x^*(\omega)$, then the Fourier transform of $x(t)/t$ is given by

$$(x(t)/t)^* = i\omega \int_\omega^\infty x(a)\,da. \tag{5.126}$$

Finally, Table 5.1 gives some Laplace and Fourier transforms of often used functions. In the table the step function $H(t)$ (Heaviside function) as defined in Eq. (5.18) is used, as well as the delta function $\delta(t)$, defined as

Table 5.1 Time functions with their Fourier and Laplace transforms.

Original function	Fourier transform	Laplace transform
$\delta(t)$	1	1
$H(t)$		$1/s$
$t\,H(t)$		$1/s^2$
$e^{-bt}H(t)$	$1/(i\omega + b)$	$1/(s+b)$
$e^{i\omega_0}$	$\delta(\omega - \omega_0)$	
$\sin(at)\,H(t)$	$a/(a^2 - \omega^2)$	$a/(s^2 + a^2)$
$\cos(at)\,H(t)$	$i\omega/(a^2 - \omega^2)$	$s/(s^2 + a^2)$

$$\delta(t) = \begin{cases} 0 & \text{if: } t \neq 0 \\ \infty & \text{if: } t = 0 \end{cases} \tag{5.127}$$

and

$$\int_{-\infty}^{+\infty} \delta(t)\,dt = 1. \tag{5.128}$$

Exercises

5.1 A visco-elastic material is described by means of a Maxwell model. The model consists of a linear dashpot, with damping coefficient c_η, in series with a linear spring, with spring constant c (see figure below).

The numerical values for the material properties are:

$$c = 8\ 10^4 \qquad [\,Nc\,]$$
$$c_\eta = 0.8\ 10^4 \qquad [\,Ncs\,]$$

(a) Derive the differential equation for this model.
(b) Give the response for a unit step in the strain. Make a drawing of the response.
(c) The material is subjected to an harmonic strain excitation with amplitude ε_0 and an angular frequency ω. Give the complex modulus $E^*(\omega)$ and the phase shift $\phi(\omega)$ for this material.

    (d)   Give the amplitude of the force for the case $\varepsilon_0 = 0.01$ and a frequency $f = 5$ [Hz].

5.2   A material can be characterized with a standard linear model (see figure below). The material is loaded with a step in the force at time $t = 0$. At time $t = t_1$ the material is suddenly unloaded stepwise.

    (a)   Derive the strain response for this loading history.

    (b)   Make a drawing of the strain response as a function of time.

    (c)   Calculate $d\varepsilon/dt$ for $t = 0$.

5.3   In a dynamic experiment a specimen is loaded with a strain as given in the figure below. The strain is defined by:

$$\varepsilon = 0.01 \, \sin(\omega t),$$

with $\omega = 0.1$ [rad s^{-1}].

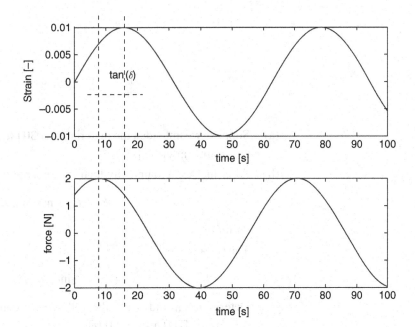

The response is also given in the figure. Let us assume that we are dealing with a material that can be described by a Maxwell model (one spring and dashpot in series).

(a) What is the phase shift if we double the angular frequency of the load?

(b) What will be the amplitude of the force for that frequency?

(c) Now we do a relaxation experiment, with the same material. We apply a step strain of $\varepsilon = 0.02$. What is the force at:
$t = 0$ [s], $t = 10$ [s] and $t = 100$ [s]?

5.4 A creep test was performed on some biological material. The applied step-wise load was $F = 20$ [N]. Assume that we are dealing with a material that can be described by a Kelvin–Voigt model. This means that the material can be modelled as a linear spring and dashpot in parallel.

(a) Suppose we unloaded the material after 30 [s]. Determine for this case the strain ε after 60 [s].

(b) After that we dynamically load the material with a force F:

$$F = F_0 \sin(\omega t),$$

with $F_0 = 10$ [N].
The response is:

$$\varepsilon = \varepsilon_0 \sin(\omega t + \delta).$$

Determine ε_0 and δ for the following frequencies $f = 0.01$ [Hz], 0.1 [Hz], 1 [Hz] and 100 [Hz].

5.5   For a material the following relaxation and creep function was found,

$$G(t) = 1 + e^{-t}$$

$$J(t) = 1 - \frac{1}{2}e^{-t/2}.$$

When we relate a force to strain the unit for $G(t)$ is [N] and the unit for $J(t)$ is [N^{-1}]

(a)   Does this material law satisfy Eq. (5.34)?

(b)   Test 1: Assume that at $t = 0$ a step in the force is applied of 1 [N]. After 1.5 [s] the step is removed. What is the strain after 3 [s]?

(c)   Test 2: At $t = 0$ a step in the strain is applied equal to 0.01. This strain is removed at $t = 1.5$ [s]. What is the value of the force after 3 [s]?

5.6   A piece of tendon material is subjected to the following strain history. At $t = 0$ [s] a step in the strain is applied of 0.01. After 4 [s] the strain is increased by an extra step of 0.01. After 8 [s] the strain is reduced by 0.005. The total strain evolution is given in the figure below.

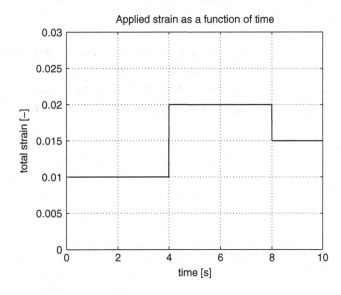

The only information that is available for the material is the result from a creep test as given in the figure below. In the creep test a tendon specimen was loaded with a mass of 500 [kg]. The gravitational constant is 10 [m s^{-2}].

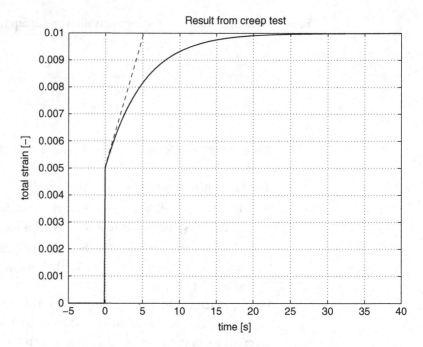

(a)   We try to describe the material with a standard linear model. The relaxation function for this material is:

$$G(t) = c_2 + c_1 e^{-t/\tau_R} .$$

Determine the value of c_1 and c_2 using the information in the figure with the result of the creep test.

(b)   Sketch the force response on the strain history as given in the first figure with the applied strain as a function of time.

6 Analysis of a one-dimensional continuous elastic medium

6.1 Introduction

In the previous chapters the global behaviour of fibres was considered, without much attention to the detailed shape of these structures. Only the length change of the fibre played a role in the analysis. In the present chapter we address in a little bit more detail the deformation of long slender structures. These can be tendons, muscles, but also long bones. The aim is, to generalize the concepts introduced in previous chapters for discrete systems (i.e. springs) to continuous systems. To simplify matters the loading and deformation of a one-dimensional elastic bar is considered.

6.2 Equilibrium in a subsection of a slender structure

Consider a straight bar as visualized in Fig. 6.1(a), loaded by an external force F at $x = L$ and fixed in space at $x = 0$. The figure shows the bar with the x-axis in the longitudinal or axial direction. It is assumed, that each cross section initially perpendicular to the axis of the bar remains perpendicular to the axis after loading. In fact it is assumed that all properties and displacements are a function of the x-coordinate only.

The objective is, to compute the displacement of each cross section of the bar due to the loading. The area of the cross section perpendicular to the central axis of the bar as well as the mechanical properties of the bar may be a function of the x-coordinate. Therefore the displacement may be a non-linear function of the axial coordinate.

To compute the displacements of the cross sections a procedure analogous to the discrete case (elastic springs) is followed. In contrast with the discrete case, the equilibrium conditions are not applied on a global scale but are applied on a local scale. For this purpose the free body diagram of an arbitrary slice of the bar, for example the grey slice in Fig. 6.1(a), is investigated. This free body diagram is depicted in Fig. 6.1(b). The left side of the slice is located at position x and the

(a) Representation of a bar

(b) Free body diagram of a slice at position x

Figure 6.1

Bar and free body diagram of a slice of the bar.

slice has a length Δx. The net force on the left side of the slice equals $N(x)$, while on the right side of the slice a force $N(x + \Delta x)$ is present. The net force on the right side of the bar may be different from the net force on the left side of the slice due to the presence of a so-called distributed volume force. A volume force Q is a force per unit of volume, and may be due to, for instance, gravity. Integration over the cross section area of the slice yields a load per unit length, called q.

If the distributed load q is assumed constant within the slice of thickness Δx, force equilibrium of the slice implies that

$$N(x) = N(x + \Delta x) + q\Delta x. \tag{6.1}$$

This may also be written as

$$\frac{N(x + \Delta x) - N(x)}{\Delta x} + q = 0. \tag{6.2}$$

If the length of the slice Δx approaches zero, we can write

$$\lim_{\Delta x \to 0} \frac{N(x + \Delta x) - N(x)}{\Delta x} = \frac{dN}{dx}, \tag{6.3}$$

where dN/dx denotes the derivative of $N(x)$ with respect to x. The transition expressed in Eq. (6.3) is illustrated in Fig. 6.2. In this graph a function $N(x)$ is sketched. The function $N(x)$ is evaluated at x and $x + \Delta x$, while Δx is small. When moving from x to $x + \Delta x$ the function $N(x)$ changes a small amount: from $N(x)$ to $N(x + \Delta x)$. If Δx is sufficiently small the ratio $\Delta N/\Delta x$ defines the tangent line to the function $N(x)$ at point x, and hence equals the derivative of $N(x)$ with respect to x. Notice that this implies that for sufficiently small Δx:

Figure 6.2

First-order derivative of a function $N(x)$.

$$N(x + \Delta x) \approx N(x) + \frac{dN}{dx}\Delta x. \qquad (6.4)$$

The result of this transformation is that the force equilibrium relation Eq. (6.2) may be written as

$$\frac{dN}{dx} + q = 0. \qquad (6.5)$$

If the load per unit length q equals zero, then the equilibrium equation reduces to

$$\frac{dN}{dx} = 0, \qquad (6.6)$$

which means that the force N is constant throughout the bar. It actually has to be equal to the force F applied to the right end of the bar see Fig. 6.1(a). Consequently, if the slice of Fig. 6.1(b) is considered the force $N(x)$ equals $N(x + \Delta x)$. In other words, the force in the bar can only be non-constant if q can be neglected.

6.3   Stress and strain

The equilibrium equation (6.5) derived in the previous section does not give information about the deformation of the bar. For this purpose a relation between force and strain or strain rate must be defined, similar to the force-strain relation for an elastic spring, discussed in Chapter 4. For continuous media it is more appropriate to formulate a relation between force per unit area (stress) and a deformation measure, such as strain or strain rate. The concepts of stress and strain in continuous media are introduced in this section.

In the one-dimensional case discussed in this chapter, the force N acting on a cross section of the bar is assumed to be homogeneously distributed over the

(a) Force acting on a surface having area A

(b) Stress σ if N is homogeneously distributed over the area A

$\sigma = \dfrac{N}{A}$

Figure 6.3

Stress σ as a homogeneously distributed force N over an area A.

Figure 6.4

Inhomogeneous distribution of the force N.

surface of this cross section. If A denotes the area of the cross section, the **stress** σ is defined as

$$\sigma = \frac{N}{A}. \tag{6.7}$$

So, the stress is a force per unit area. If the force is not homogeneously distributed over a surface, see Fig. 6.4, we must consider a small part of the surface, ΔA. Actually an infinitesimally small area ΔA is considered. This surface area only carries a part ΔN of the total force N. Then, the stress σ is, formally, defined as

$$\sigma = \lim_{\Delta A \to 0} \frac{\Delta N}{\Delta A}. \tag{6.8}$$

This means that the stress is defined by the ratio of an infinitesimal amount of force ΔN over an infinitesimal amount of area ΔA, while the infinitesimal area ΔA approaches zero. In the following it is assumed that Eq. (6.7) can be applied.

Figure 6.5

Displacements of a thin slice within a continuous bar.

Consider a slice of the bar having length Δx, as depicted in Fig. 6.5. The linear **strain** ε is expressed in the stretch λ of the slice by

$$\varepsilon = \lambda - 1, \tag{6.9}$$

where the stretch is the ratio of the deformed length of the slice and the initial length. At position x the displacement of the cross section of the bar equals $u(x)$, while at $x + \Delta x$ the displacement equals $u(x + \Delta x)$. The initial length of the slice equals:

$$\ell_0 = \Delta x, \tag{6.10}$$

while the current length is given by

$$\ell = \Delta x + u(x + \Delta x) - u(x). \tag{6.11}$$

Therefore, the stretch, that is the ratio of the deformed length over the initial length is given by:

$$\lambda = \frac{\Delta x + u(x + \Delta x) - u(x)}{\Delta x}. \tag{6.12}$$

Consequently, if the width of the slice Δx approaches zero, the strain is computed from

$$\begin{aligned}
\varepsilon &= \lim_{\Delta x \to 0} \frac{\Delta x + u(x + \Delta x) - u(x)}{\Delta x} - 1 \\
&= \lim_{\Delta x \to 0} \frac{u(x + \Delta x) - u(x)}{\Delta x}.
\end{aligned} \tag{6.13}$$

Using the definition of the derivative this yields

$$\varepsilon = \frac{du}{dx}. \tag{6.14}$$

In conclusion, the strain is defined as the derivative of the displacement field u with respect to the coordinate x.

6.4 Elastic stress–strain relation

Recall, that the force–strain relation for an elastic spring at small, infinitesimal displacements is given by

$$\vec{F} = c \underbrace{\frac{\vec{a} \cdot (\vec{u}_B - \vec{u}_A)}{l_0}}_{\text{fibre strain}} \vec{a}. \tag{6.15}$$

Here, c represents the stiffness of the spring, while the unit vector $\vec{a}$ denotes the orientation of the spring in space. In analogy with this, the (one-dimensional) stress–strain relation for linearly elastic materials is defined as

$$\sigma = E\varepsilon, \tag{6.16}$$

where E is the so-called **Young's modulus**. Using the definition of the strain in terms of the derivative of the displacement field, this may also be written as

$$\sigma = E\frac{du}{dx}. \tag{6.17}$$

Example 6.1 For a given displacement field $u(x)$, the stress field can be computed. Suppose, for instance, that the Young's modulus is constant and that u is given by a polynomial expression, say:

$$u = a_1 x + 2a_2 x^2 + 5a_3 x^3,$$

with a_1, a_2 and a_3 known coefficients. Then the stress will be

$$\sigma = E\frac{du}{dx} = E(a_1 + 4a_2 x + 15a_3 x^2).$$

6.5 Deformation of an inhomogeneous bar

In case of a one-dimensional bar, the stress at each cross section is uniquely defined according to

$$\sigma = \frac{N}{A}. \tag{6.18}$$

Substitution of $N = A\sigma$ into the equilibrium equation Eq. (6.5) yields

$$\frac{d(A\sigma)}{dx} + q = 0. \tag{6.19}$$

Subsequently, the stress–strain relation Eq. (6.17) is substituted such that the following second-order differential equation in terms of the displacement field $u(x)$ is obtained:

$$\frac{d}{dx}\left(EA\frac{du}{dx}\right) + q = 0. \tag{6.20}$$

In the absence of a force per unit length, $q = 0$, the force in the bar must be constant. The stress σ does not have to be constant, because the cross section area A may be a function of the coordinate x.

Suppose that both the force and the stress are constant (this can only occur if $q = 0$ and A is constant). Then it follows from Eq. (6.20) that

$$EA\frac{du}{dx} = c, \tag{6.21}$$

with c a constant. Nevertheless, the strain $\varepsilon = du/dx$ may be a function of the coordinate x if the Young's modulus is non-constant.

The solution of the differential Eq. (6.20) yields the displacement as a function of x, and once $u(x)$ is known the strain $\varepsilon(x)$ and the stress $\sigma(x)$ in the bar can be retrieved. However, this differential equation can only be solved if two appropriate boundary conditions are specified. Two types of boundary conditions are distinguished. Firstly, **essential** boundary conditions, formulated in terms of specified boundary displacements. The displacement $u(x)$ must at least be specified at one end point, and depending on the problem at hand, possibly at two. This is required to uniquely determine $u(x)$ and may be understood as follows. Suppose that $\hat{u}$ satisfies the equilibrium equation Eq. (6.20). Then, if the displacement $u(x)$ is not specified at, at least, one end point, an arbitrary constant displacement c may be added to $\hat{u}(x)$, while Eq. (6.20) for this modified displacement field $\hat{u}(x) + c$ is still satisfied, since the strain is given by

$$\varepsilon = \frac{d(\hat{u} + c)}{dx} = \frac{d\hat{u}}{dx} + \underbrace{\frac{dc}{dx}}_{=0} = \frac{d\hat{u}}{dx} \tag{6.22}$$

for any constant c. Such a constant c would correspond to a rigid body translation of the bar. So, in conclusion, at least one essential boundary condition **must** be specified.

Secondly, **natural** boundary conditions, formulated in terms of external boundary loads, may be specified, depending on the problem at hand. In the configuration visualized in Fig. 6.1(a) the bar is loaded by an external load F at the right end of the bar, at $x = L$, with L the length of the bar. At this boundary, the force equals

$$N(x = L) = F = \sigma A. \tag{6.23}$$

Since $\sigma = E\,du/dx$, the natural boundary condition at $x = L$ reads

$$EA\frac{du}{dx} = F. \qquad (6.24)$$

Because the equilibrium equation Eq. (6.20) is a second-order differential equation, **two** boundary conditions must be specified, one must be an essential boundary condition (the displacement must be specified at least at one point to avoid rigid body displacement) and the other may either be an essential or natural boundary condition. The combination of the equilibrium equation with appropriate boundary conditions is called a (determinate) **boundary value** problem.

Example 6.2

As a first example the solution of a well-defined boundary value problem for a homogeneous bar without a distributed load is analysed. Consider a bar of length L that has a uniform cross section, hence A is constant, and with constant Young's modulus E. There is no volume load present, hence $q = 0$. At $x = 0$ the displacement is suppressed, hence $u = 0$ at $x = 0$, while at the other end of the bar a force F is applied. Then, the boundary value problem is fully described by the following set of equations:

$$\frac{d}{dx}\left(EA\frac{du}{dx}\right) = 0 \quad \text{for} \quad 0 < x < L$$
$$u = 0 \quad \text{at} \quad x = 0$$
$$EA\frac{du}{dx} = F \quad \text{at} \quad x = L.$$

Integrating the equilibrium equation once yields

$$EA\frac{du}{dx} = c,$$

where c denotes an integration constant. This may also be written as

$$\frac{du}{dx} = \frac{c}{EA}.$$

Because both the Young's modulus E and the cross section area A are constant, integration of this relation gives

$$u = \frac{c}{EA}x + d,$$

with d yet another integration constant. So, the solution $u(x)$ is known provided that the integration constants c and d can be determined. For this purpose the boundary conditions at both ends of the bar are used. First, since at $x = 0$ the displacement $u = 0$, the integration constant d must be zero, hence

$$u = \frac{c}{EA}x.$$

Second, at $x = L$ the force is known, such that

$$F = EA\frac{du}{dx}\Big|_{x=L} = EA\frac{c}{EA} = c.$$

So, the (unique) solution to the boundary value problem reads

$$u = \frac{F}{EA}x.$$

The strain ε is directly obtained via

$$\varepsilon = \frac{du}{dx} = \frac{F}{EA},$$

while the stress σ follows from

$$\sigma = E\varepsilon = \frac{F}{A}$$

as expected.

Example 6.3 Consider, as before, a bar of length L, clamped at one end and loaded by a force F at the other end of the bar. The Young's modulus E is constant throughout the bar, but the cross section varies along the axis of the bar. Let the cross section $A(x)$ be given by

$$A = A_0\left(1 + \frac{x}{3L}\right),$$

with A_0 a constant, clearly representing the cross section area at $x = 0$.

The boundary value problem is defined by the same set of equations as in the previous example:

$$\frac{d}{dx}\left(EA\frac{du}{dx}\right) = 0 \quad \text{for} \quad 0 < x < L$$

$$u = 0 \quad \text{at} \quad x = 0$$

$$EA\frac{du}{dx} = F \quad \text{at} \quad x = L.$$

Integration of the equilibrium equation yields

$$\frac{du}{dx} = \frac{c}{EA} = \frac{c}{EA_0(1 + \frac{x}{3L})},$$

with c an integration constant that needs to be identified. Integration of this result gives

$$u = \frac{3cL}{EA_0} \ln\left(1 + \frac{x}{3L}\right) + d,$$

with d another integration constant. The integration constants c and d can be determined by application of the boundary conditions at $x = 0$ and $x = L$. Since at $x = 0$ the displacement $u = 0$ it follows that

$$u(0) = 0 = \frac{3cL}{EA_0} \ln(1 + 0) + d = d,$$

hence $d = 0$. At $x = L$ the force F is given, such that

$$EA\frac{du}{dx}\bigg|_{x=L} = c = F.$$

Consequently

$$u = \frac{3FL}{EA_0} \ln\left(1 + \frac{x}{3L}\right).$$

The stress σ in the bar is computed directly from

$$\sigma = E\frac{du}{dx} = \frac{F}{A_0(1 + \frac{x}{3L})}.$$

Example 6.4 Consider a bar of length L with constant Young's modulus E and cross section A. As before, the bar is fixed at $x = 0$ while at $x = L$ a force F is applied. Furthermore, the bar is loaded by a constant distributed load q.

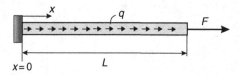

Accordingly, the boundary value problem is given by

$$\frac{d}{dx}\left(EA\frac{du}{dx}\right) = -q \quad \text{for } 0 < x < L$$

$$u = 0 \quad \text{at } x = 0$$

$$EA\frac{du}{dx} = F \quad \text{at } x = L.$$

Integration of the equilibrium equation yields

$$\frac{du}{dx} = -\frac{1}{EA}qx + \frac{c}{EA},$$

where c is an integration constant. Integration of this result yields

$$u = -\frac{1}{2EA}qx^2 + \frac{c}{EA}x + d.$$

Since at $x = 0$ the displacement $u = 0$, it follows immediately that $d = 0$. At $x = L$ the force F is prescribed, hence

$$EA\frac{du}{dx}\bigg|_{x=L} = -qL + c = F,$$

which implies that

$$c = F + qL.$$

Consequently, the displacement u is given by

$$u = -\frac{1}{2EA}qx^2 + \frac{F + qL}{EA}x.$$

Example 6.5 Consider a system of two bars as depicted in Fig. 6.6(a). The Young's modulus and the cross section of both bars are constant and are given by E_1, A_1 and E_2, A_2 for the left and right bar, respectively. The length of the left bar equals L_1 while the right bar has length L_2. The boundary conditions are as depicted in the figure: fixation at both $x = 0$ and $x = L_1 + L_2$. Besides a concentrated force F is applied at $x = L_1$. Our goal is to determine the displacement of the point $x = L_1$.

To solve this problem, we recognize two difficulties:

- We have to find a way to deal with the concentrated force F.
- The problem is statically indeterminate. Reaction forces cannot be uniquely determined from force equilibrium alone.

The concentrated force has to be incorporated according to the following procedure. Three free body diagrams have to be created by virtually cutting the bar just left of the point where the force is applied and just right of that point. Thus three free bodies can be distinguished: left bar, the right bar and a very thin slice around the concentrated force. The free body diagrams are shown in Fig. 6.6(b).

The boundary value problem for the first bar is given by

$$\frac{d}{dx}\left(E_1 A_1 \frac{du_1}{dx}\right) = 0 \quad \text{for } 0 < x < L_1$$

$$u_1 = 0 \quad \text{at } x = 0$$

$$E_1 A_1 \frac{du_1}{dx} = N_1 \quad \text{at } x = L_1.$$

Figure 6.6

A two bar system and associated free body diagram.

Clearly, based on the previous examples, for the first bar the displacement field is given by

$$u_1 = \frac{N_1}{E_1 A_1} x.$$

For the second bar the following boundary value problem holds:

$$\frac{d}{dx}\left(E_2 A_2 \frac{du_2}{dx}\right) = 0 \quad \text{for } L_1 < x < L_1 + L_2$$

$$u_2 = 0 \quad \text{at } x = L_1 + L_2$$

$$E_2 A_2 \frac{du_2}{dx} = N_2 \quad \text{at } x = L_1.$$

The solution of this system is given by

$$u_2 = \frac{N_2}{E_2 A_2} x - \frac{N_2}{E_2 A_2}(L_1 + L_2).$$

We should realize that neither N_1 nor N_2 is known so far. However, there are two additional equations that have to be satisfied. Force equilibrium of the slice (see Fig. 6.6) requires that

$$-N_1 + N_2 + F = 0,$$

while the displacement field must be continuous at $x = L_1$: the two bars must remain fitting together, hence

$$u_1(L_1) = u_2(L_1). \tag{6.25}$$

Based upon the solution for u_1 and u_2 it follows that

$$u_1(L_1) = \frac{N_1}{E_1 A_1} L_1,$$

$$u_2(L_1) = -\frac{N_2}{E_2 A_2} L_2.$$

Because the bars must fit together at $x = L_1$ we find

$$N_1 = -N_2 \frac{E_1 A_1}{L_1} \frac{L_2}{E_2 A_2}.$$

Use of force equilibrium of the slice yields

$$N_2 \frac{E_1 A_1}{L_1} \frac{L_2}{E_2 A_2} + N_2 + F = 0,$$

hence

$$N_2 = \frac{-F}{1 + \frac{E_1 A_1}{L_1} \frac{L_2}{E_2 A_2}}.$$

Now N_1 and N_2 have been determined, it is possible to find an expression for the displacement of the bar at point $x = L_1$ as a function of the force F and the material and geometrical properties of both bars:

$$u(L_1) = \frac{F L_1 L_2}{E_2 A_2 L_1 + E_1 A_1 L_2}.$$

Exercises

6.1   A displacement field as a function of the coordinate x is given as: $u(x) = ax^2 + bx + c$, with a, b and c constant coefficients. Determine the strain field as a function of x.

6.2   A strain field as a function of coordinate x is given as $\varepsilon(x) = ax^2 + bx + c$, with a, b and c constant coefficients. Determine the displacement field $u(x)$ satisfying $u(0) = 0$.

6.3   A bar with Young's modulus E and length ℓ is clamped at $x = 0$ and loaded with a force F at $x = \ell$. The cross section is a function of x ($0 \leq x \leq \ell$), according to:

$$A(x) = A_0\, e^{-\beta x}$$

with A_0 the cross section area at $x = 0$ and with β a positive constant.

(a)   Determine the stress field $\sigma(x)$.
(b)   Determine the displacement field $u(x)$.

6.4 A bar with Young's modulus E, cross section A and length ℓ is clamped at $x=0$. A distributed load $q(x)$ is applied in x-direction on the bar, according to:

$$q(x) = \alpha\, e^{\beta x},$$

with α and β constant load parameters.
(a) Determine the stress field $\sigma(x)$.
(b) Determine the displacement field $u(x)$.

6.5 A bar with Young's modulus E, density ρ, cross section A and length ℓ is hanging by its own weight. The gravitation acceleration is g. The point where the bar is clamped is located at $x = 0$. Determine the displacement field $u(x)$ and the length change of the bar, due to gravity.

6.6 The central part of a femur is modelled as a straight tube with outer diameter D, inner diameter d and length ℓ. Assume that the cortical bone behaves like a linear elastic material with Young's modulus E. In addition, assume that the bone is loaded with an axial compressive force P.

(a)   Determine the stress field $\sigma(x)$.

(b)   Determine the displacement field $u(x)$. Assume that at $x = \ell$ the displacement $u(\ell)$ satisfies $u(\ell) = 0$.

6.7   Consider a muscle/tendon complex as shown in the figure. To find out how much the tendon and the muscle are extended when the complex as a whole is loaded with a force F, a very crude two bar model can be used. The muscle is modelled as a bar with length ℓ_1, Young's modulus E_1 and cross section A_1. At point B the muscle is attached to the tendon, which is modelled as a second bar with length ℓ_2, Young's modulus E_2 and cross section A_2. At point A the muscle is attached to the bone, which we consider as a rigid fixation. At point C a force F is applied in the direction of the bar (tendon).

(a)   Determine the internal force in a cross section between A and B and in a cross section between B and C.

(b)   Determine the stress σ in a cross section between A and B and a cross section between B and C.

(c)   What happens with the calculated forces and stresses if the Young's moduli of both muscle and tendon are reduced to half of their original value?

(d)   Determine the displacements at point B and C as a result of the applied load F at point C.

7 Biological materials and continuum mechanics

7.1 Introduction

Up to this point all treated problems were in a certain way one-dimensional. Indeed, in Chapter 3 we have discussed equilibrium of two- and three-dimensional bodies and in Chapters 4 the fibres were allowed to have some arbitrary orientation in three-dimensional space. But, when deformations were involved, the focus was on fibres and bars, dealing with one-dimensional force/strain relationships. Only one-dimensional equations have been solved. In the following chapters, the theory will be extended to the description of three-dimensional bodies and it is opportune to spend some time looking at the concept of a continuum.

Consider a certain amount of solid and/or fluid material in a three-dimensional space. Although in reality for neighbouring points in space the (physical) character and behaviour of the residing material may be completely different (because of discontinuities at the microscopic level, becoming clearer by reducing the scale of observation) it is common practice that a less detailed description (at a macroscopic level) with a more gradual change of physical properties is used. The discontinuous heterogeneous reality is **homogenized** and modelled as a continuum. To make this clearer, consider the bone in Fig. 7.1. Although one might conceive the bone at a macroscopic level, as depicted in Fig. 7.1(a), as a massive structure filling all the volume that it occupies in space, it is clear from Fig. 7.1(b) that at a smaller scale the bone is a discrete structure with open spaces in between (although the spaces can be filled with a softer material or a liquid). At an even smaller scale, at the level of one single trabecula individual cells can be recognized. This phenomenon is typical for biological materials at any scale, whether it is at the organ level, the tissue level or cellular level or even smaller. On each level a very complex structure can be observed and it is not feasible to take into account every detail of this structure. That is why a homogenization of the structure is necessary and depending on the objectives of the modelling (which is the question that has to be answered in the context of the selection of the theoretical model) a certain level of homogenization is chosen. It is always necessary to

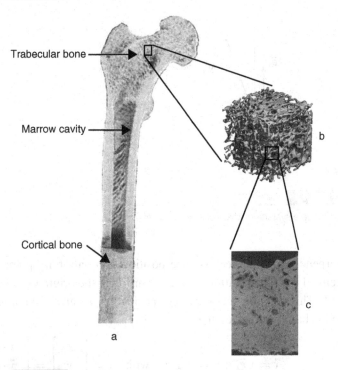

Trabecular bone

Marrow cavity

Cortical bone

b

c

a

Figure 7.1

Structure of bone at different length scales (Images courtesy of Mr Bert van Rietbergen).

critically evaluate whether or not the chosen homogenization is allowed and to be aware of the limitations of the model at hand. In Section 7.3 we will discuss when a continuum approach is allowed.

To model a continuum, physical variables are formulated as continuous functions of the position in space. Related to this, some attention will be given in this chapter to the visualization of physical fields. In addition, derivatives of variables will be discussed. The gradient operator, important in all kinds of theoretical derivations, constitutes a central part of this. The chapter ends with a section on the properties of second-order tensors, which form indispensable tools in continuum mechanics.

7.2 Orientation in space

For an orientation in three-dimensional space a Cartesian xyz-coordinate system is defined with origin O, see Fig. 7.2. The orientation of the coordinate system is laid down by means of the Cartesian vector basis $\{\vec{e}_x, \vec{e}_y, \vec{e}_z\}$, containing mutually

Figure 7.2

Coordinate system, position of a point P in space.

perpendicular unit vectors. The position of an arbitrary point P in space can be defined with the position vector $\vec{x}$, starting at the origin O and with the end point in P. This position vector can be specified by means of components with respect to the basis, defined earlier:

$$\vec{x} = x\,\vec{e}_x + y\,\vec{e}_y + z\,\vec{e}_z \quad \text{with} \quad \underset{\sim}{x} = \begin{bmatrix} x \\ y \\ z \end{bmatrix} = \begin{bmatrix} \vec{x}\cdot\vec{e}_x \\ \vec{x}\cdot\vec{e}_y \\ \vec{x}\cdot\vec{e}_z \end{bmatrix}. \tag{7.1}$$

The column $\underset{\sim}{x}$ is in fact a representation of the vector $\vec{x}$ with respect to the chosen basis vectors $\vec{e}_x$, $\vec{e}_y$ and $\vec{e}_z$ only. Nevertheless, the position vector at hand will sometimes also be indicated with $\underset{\sim}{x}$. This will not jeopardize uniqueness, because in the present context only one single set of basis vectors will be used.

The geometry of a curve (a set of points joined together) in three-dimensional space can be defined by means of a parameter description $\vec{x} = \vec{x}(\xi)$, in components $\underset{\sim}{x} = \underset{\sim}{x}(\xi)$, where ξ is specified within a certain interval, see Fig. 7.3. The tangent vector, with unit length, at an arbitrary point of this curve is depicted by $\vec{e}$, satisfying

$$\vec{e} = \frac{1}{\ell}\frac{d\vec{x}}{d\xi} \quad \text{with} \quad \ell = \sqrt{\left(\frac{d\vec{x}}{d\xi}\right)\cdot\left(\frac{d\vec{x}}{d\xi}\right)} \tag{7.2}$$

and after a transition to component format:

$$\underset{\sim}{e} = \frac{1}{\ell}\frac{d\underset{\sim}{x}}{d\xi} \quad \text{with} \quad \ell = \sqrt{\left(\frac{d\underset{\sim}{x}}{d\xi}\right)^{\text{T}}\left(\frac{d\underset{\sim}{x}}{d\xi}\right)}. \tag{7.3}$$

The parameter ξ is distance measuring if ℓ equals 1; in that special case the length of an (infinitesimally small) line piece $d\vec{x}$ (with components $d\underset{\sim}{x}$) of the curve is exactly the same as the accompanying change $d\xi$.

Figure 7.3

A parameterized curve in space.

Figure 7.4

A subvolume $\Delta V = \Delta x \Delta y \Delta z$ of *V* in space.

A bound volume in three-dimensional space is denoted by V. That volume can be regarded as a continuous set of geometric points, of which the position vectors $\vec{x}$ have their end points within V. A subvolume is indicated by ΔV. That subvolume may have the shape of a (rectangular) block, with edges in the same direction as the coordinate axes, $\Delta V = \Delta x \Delta y \Delta z$ (see Fig. 7.4). An infinitesimally small volume element in the shape of a rectangular block is written as $dV = dxdydz$.

7.3 Mass within the volume V

Assume that the volume V is filled with a material with a certain mass. Sometimes different fractions of materials can be distinguished, which usually physically interact with each other. In the following we consider one such fraction, that is not necessarily homogeneous when observed in detail (at a 'microscopic' scale). Observation at a microscopic scale, in the current context, means a scale that is

Figure 7.5

Cross section of a heterogeneous structure.

much smaller than the global dimensions of the considered volume. To support this visually one might imagine a kind of base material containing very small substructures, that can be visualized under a microscope (individual muscle cells in a skeletal muscle for example, see Fig. 7.5).

For the substructures (cells in the example) λ may be regarded as a relevant characteristic length scale for the size, as well as for the mutual stacking. The characteristic macroscopic length scale for the volume V is depicted by L. It is assumed that the inequality $L \gg \lambda$ is satisfied, meaning that on a macroscopic scale the heterogeneity at microscopic level is no longer recognizable. Attention is now focussed on physical properties, which are coupled to the material. As an illustration let us take the (mass) density. The local density ρ_ℓ in a spatial point with position vector $\vec{x}$ is defined by

$$\rho_\ell = \frac{dm}{dV} \quad \text{with:} \quad dV = dxdydz \rightarrow 0, \tag{7.4}$$

where dm is the mass of the considered fraction (the cells) in the volume dV. Ignoring some of the mathematical subtleties, a discontinuous field $\rho_\ell(\vec{x})$ results for the local density, with λ as the normative measure for the mutual distance of the discontinuities. On the microscopic level such a variation can be expected, but on the macroscopic level it is often (not) observable, (often) not interesting and not manageable.

Let us define the (locally averaged) density ρ in the spatial point $\vec{x}$ by

$$\rho = \frac{\Delta m}{\Delta V} \quad \text{with:} \quad \Delta V = \Delta x \Delta y \Delta z \quad \text{and:} \quad \Delta m = \int_{\Delta V} \rho_\ell \, dV. \tag{7.5}$$

In this case it is assumed that the dimensions of ΔV are much larger than λ and at the same time much smaller than L. Provided, that it is possible to specify such a subvolume ΔV this results in a continuous density field $\rho(\vec{x})$, from which the microscopic deviations have disappeared, however, which is still containing macroscopic variations: the real density field, according to Eq. (7.4), is **homogenized**. Continuum theory deals with physical properties that, in a way as described above, are made continuous. The results cannot be used on a microscopic level with a length scale λ, but on a much more global level.

In the above reasoning we talk about two scales: a macroscopic and a microscopic scale. However, in many applications a number of intermediate steps from large to small can be of interest. This is often seen in technical applications, but is especially of importance in biological materials. In such a case the characteristic measure λ of the components could suddenly become the relevant macroscopic length scale!

Example 7.1   Figure 7.6 illustrates some of the scales that can be distinguished in biomechanical modelling. Figure 7.6(a) could be a model to study a high-jumper. Typically, this kind of model would be used to study the coordination of muscles, describing the human body as a whole and examining how it moves. The macroscopic length

(a)   (b)

(c)   (d)

Figure 7.6

Examples of models at different length scales (a) model of a leg (b) model of a skeletal muscle (c) representative volume element of a cross section of a muscle (d) model of the cytoskeleton of a single cell.

scale L would be in the order of 1 [m]. In this case the user is often interested in stresses that are found in the bones and joints. The length scale λ at which these stresses and strains are studied is in the order of centimetres. Muscles are often described with one-dimensional fibre-like models as discussed in Chapter 4.

Figure 7.6(b) shows an MRI-image of the tibialis anterior of a rat and a schematic of the theoretical Finite Element Model that it was based on to describe its mechanical behaviour ([15]). This was a model designed to study in detail the force transmission through the tibialis anterior. The muscle was modelled as a continuum and the fibre lay-out was included, as well as a detailed description of the active and passive mechanical behaviour of the muscle cells. For this model, L is of the order of a few centimetres and λ of the order of 100 [μm].

Figure 7.6(c) shows a microscopic cross section of a skeletal muscle together with a so-called representative volume element, that is used for a microstructural model of a skeletal muscle. This model was used to study the transport of oxygen from the capillaries to the cells and how the oxygen is distributed over cell and intercellular space. Here each substructure: cells, intercellular spaces were modelled as a continuum. In this case $L \approx 50$ [μm] and $\lambda \approx 0.1$ [μm]. One could even go another step further in modelling the single cell as shown in Fig. 7.6(d). In this case the actin skeleton that is shown could be modelled as a fibre-like structure and the cytoplasm of the cell as a continuum, filled with a liquid or a solid ($L \approx 10$ [μm], $\lambda \approx 10$ [nm]).

7.4 Scalar fields

Let us consider the variation of a scalar physical property of the material (or a fraction of it) in the volume V. An example could be the temperature field: $T = T(\vec{x})$. Visualization of such a field in a three-dimensional configuration could be done by drawing (contour) planes with constant temperature. A clearer picture is obtained, when lines of constant temperature (isotherms or contour lines) are drawn on a flat two-dimensional surface of a cross section (see Fig. 7.7). By means of this figure estimates can be made of the partial derivatives of the temperature with respect to to x (with y and z constant) and y (with x and z constant). For example at the lower boundary the partial derivatives could be approximated by

$$\frac{\partial T}{\partial x} = 2\left[{}^{o}\text{C m}^{-1}\right], \ \frac{\partial T}{\partial y} = 0\left[{}^{o}\text{C m}^{-1}\right]. \tag{7.6}$$

In a three-dimensional scalar field the partial derivatives with respect to the coordinates are assembled in a column, i.e. the column with the components of the gradient vector associated with the scalar field. The gradient operator is specified

Figure 7.7

Isotherms (°C) in a cross section with constant z.

by $\vec{\nabla}$ when vector notation is used and with $\underset{\sim}{\nabla}$ when components are used. In the example of the temperature field this yields

$$\underset{\sim}{\nabla} T = \begin{bmatrix} \dfrac{\partial T}{\partial x} \\[2mm] \dfrac{\partial T}{\partial y} \\[2mm] \dfrac{\partial T}{\partial z} \end{bmatrix} \quad \text{with} \quad \underset{\sim}{\nabla} = \begin{bmatrix} \dfrac{\partial}{\partial x} \\[2mm] \dfrac{\partial}{\partial y} \\[2mm] \dfrac{\partial}{\partial z} \end{bmatrix}, \tag{7.7}$$

and also

$$\vec{\nabla} T = \vec{e}_x \frac{\partial T}{\partial x} + \vec{e}_y \frac{\partial T}{\partial y} + \vec{e}_z \frac{\partial T}{\partial z} \quad \text{with} \quad \vec{\nabla} = \vec{e}_x \frac{\partial}{\partial x} + \vec{e}_y \frac{\partial}{\partial y} + \vec{e}_z \frac{\partial}{\partial z}. \tag{7.8}$$

The gradient of a certain property is often a measure for the intensity of a physical transport phenomenon; the gradient of the temperature for example is directly related to the heat flux. Having the gradient of a certain property (T), the derivative of that property along a spatial curve (given in a parameter description, see Fig. 7.3), $\vec{x} = \vec{x}(\xi)$, can be determined (by using the chain rule for differentiation):

$$\frac{dT}{d\xi} = \frac{\partial T}{\partial x}\frac{dx}{d\xi} + \frac{\partial T}{\partial y}\frac{dy}{d\xi} + \frac{\partial T}{\partial z}\frac{dz}{d\xi}$$

$$= (\underset{\sim}{\nabla} T)^{\mathrm{T}} \frac{d\underset{\sim}{x}}{d\xi} = (\underset{\sim}{\nabla}^{\mathrm{T}} T) \frac{d\underset{\sim}{x}}{d\xi}$$

$$= \left(\frac{d\underset{\sim}{x}}{d\xi} \right)^{\mathrm{T}} \underset{\sim}{\nabla} T. \tag{7.9}$$

On the right-hand side of the equation the inner product of the (unnormalized) tangent vector to the curve and the gradient vector can be recognized. Eq. (7.9) can also be written as

$$\frac{dT}{d\xi} = \frac{d\vec{x}}{d\xi} \cdot \vec{\nabla}T. \tag{7.10}$$

Along a contour line (isotherm) in a two-dimensional cross section $dT/d\xi = 0$, so the gradient vector $\vec{\nabla}T$ (with components $\underset{\sim}{\nabla}T$) must be perpendicular to the contour line in each arbitrary point. In a three-dimensional configuration it can be said that the gradient vector of a certain scalar property in a point of V will be perpendicular to the contour plane through that point. The directional derivative dT/de of the temperature (the increase of the temperature per unit length along the curve) in the direction of the unit vector $\vec{e}$, with components $\underset{\sim}{e}$ can be written as

$$\frac{dT}{de} = \underset{\sim}{e}^{\mathrm{T}}\underset{\sim}{\nabla}T \quad \text{and also} \quad \frac{dT}{de} = \vec{e} \cdot \vec{\nabla}T. \tag{7.11}$$

7.5 Vector fields

In the previous section, as an example, a scalar temperature field was considered. Departing from a given temperature field, the associated (temperature) gradient field can be determined. This gradient field can be regarded as a vector field (in every point of the volume V the components of the associated vector can be determined). Considering the special meaning of such a gradient field, we will not use that field as a typical example in the current section. Instead, the (momentary) velocity field $\vec{v} = \vec{v}(\vec{x})$ of the material in the volume V will be used as an illustration. With respect to the Cartesian xyz-coordinate system the velocity vector $\vec{v}$ and the associated column $\underset{\sim}{v}$ can be written as

$$\vec{v} = v_x\vec{e}_x + v_y\vec{e}_y + v_z\vec{e}_z, \quad \underset{\sim}{v} = \begin{bmatrix} v_x \\ v_y \\ v_z \end{bmatrix}. \tag{7.12}$$

The possibilities for a clear graphical illustration of a vector field (velocity) in a three-dimensional configuration are limited. For a problem with a flat fluid flow pattern, for example:

$$\begin{aligned} v_x &= v_x(x, y) \\ v_y &= v_y(x, y) \\ v_z &= 0, \end{aligned} \tag{7.13}$$

a representation like the one given in Fig. 7.8 can be used. The arrows represent the velocity vector (magnitude and orientation) of the fluid at the tail of the arrow.

At a given velocity $\vec{v}$ in a certain point of V, the velocity component v in the direction of an arbitrary unit vector $\vec{e}$ (components in the column $\underset{\sim}{e}$) can be calculated with

Figure 7.8

Velocity field in a cross section with constant z.

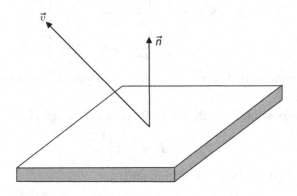

Figure 7.9

Flux of material through a small surface area.

$$v = v(\vec{e}) = \vec{v} \cdot \vec{e} \quad \text{and also} \quad v = v(\underset{\sim}{e}) = \underset{\sim}{v}^T \underset{\sim}{e}. \tag{7.14}$$

In Fig. 7.9 a small plane is drawn at a certain point in V. The plane has an outward (unit) normal $\vec{n}$. The local velocity of the material is specified with $\vec{v}$. Thus, the amount of mass per unit of surface and time flowing out of the plane in the figure is given by

$$\rho\, \vec{v} \cdot \vec{n} = \rho\, \underset{\sim}{v}^T \underset{\sim}{n}. \tag{7.15}$$

This is called the mass flux.

Also for vector fields the associated gradient can be determined. The gradient of a vector field results in a tensor field. For example, for the gradient of the velocity field we find

$$\vec{\nabla}\vec{v} = \left(\vec{e}_x\frac{\partial}{\partial x} + \vec{e}_y\frac{\partial}{\partial y} + \vec{e}_z\frac{\partial}{\partial z}\right)(v_x\vec{e}_x + v_y\vec{e}_y + v_z\vec{e}_z)$$

$$= \frac{\partial v_x}{\partial x}\vec{e}_x\vec{e}_x + \frac{\partial v_y}{\partial x}\vec{e}_x\vec{e}_y + \frac{\partial v_z}{\partial x}\vec{e}_x\vec{e}_z$$

$$+ \frac{\partial v_x}{\partial y}\vec{e}_y\vec{e}_x + \frac{\partial v_y}{\partial y}\vec{e}_y\vec{e}_y + \frac{\partial v_z}{\partial y}\vec{e}_y\vec{e}_z$$

$$+ \frac{\partial v_x}{\partial z}\vec{e}_z\vec{e}_x + \frac{\partial v_y}{\partial z}\vec{e}_z\vec{e}_y + \frac{\partial v_z}{\partial z}\vec{e}_z\vec{e}_z. \tag{7.16}$$

The associated matrix representation can be written as

$$\underset{\sim}{\nabla}\, \underset{\sim}{v}^{\mathrm{T}} = \begin{bmatrix} \dfrac{\partial v_x}{\partial x} & \dfrac{\partial v_y}{\partial x} & \dfrac{\partial v_z}{\partial x} \\[2ex] \dfrac{\partial v_x}{\partial y} & \dfrac{\partial v_y}{\partial y} & \dfrac{\partial v_z}{\partial y} \\[2ex] \dfrac{\partial v_x}{\partial z} & \dfrac{\partial v_y}{\partial z} & \dfrac{\partial v_z}{\partial z} \end{bmatrix}. \tag{7.17}$$

It should be noticed, that with the current notation an operation of the type $\underset{\sim}{\nabla}\underset{\sim}{v}$ is not allowed.

The transposed form of the tensor $\vec{\nabla}\vec{v}$ is called the velocity gradient tensor $\boldsymbol{L}$:

$$\boldsymbol{L} = \left(\vec{\nabla}\vec{v}\right)^{\mathrm{T}}. \tag{7.18}$$

The associated matrix representation reads:

$$\underline{L} = \left(\underset{\sim}{\nabla}\underset{\sim}{v}^{\mathrm{T}}\right)^{\mathrm{T}}. \tag{7.19}$$

Fig. 7.10 gives an illustration of the application of this tensor (matrix). Consider an infinitesimally small material line element $d\vec{x}$ (with components dx) in volume V at time t. If, at the tail of the vector $d\vec{x}$ the velocity is $\vec{v}$, then at the head of the vector $d\vec{x}$ the velocity will be: $\vec{v} + d\vec{x} \cdot \vec{\nabla}\vec{v} = \vec{v} + (\vec{\nabla}\vec{v})^{\mathrm{T}} \cdot d\vec{x} = \vec{v} + \boldsymbol{L} \cdot d\vec{x}$.

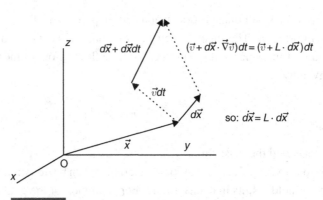

Figure 7.10

Change of a material line element $d\vec{x}$ after a time increment dt.

After an infinitesimally small increase in time dt the material line element $d\vec{x}$ will change into $d\vec{x} + \dot{d\vec{x}}dt$. By means of the figure it can be proven that:

$$\dot{d\vec{x}} = \boldsymbol{L} \cdot d\vec{x} \quad \text{and also} \quad \dot{d\underset{\sim}{x}} = \underline{L}\, d\underset{\sim}{x}, \tag{7.20}$$

and so the tensor $\boldsymbol{L}$ (with matrix representation $\underline{L}$) is a measure for the current change (per unit of time) of material line elements. This holds for the length as well as for the orientation.

The divergence (an operator that is often used for vector fields) of the velocity, div($\vec{v}$), defined as $\vec{\nabla} \cdot \vec{v}$ can be written as

$$\begin{aligned}
\text{div}(\vec{v}) &= \left(\vec{e}_x \frac{\partial}{\partial x} + \vec{e}_y \frac{\partial}{\partial y} + \vec{e}_z \frac{\partial}{\partial z} \right) \cdot (v_x \vec{e}_x + v_y \vec{e}_y + v_z \vec{e}_z) \\
&= \frac{\partial v_x}{\partial x} + \frac{\partial v_y}{\partial y} + \frac{\partial v_z}{\partial z} \\
&= \text{tr}(\boldsymbol{L}),
\end{aligned} \tag{7.21}$$

and also

$$\text{div}(\vec{v}) = \frac{\partial v_x}{\partial x} + \frac{\partial v_y}{\partial y} + \frac{\partial v_z}{\partial z} = \text{tr}(\underset{\sim}{\nabla}\,\underset{\sim}{v}^{\mathrm{T}}) = \underset{\sim}{\nabla}^{\mathrm{T}} \underset{\sim}{v} = \text{tr}(\underline{L}). \tag{7.22}$$

For the sake of completeness the following rather trivial result for the gradient $\vec{\nabla}$ applied to the position vector $\vec{x}$ is given:

$$\begin{aligned}
\vec{\nabla}\vec{x} &= \frac{\partial x}{\partial x}\vec{e}_x\vec{e}_x + \frac{\partial y}{\partial x}\vec{e}_x\vec{e}_y + \frac{\partial z}{\partial x}\vec{e}_x\vec{e}_z \\
&+ \frac{\partial x}{\partial y}\vec{e}_y\vec{e}_x + \frac{\partial y}{\partial y}\vec{e}_y\vec{e}_y + \frac{\partial z}{\partial y}\vec{e}_y\vec{e}_z \\
&+ \frac{\partial x}{\partial z}\vec{e}_z\vec{e}_x + \frac{\partial y}{\partial z}\vec{e}_z\vec{e}_y + \frac{\partial z}{\partial z}\vec{e}_z\vec{e}_z \\
&= \vec{e}_x\vec{e}_x + \vec{e}_y\vec{e}_y + \vec{e}_z\vec{e}_z = \boldsymbol{I},
\end{aligned} \tag{7.23}$$

and the according matrix formulation yields

$$\underset{\sim}{\nabla}\, \underset{\sim}{x}^{\mathrm{T}} = \begin{bmatrix} \dfrac{\partial x}{\partial x} & \dfrac{\partial y}{\partial x} & \dfrac{\partial z}{\partial x} \\[2mm] \dfrac{\partial x}{\partial y} & \dfrac{\partial y}{\partial y} & \dfrac{\partial z}{\partial y} \\[2mm] \dfrac{\partial x}{\partial z} & \dfrac{\partial y}{\partial z} & \dfrac{\partial z}{\partial z} \end{bmatrix} = \underline{I}. \tag{7.24}$$

7.6 Rigid body rotation

In the present section it is assumed that the mass in the rigid volume V rotates around a fixed axis in three-dimensional space. We focus our attention on the

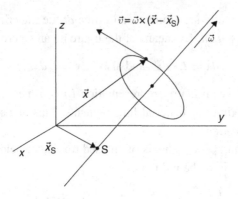

Figure 7.11

Rotation of material.

velocity field $\vec{v} = \vec{v}(\vec{x})$ and on the velocity gradient tensor $\boldsymbol{L}$ that can be derived from it. Figure 7.11 illustrates the considered problem. The axis of rotation is defined by means of a point S, fixed in space, with position vector $\vec{x}_S$ and the constant angular velocity vector $\vec{\omega}$. The associated columns with the components with respect to the Cartesian basis can be written as

$$
\underset{\sim}{x}_S = \begin{bmatrix} x_S \\ y_S \\ z_S \end{bmatrix} , \quad \underset{\sim}{\omega} = \begin{bmatrix} \omega_x \\ \omega_y \\ \omega_z \end{bmatrix} . \tag{7.25}
$$

The velocity vector $\vec{v}$ at a certain point with position vector $\vec{x}$ satisfies

$$
\vec{v} = \vec{v}(\vec{x}) = \vec{\omega} \times (\vec{x} - \vec{x}_S) . \tag{7.26}
$$

It can be derived that the components of the velocity vector $\vec{v}$ and the spatial coordinate vector $\vec{x}$, stored in the columns

$$
\underset{\sim}{v} = \begin{bmatrix} v_x \\ v_y \\ v_z \end{bmatrix} \quad \text{and} \quad \underset{\sim}{x} = \begin{bmatrix} x \\ y \\ z \end{bmatrix} , \tag{7.27}
$$

are related according to

$$\underset{\sim}{v} = \underset{\sim}{v}(\underset{\sim}{x}) = \underline{\Omega}(\underset{\sim}{x} - \underset{\sim}{x}_S) \quad \text{with} \quad \underline{\Omega} = \begin{bmatrix} 0 & -\omega_z & \omega_y \\ \omega_z & 0 & -\omega_x \\ -\omega_y & \omega_x & 0 \end{bmatrix}. \tag{7.28}$$

The spin matrix $\underline{\Omega}$ that is associated with the angular velocity vector $\vec{\omega}$ (with components $\underset{\sim}{\omega}$) is skew symmetric: $\underline{\Omega}^T = -\underline{\Omega}$. With the uniform rotation as a rigid body considered in this section, we find:

$$\underline{L} = \left(\underset{\sim}{\nabla} \underset{\sim}{v}^T\right)^T = \left(\left(\underset{\sim}{\nabla}(\underset{\sim}{x} - \underset{\sim}{x}_S)^T\right)\underline{\Omega}^T\right)^T = \left((\underline{I} + \underline{O})\underline{\Omega}^T\right)^T = \underline{\Omega} \tag{7.29}$$

and for the associated velocity gradient tensor $\boldsymbol{L}$:

$$\boldsymbol{L} = \omega_x(\vec{e}_z\vec{e}_y - \vec{e}_y\vec{e}_z) + \omega_y(\vec{e}_x\vec{e}_z - \vec{e}_z\vec{e}_x) + \omega_z(\vec{e}_y\vec{e}_x - \vec{e}_x\vec{e}_y). \tag{7.30}$$

7.7 Some mathematical preliminaries on second-order tensors

In the chapters that follow, extensive use will be made of second-order tensors. This section will summarize some of the mathematical background on this subject.

An arbitrary second-order tensor $\boldsymbol{M}$ can be written with respect to the Cartesian basis introduced earlier as

$$\begin{aligned} \boldsymbol{M} = &M_{xx}\vec{e}_x\vec{e}_x + M_{xy}\vec{e}_x\vec{e}_y + M_{xz}\vec{e}_x\vec{e}_z \\ &+ M_{yx}\vec{e}_y\vec{e}_x + M_{yy}\vec{e}_y\vec{e}_y + M_{yz}\vec{e}_y\vec{e}_z \\ &+ M_{zx}\vec{e}_z\vec{e}_x + M_{zy}\vec{e}_z\vec{e}_y + M_{zz}\vec{e}_z\vec{e}_z. \end{aligned} \tag{7.31}$$

The components of the tensor $\boldsymbol{M}$ are stored in the associated matrix $\underline{M}$ defined as

$$\underline{M} = \begin{bmatrix} M_{xx} & M_{xy} & M_{xz} \\ M_{yx} & M_{yy} & M_{yz} \\ M_{zx} & M_{zy} & M_{zz} \end{bmatrix}. \tag{7.32}$$

A tensor identifies a linear transformation. If the vector $\vec{b}$ is the result of the tensor $\boldsymbol{M}$ operating on vector $\vec{a}$, this is written as: $\vec{b} = \boldsymbol{M} \cdot \vec{a}$. In component form this leads to:

$$\begin{aligned} \vec{b} = &(M_{xx}\vec{e}_x\vec{e}_x + M_{xy}\vec{e}_x\vec{e}_y + M_{xz}\vec{e}_x\vec{e}_z \\ &+ M_{yx}\vec{e}_y\vec{e}_x + M_{yy}\vec{e}_y\vec{e}_y + M_{yz}\vec{e}_y\vec{e}_z \\ &+ M_{zx}\vec{e}_z\vec{e}_x + M_{zy}\vec{e}_z\vec{e}_y + M_{zz}\vec{e}_z\vec{e}_z) \cdot (a_x\vec{e}_x + a_y\vec{e}_y + a_z\vec{e}_z) \end{aligned}$$

$$= (M_{xx}a_x + M_{xy}a_y + M_{xz}a_z)\,\vec{e}_x$$
$$+ (M_{yx}a_x + M_{yy}a_y + M_{yz}a_z)\,\vec{e}_y$$
$$+ (M_{zx}a_x + M_{zy}a_y + M_{zz}a_z)\,\vec{e}_z$$

$$= b_x\vec{e}_x + b_y\vec{e}_y + b_z\vec{e}_z \ . \tag{7.33}$$

Using matrix notation we can write: $\underset{\sim}{b} = \underline{M}\,\underset{\sim}{a}$, in full:

$$
\begin{bmatrix} b_x \\ b_y \\ b_z \end{bmatrix}
=
\begin{bmatrix} M_{xx} & M_{xy} & M_{xz} \\ M_{yx} & M_{yy} & M_{yz} \\ M_{zx} & M_{zy} & M_{zz} \end{bmatrix}
\begin{bmatrix} a_x \\ a_y \\ a_z \end{bmatrix}
$$

$$
=
\begin{bmatrix} M_{xx}a_x + M_{xy}a_y + M_{xz}a_z \\ M_{yx}a_x + M_{yy}a_y + M_{yz}a_z \\ M_{zx}a_x + M_{zy}a_y + M_{zz}a_z \end{bmatrix} . \tag{7.34}
$$

Along with the earlier specified matrix $\underline{M}$ the transposed matrix $\underline{M}^{\mathrm{T}}$ is defined according to (taking a mirror image along the principal diagonal):

$$
\underline{M}^{\mathrm{T}} =
\begin{bmatrix} M_{xx} & M_{yx} & M_{zx} \\ M_{xy} & M_{yy} & M_{zy} \\ M_{xz} & M_{yz} & M_{zz} \end{bmatrix} . \tag{7.35}
$$

The tensor M^{T} is associated with the matrix $\underline{M}^{\mathrm{T}}$. Notice that

$$\underset{\sim}{b} = \underline{M}\,\underset{\sim}{a} \quad \text{is equivalent to} \quad \underset{\sim}{b}^{\mathrm{T}} = \underset{\sim}{a}^{\mathrm{T}}\underline{M}^{\mathrm{T}},$$
$$\vec{b} = M \cdot \vec{a} \quad \text{is equivalent to} \quad \vec{b} = \vec{a} \cdot M^{\mathrm{T}}.$$

The inverse of the tensor M is denoted by M^{-1}. By definition:

$$M \cdot M^{-1} = I, \tag{7.36}$$

with I the unit tensor, $I = \vec{e}_x\vec{e}_x + \vec{e}_y\vec{e}_y + \vec{e}_z\vec{e}_z$. The inverse of matrix $\underline{M}$ is denoted with $\underline{M}^{-1}$. By definition:

$$\underline{M}\,\underline{M}^{-1} = \underline{I} \ , \tag{7.37}$$

with $\underline{I}$ the unit matrix.

The trace of tensor M (associated matrix $\underline{M}$) is denoted as $\text{tr}(M) = \text{tr}(\underline{M})$ and given by

$$\text{tr}(M) = \text{tr}(\underline{M}) = M_{xx} + M_{yy} + M_{zz}. \tag{7.38}$$

For the determinant of the tensor M with matrix representation $\underline{M}$ it can be written:

$$\begin{aligned} \det(M) = \det(\underline{M}) = {} & M_{xx}(M_{yy}M_{zz} - M_{yz}M_{zy}) \\ & - M_{xy}(M_{yx}M_{zz} - M_{yz}M_{zx}) \\ & + M_{xz}(M_{yx}M_{zy} - M_{yy}M_{zx}). \end{aligned} \tag{7.39}$$

The deviatoric part of the tensor M is denoted by M^{d} and defined by

$$M^{\text{d}} = M - \frac{1}{3}\text{tr}(M)\,I. \tag{7.40}$$

In matrix notation this reads:

$$\underline{M}^{\text{d}} = \underline{M} - \frac{1}{3}\text{tr}(\underline{M})\,\underline{I}. \tag{7.41}$$

Let M be an arbitrary tensor. A non-zero vector $\vec{n}$ is said to be an **eigenvector** of M if a scalar λ exists such that

$$M \cdot \vec{n} = \lambda\vec{n} \qquad \text{or} \qquad (M - \lambda I) \cdot \vec{n} = \vec{0}. \tag{7.42}$$

A non-trivial solution $\vec{n}$ from Eq. (7.42) only exists if

$$\det(M - \lambda I) = 0. \tag{7.43}$$

Using the components of $\underline{M}$ and Eq. (7.39) will lead, after some elaboration, to the following equation:

$$\lambda^3 - I_1\lambda^2 + I_2\lambda - I_3 = 0, \tag{7.44}$$

with

$$\begin{aligned} I_1 &= \text{tr}(\underline{M}) \\ I_2 &= M_{xx}M_{yy} + M_{xx}M_{zz} + M_{yy}M_{zz} - M_{xy}^2 - M_{yz}^2 - M_{xz}^2 \\ I_3 &= \det(\underline{M}). \end{aligned} \tag{7.45}$$

Eq. (7.44) is called the **characteristic equation** and the scalar coefficients I_1, I_2 and I_3 are called the invariants of tensor M. In tensor form the invariants can be written as

$$I_1 = \text{tr}(\boldsymbol{M})$$

$$I_2 = \frac{1}{2}\left[(\text{tr}(\boldsymbol{M}))^2 - \text{tr}(\boldsymbol{M} \cdot \boldsymbol{M})\right]$$

$$I_3 = \det(\boldsymbol{M}).$$

$$(7.46)$$

Exercises

7.1   A drug is administered to the blood via an intravenous drug delivery system. In the figure a two-dimensional schematic is given of the blood vessel (in a fixed xy-coordinate system). The stationary concentration field $c(x,y)$ of the drug in the blood is sketched by means of iso-concentration lines. It may be assumed that the concentration is constant in the z-direction. The coordinates x and y are defined in the unit [mm], the concentration is in [mMol per litre].

Give an estimate of the gradient $\vec{\nabla}c$ of the concentration field in the point with coordinates $x = 20$ [mm], $y = 5$ [mm].

7.2   A cube ABCDEFGH (with edge a) rotates with angular velocity ω around the straight line ℓ, coinciding with the body diagonal CE. The points C and E, and consequently the rotation axis ℓ, have a fixed position in the xyz-coordinate system.

At a certain time the edges AB, AD and AE are exactly oriented in the directions of the x-, y- and z-axes, respectively (see figure). Determine for that time instant the velocity vector $\underset{\sim}{v}_B$ of the point B.

7.3 Bone mineral density is often measured, because this quantity can be related to the strength and stiffness of bone. The measurement can also be used as a diagnostic tool for osteoporosis. Consider a rectangular piece of bone in the xy-plane of a Cartesian xyz-coordinate system, with varying density $\rho = \rho(x, y)$. For the corner points ABCD of the specimen the density is given (expressed in [kg dm^{-3}]), see figure. Prove, that it is impossible for the gradient $\vec{\nabla}\rho$ of the density field in the figure to be constant.

7.4 Consider a domain in the form of a cube in three-dimensional space, given by $-\ell \leq x \leq 3\ell$, $-2\ell \leq y \leq 2\ell$ and $-2\ell \leq z \leq 2\ell$ with ℓ a constant and x, y and z the Cartesian coordinates. Within this domain a temperature field exists with known gradient $\vec{\nabla}T$ which is given by

$$\vec{\nabla}T = \theta\vec{e}_y \quad \text{with } \theta \text{ a constant.}$$

In the domain a curve is given. The points on the curve satisfy

$$(x - \ell)^2 + y^2 = \ell^2 \quad \text{and} \quad z = \ell .$$

Calculate the coordinates of the point (or points) on the curve where the derivative of the temperature in the direction of the curve is equal to zero.

7.5 In a two-dimensional xy-coordinate system (mutually perpendicular unit vectors $\vec{e}_x$ and $\vec{e}_y$ along the axes) a two-dimensional stationary fluid flow is considered. The fluid flow is caused by a fluid source in the origin. An arbitrary point in the coordinate system is given by the vector $\vec{x} = x\vec{e}_x + y\vec{e}_y$. At some distance from the origin (with $|\vec{x}| > \lambda$) the velocity field $\vec{v}(\vec{x})$ can be written as

$$\vec{v}(\vec{x}) = \alpha\frac{\vec{x}}{|\vec{x}|^2} .$$

The location of a point P is defined by the position vector $\vec{x}_P = \ell(\vec{e}_x + \vec{e}_y)$ with $\ell > \lambda$. Determine the velocity gradient tensor $\mathbf{L}$ in point P.

8 Stress in three-dimensional continuous media

In this chapter the concepts introduced in Chapter 6 for a one-dimensional continuous system are generalized to two-dimensional configurations. Extension to three-dimensional problems is briefly discussed. First the equilibrium conditions in a two- or three-dimensional body are derived from force equilibrium of an infinitesimally small volume element. Thereafter, the concept of a stress tensor, as a sum of dyads, is introduced to compute the stress vector acting on an arbitrary surface in a material point of the body.

8.1 Stress vector

Before examining the equilibrium conditions in a two-dimensional body, the concept of a stress vector is introduced. For this purpose we consider an infinitesimally small surface element having area ΔA, see Fig. 8.1.

On this surface an infinitesimally small force vector $\Delta \vec{F}$ is applied with components in the x-, y- and z-direction: $\Delta \vec{F} = \Delta F_x \vec{e}_x + \Delta F_y \vec{e}_y + \Delta F_z \vec{e}_z$. Following the definition of stress, Eq. (6.8), three stresses may be defined:

$$s_x = \lim_{\Delta A \to 0} \frac{\Delta F_x}{\Delta A},\tag{8.1}$$

that acts in the x-direction, and

$$s_y = \lim_{\Delta A \to 0} \frac{\Delta F_y}{\Delta A},\tag{8.2}$$

that acts in the y-direction, and

$$s_z = \lim_{\Delta A \to 0} \frac{\Delta F_z}{\Delta A},\tag{8.3}$$

that acts in the z-direction. Hence, a **stress vector** may be defined:

$$\vec{s} = s_x \vec{e}_x + s_y \vec{e}_y + s_z \vec{e}_z.\tag{8.4}$$

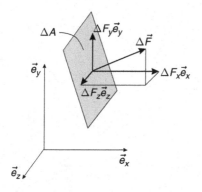

Figure 8.1

Force $\Delta \vec{F}$ on an infinitesimal surface element having area ΔA.

8.2 From stress to force

Suppose, that a free body diagram is created by means of an imaginary cutting plane through a body. The cutting plane is chosen in such a way that it coincides with the xy-plane (Fig. 8.2). On the imaginary cutting plane a stress vector $\vec{s}$ is given as a function of x and y. How can the total force vector acting on that plane be computed based on this stress vector? The complete answer to this question is somewhat beyond the scope of this course because it requires the integration of a multi-variable function. However, the more simple case where the stress vector is a function x (or y) only, while it acts on a rectangular plane at constant z is more easy to answer and sufficiently general to be useful in the remainder of this chapter. Therefore, suppose that the stress vector is a function of x such that it may be written as

$$\vec{s} = s_x(x)\,\vec{e}_x + s_y(x)\,\vec{e}_y + s_z(x)\,\vec{e}_z. \tag{8.5}$$

Let this stress vector act on a plane $z = 0$, that spans the range $0 < x < L$ and that has a width h in the y-direction. The resulting force vector on the plane considered due to this stress vector is denoted by $\vec{F} = F_x\vec{e}_x + F_y\vec{e}_y + F_z\vec{e}_z$. If s_x, s_y and s_z are **constant**, the net force is simply computed by multiplication of the stress vector components with the surface area hL in this case:

$$\vec{F} = s_x hL\vec{e}_x + s_y hL\vec{e}_y + s_y hL\vec{e}_z. \tag{8.6}$$

For **non-constant** stress vector components (e.g. as visualized in Fig. 8.3), the force components in the x-, y- and z-direction due to the stress vector $\vec{s}$ are obtained via integration of these components over the domain in the x-direction and multiplication with the width h of the plane (which is allowed because the stress components are constant in the y-direction):

Figure 8.2

The stress vector $\vec{s}$ working on a cut section of a body.

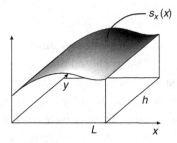

Figure 8.3

A stress (s_x) distribution where s_x is a function of x only.

$$F_x = h \int_0^L s_x(x) \ dx, \tag{8.7}$$

$$F_y = h \int_0^L s_y(x) \ dx, \tag{8.8}$$

$$F_z = h \int_0^L s_z(x) \ dx, \tag{8.9}$$

respectively. In conclusion, the force vector is found by integration of the stress vector over the plane on which it acts.

8.3 Equilibrium

In Chapter 6 the equilibrium equation for the one-dimensional case has been formulated by demanding the balance of forces of an isolated thin slice of a continuous body. In analogy with this, we consider a continuous, arbitrary material

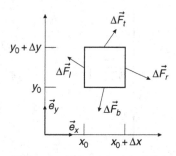

(a) An infinitesimal volume element in a continuous body

(b) Free body diagram of the infinitesimal volume element with the lower left corner located at $x = x_0$ and $y = y_0$

Figure 8.4

Free body diagram of an infinitesimal volume element of a continuous body.

volume, in particular a cross section in the xy-plane as depicted in Fig. 8.4(a). In the direction perpendicular to the drawing, hence in the $\vec{e}_z$ direction, the body has a thickness h. It is assumed that all stress components are constant across the thickness. The free body diagram of an infinitesimal volume element, which in this case is a rectangular prism having dimension $\Delta x \times \Delta y \times h$, the cross section of which is also shown in Fig. 8.4(a), is examined. The lower left corner of the rectangle as depicted in Fig. 8.4(b) is located at $x = x_0$ and $y = y_0$. Forces have been introduced on all faces of the prism in the xz- or yz-plane, see Fig. 8.4(b). These forces are a consequence of the interaction of the prism with its surroundings.

The force vectors $\Delta \vec{F}_i$, $i = l, b, t, r$, represent forces that are acting on the left, bottom, top and right face of the prism. These forces are infinitesimally small because they act on an infinitesimally small surface element (the faces of the prism) that experiences only a small part of the total force that is exerted on the body. Moreover, it is assumed that the infinitesimal volume element is sufficiently small, such that the individual force vectors can be transformed to stress vectors in the usual way, see Fig. 8.5.

Each of the stress vectors may be additively decomposed into a component acting in the $\vec{e}_x$-direction and a component acting in the $\vec{e}_y$-direction. These components are sketched in Fig. 8.6. The double subscript notation is interpreted as follows: the second subscript indicates the direction of the normal to the plane or face on which the stress component acts. The first subscript relates to the direction of the stress itself. There is also a sign convention. When both the outer normal and the stress component are oriented in positive direction relative to the coordinate axes, the stress is positive also, which is required because of Newton's law of

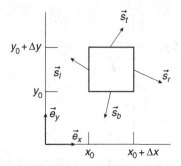

Figure 8.5

Free body diagram of the infinitesimal volume element.

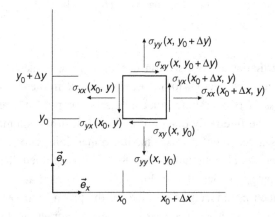

Figure 8.6

Stresses acting on the surfaces of an infinitesimal volume element.

action and reaction. When both the outer normal and the stress component are oriented in a negative direction relative to the coordinate axes, the stress is positive. When the normal points in a positive direction while the stress points in a negative direction (or vice versa), the stress is negative.

Clearly, all stress components are a function of the position in space, hence of the x- and y-coordinate (but not of the z-coordinate because all quantities are assumed constant in the z-direction):

$$\sigma_{xx} = \sigma_{xx}(x, y)$$
$$\sigma_{yy} = \sigma_{yy}(x, y)$$
$$\sigma_{xy} = \sigma_{xy}(x, y)$$
$$\sigma_{yx} = \sigma_{yx}(x, y). \tag{8.10}$$

Using the notation from Fig. 8.6 it holds that

$$\vec{s}_l = -\sigma_{xx}(x_0, y)\,\vec{e}_x - \sigma_{yx}(x_0, y)\,\vec{e}_y$$
$$\vec{s}_b = -\sigma_{xy}(x, y_0)\,\vec{e}_x - \sigma_{yy}(x, y_0)\,\vec{e}_y$$
$$\vec{s}_t = \sigma_{xy}(x, y_0 + \Delta y)\,\vec{e}_x + \sigma_{yy}(x, y_0 + \Delta y)\,\vec{e}_y \qquad (8.11)$$
$$\vec{s}_r = \sigma_{xx}(x_0 + \Delta x, y)\,\vec{e}_x + \sigma_{yx}(x_0 + \Delta x, y)\,\vec{e}_y.$$

Notice that the stress components of the vector $\vec{s}_l$ that act on the left face are taken at a constant value of x ($x = x_0$), but that these stress components may be a function of y. Likewise, the stress components on the right face are also taken at a constant x-coordinate ($x = x_0 + \Delta x$) and are also a function of y, while the stress components on the bottom and top faces are both a function of x but are taken at $y = y_0$ and $y = y_0 + \Delta y$, respectively.

Force equilibrium of the infinitesimal volume element depicted in Fig. 8.6 is established next. The force components acting on the faces of the prism have been identified in Fig. 8.7.

The net force in the negative x-direction acting on the left face is denoted by ΔF_{lx} and is obtained by integration of the stress σ_{xx} at constant value of x (i.e. $x = x_0$), from y_0 to $y_0 + \Delta y$:

$$\Delta F_{lx} = h \int_{y_0}^{y_0 + \Delta y} \sigma_{xx}(x_0, y)\ dy, \qquad (8.12)$$

Figure 8.7

Forces acting on surfaces of infinitesimal volume element.

where h is the thickness of the cube. The stress field $\sigma_{xx}(x_0, y)$ may be approximated by

$$\sigma_{xx}(x_0, y) \approx \sigma_{xx}(x_0, y_0) + \frac{\partial \sigma_{xx}}{\partial y}\Big|_{x=x_0,\, y=y_0} (y - y_0), \qquad (8.13)$$

which is allowed if Δy is sufficiently small. Notice that the partial derivative of σ_{xx} in the y-direction is taken at a **fixed** value of x and y, i.e. at $x = x_0$ and $y = y_0$ (for readability reasons, in the following the addition $|_{x=x_0,y=y_0}$ to the partial derivatives is omitted from the equations). Consequently, the force ΔF_{lx} can be written as

$$\Delta F_{lx} = h \int_{y_0}^{y_0+\Delta y} \left(\sigma_{xx}(x_0, y_0) + \frac{\partial \sigma_{xx}}{\partial y}(y - y_0) \right) dy. \qquad (8.14)$$

To compute this integral it must be realized that $\sigma_{xx}(x_0, y_0)$, $\partial \sigma_{xx}/\partial y$ and y_0 denote quantities that are not a function of the integration parameter y. Bearing this in mind, it is straightforward to show that

$$\Delta F_{lx} = \sigma_{xx}(x_0, y_0)\, h\, \Delta y + \frac{\partial \sigma_{xx}}{\partial y}\, h\, \frac{\Delta y^2}{2}. \qquad (8.15)$$

The stress on the right face of the cube, $\sigma_{xx}(x_0 + \Delta x, y)$ may be integrated to give the force in the (positive) x-direction on this face:

$$\Delta F_{rx} = h \int_{y_0}^{y_0+\Delta y} \sigma_{xx}(x_0 + \Delta x, y)\, dy. \qquad (8.16)$$

Clearly, the stress component σ_{xx} on the right face, hence at constant $x_0 + \Delta x$, may be approximated by

$$\sigma_{xx}(x_0 + \Delta x, y) \approx \sigma_{xx}(x_0, y_0) + \frac{\partial \sigma_{xx}}{\partial x}\Delta x + \frac{\partial \sigma_{xx}}{\partial y}(y - y_0). \qquad (8.17)$$

Therefore the force ΔF_{rx} can be computed as

$$\Delta F_{rx} = \sigma_{xx}(x_0, y_0)\, h\Delta y + \frac{\partial \sigma_{xx}}{\partial x}h\Delta x\Delta y + \frac{\partial \sigma_{xx}}{\partial y}\, h\, \frac{\Delta y^2}{2}. \qquad (8.18)$$

A similar exercise can be performed with respect to the forces in the x-direction on the top and bottom faces, giving

$$\Delta F_{tx} = \sigma_{xy}(x_0, y_0)\, h\Delta x + \frac{\partial \sigma_{xy}}{\partial y}h\Delta x\Delta y + \frac{\partial \sigma_{xy}}{\partial x}\, h\, \frac{\Delta x^2}{2}, \qquad (8.19)$$

while

$$\Delta F_{bx} = \sigma_{xy}(x_0, y_0)\, h\Delta x + \frac{\partial \sigma_{xy}}{\partial x}\, h\, \frac{\Delta x^2}{2}. \qquad (8.20)$$

Force equilibrium in the x-direction now yields

$$- \Delta F_{lx} - \Delta F_{bx} + \Delta F_{rx} + \Delta F_{tx} = 0. \tag{8.21}$$

Use of the above results for the force components gives

$$- \sigma_{xx}(x_0, y_0) \, h\Delta y - \frac{\partial \sigma_{xx}}{\partial y} \frac{h}{2} \Delta y^2 - \sigma_{xy}(x_0, y_0) \, h\Delta x$$

$$- \frac{\partial \sigma_{xy}}{\partial x} \frac{h}{2} \Delta x^2 + \sigma_{xx}(x_0, y_0) \, h\Delta y + \frac{\partial \sigma_{xx}}{\partial x} h\Delta x\Delta y + \frac{\partial \sigma_{xx}}{\partial y} \frac{h}{2} \Delta y^2$$

$$+ \sigma_{xy}(x_0, y_0) \, h\Delta x + \frac{\partial \sigma_{xy}}{\partial y} h\Delta x\Delta y + \frac{\partial \sigma_{xy}}{\partial x} \frac{h}{2} \Delta x^2 = 0, \tag{8.22}$$

which implies that

$$\frac{\partial \sigma_{xx}}{\partial x} + \frac{\partial \sigma_{xy}}{\partial y} = 0. \tag{8.23}$$

This should hold for any position in space of the prism, hence for all values of x and y. Performing a similar exercise in the y-direction yields the full set of partial differential equations, known as the local equilibrium equations:

$$\frac{\partial \sigma_{xx}}{\partial x} + \frac{\partial \sigma_{xy}}{\partial y} = 0 \tag{8.24}$$

$$\frac{\partial \sigma_{yx}}{\partial x} + \frac{\partial \sigma_{yy}}{\partial y} = 0. \tag{8.25}$$

Two different shear stresses are present: σ_{xy} and σ_{yx}. However, by using **equilibrium of moment** it can be proven that

$$\sigma_{xy} = \sigma_{yx}. \tag{8.26}$$

This result is revealed by considering the sum of moments with respect to the midpoint of the infinitesimal cube of Fig. 8.7. With respect to the midpoint, the moments due to the normal forces, ΔF_{lx}, ΔF_{rx}, ΔF_{ty} and ΔF_{by} are equal to zero, while the shear forces generate a moment, hence enforcing the sum of moments with respect to the midpoint to be equal to zero gives

$$- \frac{\Delta x}{2}(\Delta F_{ly} + \Delta F_{ry}) + \frac{\Delta y}{2}(\Delta F_{tx} + \Delta F_{bx}) = 0. \tag{8.27}$$

Using the above results for the force components it follows that

$$-\frac{\Delta x}{2}\left(\sigma_{yx}(x_0,y_0)\,h\Delta y + \frac{\partial \sigma_{yx}}{\partial y}\,h\,\frac{\Delta y^2}{2} + \sigma_{yx}(x_0,y_0)\,h\Delta y + \frac{\partial \sigma_{yx}}{\partial y}\,h\,\frac{\Delta y^2}{2}\right.$$

$$\left.+\frac{\partial \sigma_{yx}}{\partial x}h\Delta x\Delta y\right) = \frac{\Delta y}{2}\left(\sigma_{xy}(x_0,y_0)\,h\Delta x + \frac{\partial \sigma_{xy}}{\partial x}\,h\,\frac{\Delta x^2}{2} + \sigma_{xy}(x_0,y_0)\,h\Delta x\right.$$

$$\left.+\frac{\partial \sigma_{xy}}{\partial y}\,h\,\frac{\Delta x^2}{2} + \frac{\partial \sigma_{xy}}{\partial x}h\Delta x\Delta y\right) = 0. \tag{8.28}$$

Neglecting terms of order Δx^3, Δy^3 etc., reveals immediately that

$$\sigma_{yx} = \sigma_{xy}. \tag{8.29}$$

Based on this result, the equilibrium equations Eqs. (8.24) and (8.25) may be rewritten as

$$\frac{\partial \sigma_{xx}}{\partial x} + \frac{\partial \sigma_{xy}}{\partial y} = 0 \tag{8.30}$$

$$\frac{\partial \sigma_{xy}}{\partial x} + \frac{\partial \sigma_{yy}}{\partial y} = 0. \tag{8.31}$$

Note, that strictly speaking the resulting forces on the faces of the prism as visualized in Fig. 8.7 are not exactly located in the midpoints of the faces. This should be accounted for in the equilibrium of moment. This would complicate the derivations considerably, but it would lead to the same conclusions.

In the three-dimensional case a number of additional stress components is present, see Fig. 8.8. In total there are six independent stress components: σ_{xx}, σ_{xy}, σ_{xz}, σ_{yy}, σ_{yz} and σ_{zz}. As due to moment equilibrium it can be derived that

$$\sigma_{yx} = \sigma_{xy}, \qquad \sigma_{zx} = \sigma_{xz}, \qquad \sigma_{zy} = \sigma_{yz}. \tag{8.32}$$

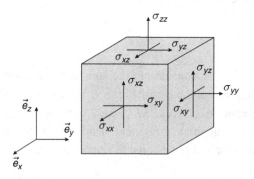

Figure 8.8

Stresses in three dimensions.

The equilibrium equations Eqs. (8.30 and (8.31) can be generalized to three dimensions as

$$\frac{\partial \sigma_{xx}}{\partial x} + \frac{\partial \sigma_{xy}}{\partial y} + \frac{\partial \sigma_{xz}}{\partial z} = 0 \tag{8.33}$$

$$\frac{\partial \sigma_{xy}}{\partial x} + \frac{\partial \sigma_{yy}}{\partial y} + \frac{\partial \sigma_{yz}}{\partial z} = 0 \tag{8.34}$$

$$\frac{\partial \sigma_{xz}}{\partial x} + \frac{\partial \sigma_{yz}}{\partial y} + \frac{\partial \sigma_{zz}}{\partial z} = 0. \tag{8.35}$$

Some interpretation of the equilibrium equations is given by considering a number of special cases.

Example 8.1 If all the individual partial derivatives appearing in Eqs. (8.30) and (8.31) are zero, that is if

$$\frac{\partial \sigma_{xx}}{\partial x} = 0, \quad \frac{\partial \sigma_{xy}}{\partial y} = 0, \quad \frac{\partial \sigma_{yy}}{\partial y} = 0, \quad \frac{\partial \sigma_{xy}}{\partial x} = 0,$$

the stresses on the faces of the cube are shown in Fig. 8.9. Clearly the forces on opposing faces are in equilibrium, as demanded by the equilibrium equations if the individual partial derivatives are zero.

Example 8.2 In the absence of shear stresses ($\sigma_{xy} = 0$), or if the shear stresses are constant ($\sigma_{xy} = c$) it follows that

$$\frac{\partial \sigma_{xy}}{\partial x} = \frac{\partial \sigma_{xy}}{\partial y} = 0.$$

Hence the equilibrium equations reduce to:

$$\frac{\partial \sigma_{xx}}{\partial x} = 0, \quad \frac{\partial \sigma_{yy}}{\partial y} = 0.$$

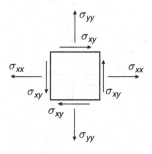

Figure 8.9

Stresses on the faces.

This means that σ_{xx} may only be a function of $y : \sigma_{xx} = \sigma_{xx}(y)$. Likewise, σ_{yy} may only be a function of x.

8.4   Stress tensor

Suppose that, in a two-dimensional configuration, all three stress components (σ_{xx}, σ_{xy} and σ_{yy}) are known. How can the resulting stress vector acting on an arbitrary cross section of a body be computed based on these stress components? To answer this question, consider an arbitrary, but infinitesimally small prism, having a triangular cross section as depicted in Fig. 8.10(a). Two faces of the prism are parallel to $\vec{e}_x$ and $\vec{e}_y$, respectively, while the third face, having length Δl, is oriented at some angle α with respect to $\vec{e}_x$. The orientation in space of this face is fully characterized by the unit outward normal vector $\vec{n}$, related to α by

$$\vec{n} = n_x\vec{e}_x + n_y\vec{e}_y = \sin(\alpha)\vec{e}_x + \cos(\alpha)\vec{e}_y. \tag{8.36}$$

The face parallel to $\vec{e}_x$ has length $h_x = n_y \Delta l$, while the face parallel to $\vec{e}_y$ has length $h_y = n_x \Delta l$. This follows immediately from

$$n_x = \sin(\alpha) = \frac{h_y}{\Delta l} \rightarrow h_y = n_x\Delta l, \tag{8.37}$$

while

$$n_y = \cos(\alpha) = \frac{h_x}{\Delta l} \rightarrow h_x = n_y\Delta l. \tag{8.38}$$

The stresses acting on the left and bottom face of the triangular prism are depicted in Fig. 8.10(b). On the inclined face a stress vector $\vec{s}$ is introduced. Force equilibrium in the x-direction yields

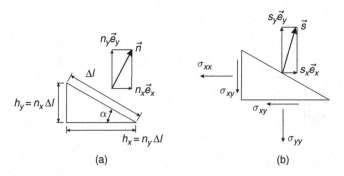

(a)   (b)

Figure 8.10

Stress vector $\vec{s}$ acting on an inclined plane with normal $\vec{n}$.

$$s_x(h\,\Delta l)\,\vec{e}_x = \sigma_{xx}(h\,\Delta l\,n_x)\,\vec{e}_x + \sigma_{xy}(h\,\Delta l\,n_y)\,\vec{e}_x, \tag{8.39}$$

where h denotes the thickness of the prism in the z-direction. Dividing by the area $h\Delta l$ yields

$$s_x = \sigma_{xx}n_x + \sigma_{xy}n_y. \tag{8.40}$$

A similar exercise in the y-direction gives

$$s_y(h\,\Delta l)\,\vec{e}_y = \sigma_{yy}(h\,\Delta l\,n_y)\,\vec{e}_y + \sigma_{xy}(h\,\Delta l\,n_x)\,\vec{e}_y, \tag{8.41}$$

hence dividing by $h\Delta l$ yields

$$s_y = \sigma_{yy}n_y + \sigma_{xy}n_x. \tag{8.42}$$

So, the stress vector $\vec{s}$ is directly related to the stress components σ_{xx}, σ_{yy} and σ_{xy} via the normal $\vec{n}$ to the infinitesimal surface element at which $\vec{s}$ acts. This can also be written in a compact form by introducing the so-called **stress tensor** σ. Let this tensor be defined according to

$$\sigma = \sigma_{xx}\vec{e}_x\vec{e}_x + \sigma_{yy}\vec{e}_y\vec{e}_y + \sigma_{xy}(\vec{e}_x\vec{e}_y + \vec{e}_y\vec{e}_x). \tag{8.43}$$

In the two-dimensional case, the stress tensor σ is the sum of four dyads. The components of the stress tensor σ can be assembled in the stress matrix $\underline{\sigma}$ according to

$$\underline{\sigma} = \begin{bmatrix} \sigma_{xx} & \sigma_{xy} \\ \sigma_{yx} & \sigma_{yy} \end{bmatrix}, \tag{8.44}$$

where σ_{yx} equals σ_{xy}, see Eq. (8.29). This stress tensor σ has been constructed such that the stress vector $\vec{s}$ (with components $\underline{s}$, acting on an infinitesimal surface element with outward unit normal $\vec{n}$ (with components $\underline{n}$) may be computed via

$$\vec{s} = \sigma \cdot \vec{n}. \tag{8.45}$$

This follows immediately from

$$\begin{aligned} \sigma \cdot \vec{n} &= (\sigma_{xx}\vec{e}_x\vec{e}_x + \sigma_{yy}\vec{e}_y\vec{e}_y + \sigma_{xy}(\vec{e}_x\vec{e}_y + \vec{e}_y\vec{e}_x)) \cdot \vec{n} \\ &= \sigma_{xx}\vec{e}_x\vec{e}_x \cdot \vec{n} + \sigma_{yy}\vec{e}_y\vec{e}_y \cdot \vec{n} + \sigma_{xy}(\vec{e}_x\vec{e}_y \cdot \vec{n} + \vec{e}_y\vec{e}_x \cdot \vec{n}) \\ &= \sigma_{xx}n_x\vec{e}_x + \sigma_{yy}n_y\vec{e}_y + \sigma_{xy}(n_y\vec{e}_x + n_x\vec{e}_y) \\ &= (\sigma_{xx}n_x + \sigma_{xy}n_y)\,\vec{e}_x + (\sigma_{yy}n_y + \sigma_{xy}n_x)\,\vec{e}_y. \end{aligned} \tag{8.46}$$

Hence, it follows immediately that, with $\vec{s} = s_x\vec{e}_x + s_y\vec{e}_y$:

$$s_x = \sigma_{xx}n_x + \sigma_{xy}n_y \tag{8.47}$$

and

$$s_y = \sigma_{xy}n_x + \sigma_{yy}n_y. \tag{8.48}$$

Figure 8.11

Stress vector $\vec{s}$ acting on an inclined plane with normal $\vec{n}$, decomposed into a stress vector normal and stress vector tangent to the plane.

Eq. (8.45) can also be written in column notation as

$$\underset{\sim}{s} = \underset{\approx}{\sigma}\, \underset{\sim}{n}. \tag{8.49}$$

The purpose of introducing the stress tensor $\boldsymbol{\sigma}$, defined as the sum of four dyads, is to compute the stress vector that acts on an infinitesimally small area that is oriented in space as defined by the normal $\vec{n}$. For any given normal $\vec{n}$ this stress vector is computed via $\vec{s} = \boldsymbol{\sigma} \cdot \vec{n}$.

At any point in a body and for any plane in that point this stress vector can be computed. This stress vector itself may be decomposed into a stress vector normal to the plane (normal stress) and a vector tangent to the plane (shear stress), see Fig. 8.11. Hence let

$$\vec{s} = \boldsymbol{\sigma} \cdot \vec{n}, \tag{8.50}$$

then the stress vector normal to the plane, $\vec{s}_n$ follows from

$$\vec{s}_n = (\vec{s} \cdot \vec{n})\vec{n}$$
$$= ((\boldsymbol{\sigma} \cdot \vec{n}) \cdot \vec{n})\vec{n} = \boldsymbol{\sigma} \cdot \vec{n} \cdot (\vec{n}\vec{n}). \tag{8.51}$$

The stress vector tangent to the plane is easily obtained via

$$\vec{s} = \vec{s}_n + \vec{s}_t, \tag{8.52}$$

hence

$$\vec{s}_t = \vec{s} - \vec{s}_n$$
$$= \boldsymbol{\sigma} \cdot \vec{n} - ((\boldsymbol{\sigma} \cdot \vec{n}) \cdot \vec{n})\vec{n}$$
$$= \boldsymbol{\sigma} \cdot \vec{n} \cdot (\boldsymbol{I} - \vec{n}\vec{n}). \tag{8.53}$$

Example 8.3   If the stress state is specified as depicted in Fig. 8.12, then

$$\boldsymbol{\sigma} = 10\vec{e}_x\vec{e}_x + 3(\vec{e}_x\vec{e}_y + \vec{e}_y\vec{e}_x).$$

Figure 8.12

Stress components.

If the normal to the plane of interest equals

$$\vec{n} = \vec{e}_x,$$

then, the stress vector on this plane follows from

$$\vec{s} = \boldsymbol{\sigma} \cdot \vec{n}$$
$$= (10\vec{e}_x\vec{e}_x + 3(\vec{e}_x\vec{e}_y + \vec{e}_y\vec{e}_x)) \cdot \vec{e}_x$$
$$= 10\vec{e}_x + 3\vec{e}_y.$$

Clearly, as expected, the operation $\boldsymbol{\sigma} \cdot \vec{e}_x$ extracts the stress components acting on the right face of the rectangle shown in Fig. 8.12. The stress vector $\vec{s}$ may be decomposed into a component normal to this face and a component tangent to this face. Clearly, the normal component should be $\vec{s}_n = 10\vec{e}_x$, while the tangent component should be $\vec{s}_t = 3\vec{e}_y$. This also follows from

$$\vec{s}_n = ((\boldsymbol{\sigma} \cdot \vec{n}) \cdot \vec{n})\vec{n}$$
$$= ((10\vec{e}_x + 3\vec{e}_y) \cdot \vec{e}_x)\vec{e}_x$$
$$= 10\vec{e}_x.$$

Generalization to three dimensions As noted before, there are six independent stress components in the three-dimensional case. These may be stored in the three-dimensional stress tensor using the sum of nine dyads:

$$\boldsymbol{\sigma} = \sigma_{xx}\vec{e}_x\vec{e}_x + \sigma_{yy}\vec{e}_y\vec{e}_y + \sigma_{zz}\vec{e}_z\vec{e}_z$$
$$+ \sigma_{xy}(\vec{e}_x\vec{e}_y + \vec{e}_y\vec{e}_x) + \sigma_{xz}(\vec{e}_x\vec{e}_z + \vec{e}_z\vec{e}_x) + \sigma_{yz}(\vec{e}_y\vec{e}_z + \vec{e}_z\vec{e}_y),$$

$$(8.54)$$

and also in the symmetric stress matrix:

$$\underline{\sigma} = \begin{bmatrix} \sigma_{xx} & \sigma_{xy} & \sigma_{xz} \\ \sigma_{yx} & \sigma_{yy} & \sigma_{yz} \\ \sigma_{zx} & \sigma_{zy} & \sigma_{zz} \end{bmatrix}. \qquad (8.55)$$

In this case the normal on a plane has three components: $\vec{n} = n_x\vec{e}_x + n_y\vec{e}_y + n_z\vec{e}_z$, therefore

$$
\begin{aligned}
\vec{s} = \boldsymbol{\sigma} \cdot \vec{n} \\
= \big[\sigma_{xx}\vec{e}_x\vec{e}_x + \sigma_{yy}\vec{e}_y\vec{e}_y + \sigma_{zz}\vec{e}_z\vec{e}_z \\
+ \sigma_{xy}(\vec{e}_x\vec{e}_y + \vec{e}_y\vec{e}_x) + \sigma_{xz}(\vec{e}_x\vec{e}_z + \vec{e}_z\vec{e}_x) \\
+ \sigma_{yz}(\vec{e}_y\vec{e}_z + \vec{e}_z\vec{e}_y)\big] \cdot (n_x\vec{e}_x + n_y\vec{e}_y + n_z\vec{e}_z) \\
= \sigma_{xx}n_x\vec{e}_x + \sigma_{yy}n_y\vec{e}_y + \sigma_{zz}n_z\vec{e}_z \\
+ \sigma_{xy}(n_y\vec{e}_x + n_x\vec{e}_y) + \sigma_{xz}(n_z\vec{e}_x + n_x\vec{e}_z) \\
+ \sigma_{yz}(n_z\vec{e}_y + n_y\vec{e}_z).
\end{aligned}
\tag{8.56}
$$

Hence the components of the stress vector $\vec{s} = s_x\vec{e}_x + s_y\vec{e}_y + s_z\vec{e}_z$ satisfy

$$
\begin{aligned}
s_x = \sigma_{xx}n_x + \sigma_{xy}n_y + \sigma_{xz}n_z \\
s_y = \sigma_{xy}n_x + \sigma_{yy}n_y + \sigma_{yz}n_z \\
s_z = \sigma_{xz}n_x + \sigma_{yz}n_y + \sigma_{zz}n_z.
\end{aligned}
\tag{8.57}
$$

Application of Eq. (8.55) and using

$$
\underline{s} = \begin{bmatrix} s_x \\ s_y \\ s_z \end{bmatrix}, \qquad
\underline{n} = \begin{bmatrix} n_x \\ n_y \\ n_z \end{bmatrix},
\tag{8.58}
$$

gives an equivalent, but much shorter, expression for Eq. (8.57):

$$
\underline{s} = \underline{\sigma}\,\underline{n}.
\tag{8.59}
$$

If all the shear stress components are zero, i.e.: $\sigma_{xy} = \sigma_{xz} = \sigma_{yz} = 0$, and all the normal stresses are equal, i.e. $\sigma_{xx} = \sigma_{yy} = \sigma_{zz}$, this normal stress is called the **pressure** p such that

$$
p = -\sigma_{xx} = -\sigma_{yy} = -\sigma_{zz},
\tag{8.60}
$$

while

$$
\boldsymbol{\sigma} = -p\,(\vec{e}_x\vec{e}_x + \vec{e}_y\vec{e}_y + \vec{e}_z\vec{e}_z) = -p\,\boldsymbol{I},
\tag{8.61}
$$

with $\boldsymbol{I}$ the unit tensor. This is illustrated in Fig. 8.13.

8.5 Principal stresses and principal stress directions

Assume, that in a certain point of the material volume the stress state is known by specification of the tensor $\boldsymbol{\sigma}$. One might ask, whether it is possible to chose the

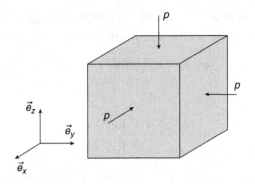

Pressure.

orientation of a surface element in such a direction, that only a normal stress acts on the surface and no shear stress. This means that we are trying to determine a vector $\vec{n}$ for which the following equation holds:

$$\boldsymbol{\sigma} \cdot \vec{n} = \lambda\, \vec{n}, \tag{8.62}$$

with λ to be interpreted as the normal stress. By shifting the right-hand side to the left, this equation can also be written as

$$(\boldsymbol{\sigma} - \lambda \boldsymbol{I}) \cdot \vec{n} = \vec{0} \ \text{ and in components as } \ (\underline{\sigma} - \lambda \underline{I})\underline{n} = \underline{0}, \tag{8.63}$$

with $\boldsymbol{I}$ the unit tensor ($\underline{I}$ the unit matrix) and $\vec{0}$ the zero vector ($\underline{0}$ the column with zeros). We can recognize an **eigenvalue problem**. The equation only has non-trivial solutions for $\vec{n}$ (so solutions with $\vec{n} \neq \vec{0}$) if

$$\det(\boldsymbol{\sigma} - \lambda \boldsymbol{I}) = 0 \ \text{ and also } \ \det(\underline{\sigma} - \lambda \underline{I}) = 0. \tag{8.64}$$

This third-order algebraic equation for λ has, because the tensor $\boldsymbol{\sigma}$ (with matrix representation $\underline{\sigma}$) is symmetric, three real solutions (sometimes coinciding), the **eigenvalues** of the stress tensor (stress matrix). For each eigenvalue it is possible to determine a normalized eigenvector. If the three eigenvalues are different, the three eigenvectors are unique and mutually perpendicular. If two (or three) eigenvalues are the same, it is still possible to determine a set of three, associated, mutually perpendicular eigenvectors of unit length, however they are no longer unique.

The three eigenvalues (solutions for λ) are called the **principal stresses** and denoted as σ_1, σ_2 and σ_3. The principal stresses are arranged in such a way that $\sigma_1 \leq \sigma_2 \leq \sigma_3$.

The associated, normalized, mutually perpendicular eigenvectors are the principal stress directions and specified by $\vec{n}_1$, $\vec{n}_2$ and $\vec{n}_3$. It will be clear that

$$\boldsymbol{\sigma} \cdot \vec{n}_i = \sigma_i \vec{n}_i \quad \text{for} \quad i = 1, 2, 3 \tag{8.65}$$

and also

$$\vec{n}_i \cdot \vec{n}_j = 1 \quad \text{for:} \quad i = j$$
$$= 0 \quad \text{for:} \quad i \neq j. \tag{8.66}$$

Based on the above it is obvious, that to every arbitrary stress cube, as depicted in Fig. 8.8, another cube can be attributed, which is differently oriented in space, upon which only normal stresses (the principal stresses) and no shear stresses are acting. Such a cube is called a principal stress cube, see Fig. 8.14. Positive principal stresses indicate extension, negative stresses indicate compression. The principal stress cube makes it easier to interpret a stress state and to identify the way a material is loaded. In the following section this will be discussed in more detail. As observed earlier, the stress state in a certain point is determined completely by the stress tensor $\boldsymbol{\sigma}$; in other words, by all stress components that act upon a cube, of which the orientation coincides with the *xyz*-coordinate system. Because, actually, the choice of the coordinate system is arbitrary, it can also be stated that the stress state is completely determined when the principal stresses and principal stress directions are known. How the principal stresses σ_i (with $i = 1, 2, 3$) and the principal stress directions $\vec{n}_i$ can be determined when the stress tensor $\boldsymbol{\sigma}$ is known is discussed above. The inverse procedure to reconstruct the

Figure 8.14

The principal stress cube.

stress tensor σ, when the principal stresses and stress directions are known, will not be outlined here.

8.6 Mohr's circles for the stress state

When the stress tensor σ is given, the stress vector $\vec{s}$ on an arbitrary oriented plane, defined by the unit outward normal $\vec{n}$ can be determined by

$$\vec{s} = \sigma \cdot \vec{n}. \tag{8.67}$$

Then, the normal stress s_n and shear stress s_t can be determined. The normal stress s_n is the inner product of the stress vector $\vec{s}$ with the unit outward normal $\vec{n}$: $s_n = \vec{s} \cdot \vec{n}$ and can be either positive or negative (extension or compression). The shear stress s_t is the magnitude of the component of $\vec{s}$ tangent to that surface, see Section 8.4. In this way, it is possible to add to each $\vec{n}$ a combination (s_n, s_t) that can be regarded as a mapping in a graph with s_n along the horizontal axis and s_t along the vertical axis, see Fig. 8.15. It can be proven that all possible combinations (s_n, s_t) are located in the shaded area between the drawn circles (the three Mohr's circles). If the principal stresses σ_1, σ_2 and σ_3 are known, Mohr's circles (the centroid is located on the s_n-axis) can be drawn at once! For this it is not necessary to determine the principal stress directions. In the $s_n s_t$-coordinate system the combinations $(\sigma_1, 0)$, $(\sigma_2, 0)$ and $(\sigma_3, 0)$ constitute the image points associated with the faces of the principal stress cube in Fig. 8.14. Based on Fig. 8.15, it can immediately be concluded that

$$(\sigma_n)_{\text{max}} = \sigma_3, \quad (\sigma_n)_{\text{min}} = \sigma_1$$
$$(\sigma_s)_{\text{max}} = (\sigma_3 - \sigma_1)/2. \tag{8.68}$$

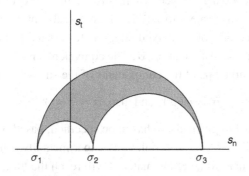

Figure 8.15

Mohr's circles for the stress.

8.7   Hydrostatic pressure and deviatoric stress

The hydrostatic pressure p is defined as the average of the normal stresses in Fig. 8.8. It is common practice to give the pressure the opposite sign of the average normal stresses, so:

$$p = -(\sigma_{xx} + \sigma_{yy} + \sigma_{zz})/3 = -\text{tr}(\boldsymbol{\sigma})/3 = -\text{tr}(\underline{\sigma})/3. \qquad (8.69)$$

It can be shown that, if expressed in principal stresses, we find

$$p = -(\sigma_1 + \sigma_2 + \sigma_3)/3. \qquad (8.70)$$

The associated hydrostatic stress tensor is denoted with $\boldsymbol{\sigma}^h$ and defined according to

$$\boldsymbol{\sigma}^h = -p\,\boldsymbol{I}. \qquad (8.71)$$

The difference of the (total) stress tensor $\boldsymbol{\sigma}$ and the hydrostatic stress tensor $\boldsymbol{\sigma}$ is called the deviatoric stress tensor $\boldsymbol{\sigma}^d$. Thus

$$\boldsymbol{\sigma} = \boldsymbol{\sigma}^h + \boldsymbol{\sigma}^d = -p\,\boldsymbol{I} + \boldsymbol{\sigma}^d \text{ and also } \underline{\sigma} = \underline{\sigma}^h + \underline{\sigma}^d = -p\,\underline{I} + \underline{\sigma}^d. \qquad (8.72)$$

Splitting the stress state in a hydrostatic part and a deviatoric part appears to be useful for the description of the material behaviour. This will be the theme of Chapter 12.

8.8   Equivalent stress

In general mechanical failure of materials is (among other items) determined by the stresses that act on the material. In principle, this means that all components of the stress tensor $\boldsymbol{\sigma}$ in one way or another may contribute to failure. It is common practice, to attribute a scalar property to the stress tensor $\boldsymbol{\sigma}$ that reflects the gravity of the stress state with respect to failure. Such a scalar property is normalized, based on a consideration of a uniaxial stress state (related to the extension of a bar) and is called an equivalent stress $\bar{\sigma}$. The equivalent stress is a scalar function of the stress tensor and thus of the components of the stress matrix

$$\bar{\sigma} = \bar{\sigma}(\boldsymbol{\sigma}) \text{ and also } \bar{\sigma} = \bar{\sigma}(\underline{\sigma}). \qquad (8.73)$$

The formal relationship given above has to be specified, based on physical understandings of the failure of the material. Only experimentally can it be assessed at which stress combinations a certain material will reach the limits of its resistance to failure. It is very well possible and obvious that one material fails by means of a completely different mechanism than another material. For example, considering technical materials, it is known that for a metal the maximum shear strain

$(\sigma_s)_{max}$ is often normative, while a ceramic material might fail because of a too high maximum extensional stress $(\sigma_n)_{max}$. Such knowledge is important for the design of hip and knee prostheses or tooth implants, where both types of materials are used. But also biological materials may have different failure mechanisms. A bone for example will often fail as a result of the maximum compression stress, but a tendon will usually fail because it is overstretched, i.e. due to the maximum extensional stress or maximum shear stress. The functional relationship $\overline{\sigma}(\sigma)$ is primarily determined by the (micro) structure of the considered material. Thus, depending on the material, different specifications of $\overline{\sigma}$ may be applied. We will limit ourselves to a few examples.

According to the equivalent stress $\overline{\sigma}_T$ ascribed to Tresca (sometimes also called Coulomb, Mohr, Guest) the maximum shear stress is held responsible for failure. The definition is

$$\overline{\sigma}_T = 2(\sigma_s)_{max} = \sigma_3 - \sigma_1, \tag{8.74}$$

which is normalized in such a way that for a uniaxially loaded bar, with an extensional stress $\sigma_{ax} > 0$, we find $\overline{\sigma}_T = \sigma_{ax}$.

The equivalent stress σ_M according to von Mises (also Hüber, Hencky) is based on the deviatoric stress tensor:

$$\overline{\sigma}_M = \sqrt{\frac{3}{2}\mathrm{tr}(\sigma^d \cdot \sigma^d)} = \sqrt{\frac{3}{2}\mathrm{tr}(\underline{\sigma}^d\underline{\sigma}^d)}. \tag{8.75}$$

This can be elaborated to

$$\overline{\sigma}_M = \sqrt{\frac{1}{2}\left[(\sigma_{xx} - \sigma_{yy})^2 + (\sigma_{yy} - \sigma_{zz})^2 + (\sigma_{zz} - \sigma_{xx})^2\right] + 3\left[\sigma_{xy}^2 + \sigma_{xz}^2 + \sigma_{yz}^2\right]}.$$

$$\tag{8.76}$$

Here also the specification is chosen in such a way that for a uniaxially loaded bar with axial stress σ_{ax}, the equivalent von Mises stress satisfies $\overline{\sigma}_M = \sigma_{ax}$. It can be proven that in terms of principal stresses:

$$\overline{\sigma}_M = \sqrt{\frac{1}{2}\left[(\sigma_1 - \sigma_2)^2 + (\sigma_2 - \sigma_3)^2 + (\sigma_3 - \sigma_1)^2\right]}. \tag{8.77}$$

In general (for arbitrary σ) the difference between the equivalent stresses according to Tresca and von Mises are relatively small.

The equivalent stress $\overline{\sigma}_R$, according to Rankine (also Galilei) expresses that, in absolute sense, the maximum principal stress determines failure. This means that

$$\overline{\sigma}_R = |\sigma_3| \quad \text{if} \quad |\sigma_3| \geq |\sigma_1|$$
$$\overline{\sigma}_R = |\sigma_1| \quad \text{if} \quad |\sigma_3| < |\sigma_1|. \tag{8.78}$$

And again for a uniaxially loaded bar $\overline{\sigma}_R = \sigma_{ax}$. For an arbitrary stress state the equivalent stress according to Rankine can be completely different from the equivalent stress according to Tresca or von Mises.

A few remarks on this subject are opportune at this point. The understanding of failure thresholds and mechanisms for biological materials involves much more than identifying a suitable equivalent stress expression. Biological materials can also fail because of a disturbance of the metabolic processes in the cells. The mechanical state in biological materials is not only determined by stresses and strains, but also by rather complicated transport processes of nutrients, oxygen and waste products and very complex biochemistry. These processes can be disturbed by mechanical deformation (for example occlusion of blood vessels, causing an ischemic state of the tissue, resulting in lack of oxygen and nutrients and accumulation of waste products). After some time this may result in cell death and thus damaged tissues. How these processes evolve and lead to tissue remodelling and/or damage is still the subject of research.

Exercises

8.1   Consider a two-dimensional plane stress state, with stress components:

$$\sigma_{xx} = ax^2 + by$$
$$\sigma_{yy} = bx^2 + ay^2 - cx.$$

Use the equilibrium equations to determine the shear stress component $\sigma_{xy}(x, y)$, satisfying $\sigma_{xy}(0, 0) = 2a$.

8.2   On an infinitesimal area segment, stress components are working as given in the figure.

(a)   Determine the stress tensor.

(b)   Determine the stress vector $\vec{s}$ acting on a plane with unit normal vector $\vec{n} = \frac{1}{2}\sqrt{2}\vec{e}_x + \frac{1}{2}\sqrt{2}\vec{e}_y$.

(c)   Determine the components of $\vec{s}$ perpendicular and parallel to the plane defined in item (b).

8.3   On an infinitesimal area segment two sets of stress components are working as shown in the figures (a) and (b).

(a)   Determine the stress tensor for both situations.

(b)   Determine the stress vector on the plane with normal $\vec{n} = \vec{e}_x \cos(\alpha) + \vec{e}_y \sin(\alpha)$. The angle α represents the angle of the normal vector with the x-axis.

(c)   Determine for both situations the length of the stress vector as a function of α.

(a)   (b)

8.4   In a material a stress state is observed that is characterized by the principal stresses (in [MPa]) and the principal stress directions (unit vectors, defined with respect to a fixed Cartesian coordinate system):

$$\sigma_1 = 0 \quad \text{with} \quad \vec{n}_1 = \vec{e}_z$$
$$\sigma_2 = 0 \quad \text{with} \quad \vec{n}_2 = -\frac{4}{5}\vec{e}_x + \frac{3}{5}\vec{e}_y$$
$$\sigma_3 = 25 \quad \text{with} \quad \vec{n}_3 = \frac{3}{5}\vec{e}_x + \frac{4}{5}\vec{e}_y.$$

Calculate the associated stress tensor σ with respect to the basis vectors $\vec{e}_x$, $\vec{e}_y$ and $\vec{e}_z$.

8.5   Consider a cube of material, with the edges oriented in the direction of the axes of a xyz-coordinate system, see the figure. In this figure also the normal and shear stresses are depicted (expressed in [MPa]) that act on the side faces of the cube.

Determine the equivalent stress $\overline{\sigma}_M$ according to von Mises.

8.6 A prismatic piece of material ABCDEF is given, see the figure.
The coordinates of the corner nodes are specified in the following table (in mm):

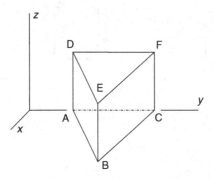

	A	B	C	D	E	F
x	0	8	0	0	8	0
y	1	7	7	1	7	7
z	0	0	0	5	5	5

For the faces of the prism that are visible in the figure, the stresses that act upon these faces (in [MPa]) are known:

$$\vec{p}_{ABED} = -5\vec{e}_x + 2\vec{e}_y + 6\vec{e}_z$$
$$\vec{p}_{BCFE} = 10\vec{e}_x + 5\vec{e}_y$$
$$\vec{p}_{DEF} = 10\vec{e}_x.$$

Calculate the associated stress tensor σ under the assumption that the stress state in the considered piece of material is homogeneous.

8.7   Consider a material cube ABCDEFGH. The edges of the cube are oriented in the direction of the axes of a Cartesian xyz-coordinate system, see the figure.

In addition, it is given that the relevant stress matrix $\underline{\sigma}$ for the particle is

$$\underline{\sigma} = \begin{bmatrix} 5 & 2 & 5 \\ 2 & 12 & 10 \\ 5 & 10 & 3 \end{bmatrix} \qquad \text{expressed in [MPa]}.$$

Determine the magnitude σ_t of the shear stress vector $\vec{s}_t$ acting on the diagonal plane BDE.

9 Motion: the time as an extra dimension

9.1 Introduction

Let us consider the geometrical change in time (the deformation and movement in three-dimensional space) of a coherent amount of material or material fraction, for which a continuum modelling approach is allowed. In case more fractions are involved, it is in principle possible to describe the behaviour of each fraction separately, as if it was isolated from the other fractions (however it wil be necessary to include interactions between fractions). The present chapter is focussed on a detailed description of motion. In addition, the consequences of configuration changes for the formulation of physical fields will be discussed. There will be no attention to the possible causes of the motion. In the present chapter, an approach will be followed that is common practice in the continuum description of solids (although it can also be applied to fluids). The specific aspects relevant for fluids will be treated at the end of the chapter.

9.2 Geometrical description of the material configuration

Consider a coherent amount of material in a completely defined geometrical state (the reference configuration). From each material point P, that can be allocated, the position vector $\vec{x}_0$ (with components stored in the column $\underset{\sim}{x}_0$) is known. In the following, this position vector will be used to identify the material point. The vector $\vec{x}_0$ is inextricably bound to the material point P, as if it were an attached label. With respect to a Cartesian xyz-coordinate system, $\vec{x}_0$ can be written as

$$\vec{x}_0 = x_0\, \vec{e}_x + y_0\, \vec{e}_y + z_0\, \vec{e}_z \text{ and also } \underset{\sim}{x}_0 = \begin{bmatrix} x_0 \\ y_0 \\ z_0 \end{bmatrix}. \tag{9.1}$$

Because $\vec{x}_0$ is uniquely coupled to a material point, the components $\underset{\sim}{x}_0$ of $\vec{x}_0$ are called material coordinates. The set of position vectors $\vec{x}_0$ that address all the material points in the configuration comprise the reference volume V_0, see

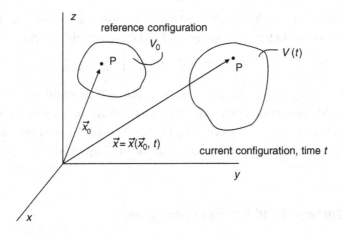

Figure 9.1

The position vector of a material point P.

Fig. 9.1. In an arbitrary current state, at time t, the position vector of the point P is specified by $\vec{x}$ and can be written as

$$\vec{x} = x\,\vec{e}_x + y\,\vec{e}_y + z\,\vec{e}_z \quad \text{and also} \quad \underset{\sim}{x} = \begin{bmatrix} x \\ y \\ z \end{bmatrix}. \tag{9.2}$$

With the attention focussed on a certain material point, it can be stated that

$$\vec{x} = \vec{x}(\vec{x}_0, t). \tag{9.3}$$

This functional relation expresses that the current position $\vec{x}$ of a material point is determined by the material identification $\vec{x}_0$ in V_0 of that point and by the current time t. When $\vec{x}_0$ is constant and with t passing through a certain time interval, $\vec{x} = \vec{x}(\vec{x}_0, t)$ can be considered to be a parameter description (with parameter t) of the trajectory of a material point (defined by $\vec{x}_0$) through three-dimensional space: the path of the particle.

Differentiation of the relation $\vec{x} = \vec{x}(\vec{x}_0, t)$ to the time t, with $\vec{x}_0$ taken constant (partial differentiation), results in the velocity vector $\vec{v}$ of the material point under consideration. It can be written:

$$\vec{v} = \dot{\vec{x}} \qquad \text{and also} \qquad \underset{\sim}{v} = \dot{\underset{\sim}{x}}, \tag{9.4}$$

with ($\dot{}$) the material time derivative: partial differentiation with respect to the time t with constant $\vec{x}_0$. For the acceleration vector $\vec{a}$ it follows:

$$\vec{a} = \dot{\vec{v}} = \ddot{\vec{x}} \qquad \text{and also} \qquad \underset{\sim}{a} = \dot{\underset{\sim}{v}} = \ddot{\underset{\sim}{x}}. \tag{9.5}$$

It will be clear that, in the context of the discussion above, the following formal relations hold for the velocity field and the acceleration field:

$$\vec{v} = \vec{v}(\vec{x}_0, t) \quad \text{and} \quad \vec{a} = \vec{a}(\vec{x}_0, t). \tag{9.6}$$

The configuration change of the material in a certain time interval can be associated with deformation. We deal with deformation when the mutual distances between material points change. The mathematical description of deformation and deformation velocity is the major theme of Chapter 10.

9.3 Lagrangian and Eulerian description

In the previous section the velocity and acceleration of the material are formally written as functions of the material identification $\vec{x}_0$ with material coordinates x_0 in V_0 and the time t. Obviously, this can also be done with other physical properties associated with the material, for example the temperature T. For a (time-dependent) temperature field it can be written

$$T = T(\vec{x}_0, t). \tag{9.7}$$

The temperature field in the current configuration $V(t)$ is mapped on the reference configuration. Such a description is referred to as **Lagrangian**. Partial differentiation to time t at constant $\vec{x}_0$ results in the **material time derivative** of the temperature, $\dot{T}$:

$$\dot{T} = \left[\frac{\partial T}{\partial t} \right]_{\vec{x}_0 \text{ constant}}. \tag{9.8}$$

This variable $\dot{T}$ has to be interpreted as the change (per unit time) of the temperature at a material point (moving through space) identified by $\vec{x}_0$.

Another approach concentrates on a fixed point in three-dimensional space. At every point in time a different material particle may be arrived at this location. For the (time-dependent) temperature field it can be written:

$$T = T(\vec{x}, t), \tag{9.9}$$

indicating the temperature of the material being present at time t in the spatial point $\vec{x}$ in $V(t)$. This alternative field specification is called **Eulerian**. When the partial derivative with respect to time t of the temperature field in the Eulerian description is determined, **the spatial time derivative $\delta T / \delta t$ is obtained**:

$$\frac{\delta T}{\delta t} = \left[\frac{\partial T}{\partial t} \right]_{\vec{x} \text{ constant}}. \tag{9.10}$$

This result $\delta T/\delta t$ should be interpreted as the change (per unit time) of the temperature at a fixed point in space $\vec{x}$, in which at consecutive time values t different material points will be found.

The temperature field at time t, used as an example above, can be written in a Lagrangian description: $T = T(\vec{x}_0, t)$ and thus be mapped on the reference configuration with the domain V_0. The field can also be written in an Eulerian description: $T = T(\vec{x}, t)$ and be associated with the current configuration with domain $V(t)$. It should be noticed that a graphical representation of such a field in both cases can be very different, especially in the case of large deformations and large rotations (both quite common in biological applications).

In Section 9.4 we focus on the relation between the time derivatives discussed above. In Section 9.6 the relation between gradient operators applied to both descriptions will be discussed.

9.4 The relation between the material and spatial time derivative

For the derivation of the relation between the material and spatial time derivative of, for example, the temperature (as an arbitrary physical state variable, associated with the material) we start with the Eulerian description of the temperature field $T = T(\vec{x}, t)$, in components formulated as $T = T(\underset{\sim}{x}, t) = T(x, y, z, t)$. For the total differential dT it can be written:

$$
dT = \left[\frac{\partial T}{\partial x}\right]_{y,z,t \text{ constant}} dx + \left[\frac{\partial T}{\partial y}\right]_{x,z,t \text{ constant}} dy
$$
$$
+ \left[\frac{\partial T}{\partial z}\right]_{x,y,t \text{ constant}} dz + \left[\frac{\partial T}{\partial t}\right]_{x,y,z \text{ constant}} dt
$$

$$(9.11)$$

and in a more compact notation, using the gradient operator (see Chapter 7):

$$
dT = d\vec{x} \cdot \vec{\nabla} T + \frac{\delta T}{\delta t} dt \quad \text{and also} \quad dT = d\underset{\sim}{x}^T \underset{\sim}{\nabla} T + \frac{\delta T}{\delta t} dt. \tag{9.12}
$$

This equation describes the change dT of T at an arbitrary, infinitesimally small change $d\vec{x}$ (with associated $d\underset{\sim}{x}$) of the location in space, combined with an infinitesimally small change dt in time.

Now the change $d\vec{x}$, in the time increment dt, is chosen in such a way that the material is followed: $d\vec{x} = \vec{v}dt$. This implies a change in temperature according to $dT = \dot{T}dt$. Substituting this special choice in Equation (9.12), directly leads to

$$
\dot{T}dt = \vec{v} \cdot \vec{\nabla} T dt + \frac{\delta T}{\delta t} dt, \tag{9.13}
$$

and thus

$$\dot{T} = \vec{v} \cdot \vec{\nabla} T + \frac{\delta T}{\delta t} \quad \text{and also} \quad \dot{T} = \underset{\sim}{v}^T \underset{\sim}{\nabla} T + \frac{\delta T}{\delta t}. \tag{9.14}$$

The first term on the right-hand side, the difference between the material derivative and the spatial derivative, is called the **convective** contribution. For an arbitrary physical variable, associated with the material the following relation between the operators has to be applied:

$$(\dot{\ }) = \vec{v} \cdot \vec{\nabla}(\ ) + \frac{\delta(\ )}{\delta t} \quad \text{and also} \quad (\dot{\ }) = \underset{\sim}{v}^T \underset{\sim}{\nabla}(\ ) + \frac{\delta(\ )}{\delta t}. \tag{9.15}$$

If $\vec{v} = \vec{0}$, in other words, if the material is not moving in three-dimensional space, there is no difference between the material and spatial time derivative.

Applying the operator to the vector $\vec{x}$ results in an identity:

$$\dot{\vec{x}} = \vec{v} \cdot \vec{\nabla}\vec{x} + \frac{\delta \vec{x}}{\delta t} \quad \rightarrow \quad \vec{v} = \vec{v} \cdot \boldsymbol{I} + \vec{0} \quad \rightarrow \quad \vec{v} = \vec{v}. \tag{9.16}$$

Of course this also holds for application to the row $\underset{\sim}{x}^T$ (application to the column $\underset{\sim}{x}$ is not allowed in the framework of the notation used):

$$\dot{\underset{\sim}{x}}^T = \underset{\sim}{v}^T \underset{\sim}{\nabla} \, \underset{\sim}{x}^T + \frac{\delta \underset{\sim}{x}^T}{\delta t} \quad \rightarrow \quad \underset{\sim}{v}^T = \underset{\sim}{v}^T \underset{\sim}{I} + \underset{\sim}{0}^T \quad \rightarrow \quad \underset{\sim}{v}^T = \underset{\sim}{v}^T. \tag{9.17}$$

Applying the operation to $\vec{v}$ (the velocity vector), leads to

$$\dot{\vec{v}} = \vec{v} \cdot \vec{\nabla}\vec{v} + \frac{\delta \vec{v}}{\delta t} \quad \rightarrow \quad \vec{a} = \vec{v} \cdot \vec{\nabla}\vec{v} + \frac{\delta \vec{v}}{\delta t}$$

$$\rightarrow \quad \vec{a} = \left(\vec{\nabla}\vec{v}\right)^T \cdot \vec{v} + \frac{\delta \vec{v}}{\delta t} \tag{9.18}$$

and application to the row $\underset{\sim}{v}^T$ with velocity components, results in

$$\dot{\underset{\sim}{v}}^T = \underset{\sim}{v}^T \underset{\sim}{\nabla} \underset{\sim}{v}^T + \frac{\delta \underset{\sim}{v}^T}{\delta t} \quad \rightarrow \quad \underset{\sim}{a}^T = \underset{\sim}{v}^T \underset{\sim}{\nabla} \underset{\sim}{v}^T + \frac{\delta \underset{\sim}{v}^T}{\delta t}$$

$$\rightarrow \quad \underset{\sim}{a} = \left(\underset{\sim}{\nabla}\underset{\sim}{v}^T\right)^T \underset{\sim}{v} + \frac{\delta \underset{\sim}{v}}{\delta t}. \tag{9.19}$$

With this equation, the acceleration field can be found if the Eulerian description of the velocity field is known. In the result the velocity gradient tensor $\boldsymbol{L}$ (with associated matrix representation $\underline{L}$), as introduced in Chapter 7 can be recognized:

$$\boldsymbol{L} = \left(\vec{\nabla}\vec{v}\right)^T \quad \text{and also} \quad \underline{L} = \left(\underset{\sim}{\nabla} \, \underset{\sim}{v}^T\right)^T. \tag{9.20}$$

Now the equation for the acceleration vector $\vec{a}$ with the components $\underset{\sim}{a}$ becomes

$$\vec{a} = \boldsymbol{L} \cdot \vec{v} + \frac{\delta \vec{v}}{\delta t} \quad \text{and also} \quad \underset{\sim}{a} = \underline{L} \, \underset{\sim}{v} + \frac{\delta \underset{\sim}{v}}{\delta t}. \tag{9.21}$$

In the case of a stationary flow, where $\vec{v} = \vec{v}(\vec{x})$ instead of $\vec{v} = \vec{v}(\vec{x}, t)$, the equation for the acceleration reduces to

$$\vec{a} = \boldsymbol{L} \cdot \vec{v} \quad \text{and also} \quad \underset{\sim}{a} = \underline{L}\,\underset{\sim}{v}. \tag{9.22}$$

9.5 The displacement vector

Consider a material point P, which is defined by the position vector $\vec{x}_0$ in the reference configuration with volume V_0. In the current configuration the position vector of that point is denoted by $\vec{x}$, see Fig. 9.2.

The displacement vector of a point P, in the current configuration with respect to the undeformed configuration, is denoted by $\vec{u}$ satisfying

$$\vec{u} = \vec{x} - \vec{x}_0 \tag{9.23}$$

and in component form:

$$\begin{bmatrix} u_x \\ u_y \\ u_z \end{bmatrix} = \begin{bmatrix} x \\ y \\ z \end{bmatrix} - \begin{bmatrix} x_0 \\ y_0 \\ z_0 \end{bmatrix}. \tag{9.24}$$

In the Lagrangian description $\vec{u}$ is considered to be a function of $\vec{x}_0$ in V_0 and t and thus

$$\vec{u} = \vec{u}(\vec{x}_0, t) = \vec{x}(\vec{x}_0, t) - \vec{x}_0. \tag{9.25}$$

Figure 9.2

The displacement of a material point P.

This relation enables, to formally calculate the displacement (in the current configuration at time t with respect to the reference configuration) of a material point, defined in the reference configuration with material identification $\vec{x}_0$. For the use of Eq. (9.25), it is assumed that $\vec{x}(\vec{x}_0, t)$ is available.

In the Eulerian description $\vec{u}$ is considered to be a function of $\vec{x}$ in $V(t)$ and t and thus

$$\vec{u} = \vec{u}(\vec{x}, t) = \vec{x} - \vec{x}_0(\vec{x}, t). \tag{9.26}$$

With Eq. (9.26), it is possible to formally calculate the displacement (in the current configuration with respect to the reference configuration) of a material point, which is actually (at time t) at position $\vec{x}$ in the three-dimensional space. Necessary for this is, that $\vec{x}_0(\vec{x}, t)$ is known, expressing which material point $\vec{x}_0$ at time t is present at the spatial point $\vec{x}$, in other words, the inverse relation of $\vec{x}(\vec{x}_0, t)$.

9.6 The gradient operator

In Chapter 7 the gradient operator with respect to the current configuration was treated extensively. In fact, the current field of a physical variable (for example the temperature T) was considered in the current configuration with domain $V(t)$ and as such defined according to an Eulerian description. The gradient of such a variable is built up from the partial derivatives with respect to the spatial coordinates, for example:

$$\vec{\nabla} T = \vec{e}_x \frac{\partial T}{\partial x} + \vec{e}_y \frac{\partial T}{\partial y} + \vec{e}_z \frac{\partial T}{\partial z} \quad \text{and also} \quad \underset{\sim}{\nabla} T = \begin{bmatrix} \frac{\partial T}{\partial x} \\[6pt] \frac{\partial T}{\partial y} \\[6pt] \frac{\partial T}{\partial z} \end{bmatrix}. \tag{9.27}$$

The current field (with respect to time t) can also be mapped onto the reference configuration with volume V_0 and thus formulated by means of a Lagrangian description. In this formulation the gradient can also be defined and is built up from partial derivatives with respect to the material coordinates:

$$\vec{\nabla}_0 T = \vec{e}_x \frac{\partial T}{\partial x_0} + \vec{e}_y \frac{\partial T}{\partial y_0} + \vec{e}_z \frac{\partial T}{\partial z_0} \quad \text{and also} \quad \underset{\sim}{\nabla}_0 T = \begin{bmatrix} \frac{\partial T}{\partial x_0} \\[6pt] \frac{\partial T}{\partial y_0} \\[6pt] \frac{\partial T}{\partial z_0} \end{bmatrix}. \tag{9.28}$$

To relate the afore mentioned gradient operators, the chain rule for differentiation is used. For a fixed time t we find

$$\frac{\partial T}{\partial x_0} = \frac{\partial T}{\partial x}\frac{\partial x}{\partial x_0} + \frac{\partial T}{\partial y}\frac{\partial y}{\partial x_0} + \frac{\partial T}{\partial z}\frac{\partial z}{\partial x_0}$$

$$\frac{\partial T}{\partial y_0} = \frac{\partial T}{\partial x}\frac{\partial x}{\partial y_0} + \frac{\partial T}{\partial y}\frac{\partial y}{\partial y_0} + \frac{\partial T}{\partial z}\frac{\partial z}{\partial y_0}$$

$$\frac{\partial T}{\partial z_0} = \frac{\partial T}{\partial x}\frac{\partial x}{\partial z_0} + \frac{\partial T}{\partial y}\frac{\partial y}{\partial z_0} + \frac{\partial T}{\partial z}\frac{\partial z}{\partial z_0}, \tag{9.29}$$

and in a more concise notation:

$$\nabla_0 T = \underline{F}^{\mathrm{T}} \nabla T \quad \text{with} \quad \underline{F}^{\mathrm{T}} = \begin{pmatrix} \frac{\partial x}{\partial x_0} & \frac{\partial y}{\partial x_0} & \frac{\partial z}{\partial x_0} \\ \frac{\partial x}{\partial y_0} & \frac{\partial y}{\partial y_0} & \frac{\partial z}{\partial y_0} \\ \frac{\partial x}{\partial z_0} & \frac{\partial y}{\partial z_0} & \frac{\partial z}{\partial z_0} \end{pmatrix}. \tag{9.30}$$

The matrix $\underline{F}$, for which the transpose is defined in Eq. (9.30), is called the **deformation matrix** or **deformation gradient matrix**, from the current configuration with respect to the reference configuration. In the next chapter this matrix will be discussed in full detail. For the deformation matrix $\underline{F}$ we can write in a more concise notation:

$$\underline{F}^{\mathrm{T}} = \nabla_0 \underline{x}^{\mathrm{T}} \quad \text{and} \quad \underline{F} = \left(\nabla_0 \underline{x}^{\mathrm{T}}\right)^{\mathrm{T}}. \tag{9.31}$$

By substituting $\underline{x} = \underline{x}_0 + \underline{u}$ into Eq. (9.31), and using $\nabla_0 \underline{x}^{\mathrm{T}} = \underline{I}$ we obtain

$$\underline{F}^{\mathrm{T}} = \underline{I} + \nabla_0 \underline{u}^{\mathrm{T}} \quad \text{and} \quad \underline{F} = \underline{I} + \left(\nabla_0 \underline{u}^{\mathrm{T}}\right)^{\mathrm{T}}. \tag{9.32}$$

In tensor notation the relation between $\nabla_0 T$ and ∇T can be written as

$$\vec{\nabla}_0 T = \boldsymbol{F}^{\mathrm{T}} \cdot \vec{\nabla} T \quad \text{with} \quad \boldsymbol{F} = \left(\vec{\nabla}_0 \vec{x}\right)^{\mathrm{T}} = \boldsymbol{I} + \left(\vec{\nabla}_0 \vec{u}\right)^{\mathrm{T}}. \tag{9.33}$$

with $\boldsymbol{F}$ the **deformation tensor** (also called **deformation gradient tensor**).

Above, a relation is derived between the gradient of a physical property (at time t) with respect to the reference configuration and the gradient of that property with respect to the current configuration. Consequently the mutual relation between the gradient operators can be written as

$$\vec{\nabla}_0( ) = \boldsymbol{F}^{\mathrm{T}} \cdot \vec{\nabla}( ) \quad \text{and also} \quad \nabla_0( ) = \underline{F}^{\mathrm{T}} \nabla( ), \tag{9.34}$$

and the inverse relation

$$\vec{\nabla}( ) = \boldsymbol{F}^{-\mathrm{T}} \cdot \vec{\nabla}_0( ) \quad \text{and also} \quad \nabla( ) = \underline{F}^{-\mathrm{T}} \nabla_0( ). \tag{9.35}$$

If the current and reference configuration are identical (in that case $\vec{u} = \vec{0}$ and $\boldsymbol{F} = \boldsymbol{I}$) the gradient operators are also identical.

9.7 Extra displacement as a rigid body

In this section the consequences of a (fictitious) extra displacement of the current configuration as a rigid body will be discussed. Consider a hypothetical current configuration, that originates by first rotating the current configuration around the origin of the xyz-coordinate system and by translating it subsequently. The rotation around the origin is defined by means of a rotation tensor P (orthogonal) with matrix representation $\underline{P}$, satisfying

$$P^{-1} = P^{\mathrm{T}} \quad \text{and} \quad \underline{P}^{-1} = \underline{P}^{\mathrm{T}} \quad \text{while} \quad \det(P) = \det(\underline{P}) = 1, \tag{9.36}$$

and the translation by a vector $\vec{\lambda}$ with components $\underset{\sim}{\lambda}$. Fig. 9.3 shows the rigid body motion. Variables associated with the extra rotated and translated configuration are indicated with the superscript *. Because of the extra displacement as a rigid body the current position of a material point will change from $\vec{x}$ to $\vec{x}^*$ according to

$$\vec{x}^* = P \cdot \vec{x} + \vec{\lambda} = P \cdot (\vec{x}_0 + \vec{u}) + \vec{\lambda}, \tag{9.37}$$

while the displacement of the virtual configuration with respect to the reference configuration can be written as

Figure 9.3

The displacement as a rigid body.

$$\vec{u}^* = \vec{x}^* - \vec{x}_0 = P \cdot \vec{x} + \vec{\lambda} - \vec{x}_0$$
$$= P \cdot (\vec{x}_0 + \vec{u}) + \vec{\lambda} - \vec{x}_0 = (P - I) \cdot \vec{x}_0 + P \cdot \vec{u} + \vec{\lambda}$$
$$= P \cdot \vec{x} + \vec{\lambda} - (\vec{x} - \vec{u}) = (P - I) \cdot \vec{x} + \vec{u} + \vec{\lambda}. \tag{9.38}$$

In the following, the attention is focussed on a fixed material point with position vector $\vec{x}_0$ in the reference configuration, the position vector $\vec{x}$ in the current configuration at time t and the position vector $\vec{x}^*$ in the extra rotated and translated virtual configuration.

For a scalar physical variable, for example the temperature T, the value will not change as a result of an extra rigid body motion, thus, with respect to the same material point: $T^* = T$.

For the gradient operator, applied to a certain physical variable connected to the material, it follows, based on the relation between $\vec{x}$ and $\vec{x}^*$, directly:

$$\left(\vec{\nabla}(\ ) \right)^* = P \cdot \vec{\nabla}(\ ) = P \cdot F^{-T} \cdot \vec{\nabla}_0(\ ). \tag{9.39}$$

Note, that the gradient is the same operator for the real as well as for the imaginary, extra displaced, current configuration. However, the effect on for example the temperature field T and the field T^* is different. Eq. (9.39) shows this difference.

For the deformation tensor of the virtual configuration with respect to the undeformed reference configuration it is found that

$$F^* = \left(\vec{\nabla}_0 \vec{x}^* \right)^{\mathrm{T}} = \left(\vec{\nabla}_0 (P \cdot \vec{x} + \vec{\lambda}) \right)^{\mathrm{T}}$$
$$= \left(\vec{\nabla}_0 (\vec{x} \cdot P^{\mathrm{T}} + \vec{\lambda}) \right)^{\mathrm{T}} = \left(F^{\mathrm{T}} \cdot P^{\mathrm{T}} \right)^{\mathrm{T}}$$
$$= P \cdot F. \tag{9.40}$$

Finally, the influence of the rigid body motion on the stress state will be determined. Assume, that the internal interaction between the material particles, with an exception for the direction, will not change because of the motion as a rigid body. For the considered material point the stress tensor σ relates the stress vector $\vec{p}$ on a surface element with the unit normal $\vec{n}$ of that element, according to: $\vec{p} = \sigma \cdot \vec{n}$. Because the imaginary configuration is rotated with respect to the current configuration, both vectors $\vec{n}^*$ and $\vec{p}^*$ can be written as

$$\vec{n}^* = P \cdot \vec{n} \quad \text{and} \quad \vec{p}^* = P \cdot \vec{p}. \tag{9.41}$$

This reveals

$$\vec{p}^* = P \cdot \vec{p} = P \cdot \sigma \cdot \vec{n} = P \cdot \sigma \cdot P^{-1} \cdot \vec{n}^*$$
$$= \left(P \cdot \sigma \cdot P^{T} \right) \cdot \vec{n}^*, \tag{9.42}$$

and so for the stress tensor σ^* and the associated matrix $\underline{\sigma}^*$ in the imaginary configuration it is found that

$$\sigma^* = \boldsymbol{P} \cdot \boldsymbol{\sigma} \cdot \boldsymbol{P}^{\mathrm{T}} \quad \text{and also} \quad \underline{\sigma}^* = \underline{P}\,\underline{\sigma}\,\underline{P}^{\mathrm{T}} \tag{9.43}$$

with $\underline{P}$ the matrix representation of the tensor $\boldsymbol{P}$.

9.8   Fluid flow

For fluids it is not common practice (and in general not very useful) to define a reference state. This implies, that the Lagrangian description (expressing properties as a function of $\vec{x}_0$ and t) is not commonly used for fluids. Related to this, derivatives with respect to $\vec{x}_0$ (the gradient operator $\vec{\nabla}_0$) and derivatives with respect to time under constant $\vec{x}_0$ will not appear in fluid mechanics. The deformation tensor $\boldsymbol{F}$ is not relevant for fluids. However, the material time derivative (for example to calculate the acceleration) is important nevertheless. For fluids an Eulerian description is used, meaning that all physical properties are considered in the current configuration, so as functions of $\vec{x}$ in the volume $V(t)$ and t.

The kinematic variables that generally play a role in fluid mechanics problems are the velocity $\vec{v} = \vec{v}(\vec{x}, t)$ and the acceleration $\vec{a} = \vec{a}(\vec{x}, t)$, both in an Eulerian description. Their relation is given by (see the end of Section 9.4)

$$\vec{a} = \left(\vec{\nabla}\vec{v}\right)^{T} \cdot \vec{v} + \frac{\delta \vec{v}}{\delta t} = \boldsymbol{L} \cdot \vec{v} + \frac{\delta \vec{v}}{\delta t}. \tag{9.44}$$

Figure 9.4

Streamlines in a model of a carotid artery bifurcation.

Based on the velocity field in $V(t)$ often streamlines are drawn. Streamlines are representative for the current (at time t) direction of the velocity: the direction of the velocity at a certain point $\vec{x}$ corresponds to the direction of the tangent to the streamline in point $\vec{x}$. Fig. 9.4 gives an example of streamlines in a flow through a constriction.

For a stationary flow, $\vec{v} = \vec{v}(\vec{x})$ and thus $\delta\vec{v}/\delta t = 0$, the streamline pattern is the same at each time point. In that case the material particles follow exactly the streamlines, i.e. the particle tracks coincide with the streamlines.

Exercises

9.1 The material points of a deforming continuum are identified with the position vectors $\underset{\sim}{x}_0$ of these points in the reference configuration at time $t = 0$. The deformation (Lagrangian approach) is described with the current position vectors $\underset{\sim}{x}$ as a function of $\underset{\sim}{x}_0$ and time t, according to

$$\underset{\sim}{x}(\underset{\sim}{x}_0, t) = \begin{bmatrix} x_0 + (a + b\,y_0)\,t \\ y_0 + a\,t \\ z_0 \end{bmatrix} \quad \text{with } a \text{ and } b \text{ constant.}$$

Determine the velocity field as a function of time in an Eulerian description, in other words, give an expression for $\underset{\sim}{v} = \underset{\sim}{v}(\underset{\sim}{x}, t)$.

9.2 Consider a fluid that flows through three-dimensional space (with an xyz-coordinate system). In a number of fixed points in space the fluid velocity $\underset{\sim}{v}$ is measured as a function of the time t. Based on these measurements it appears that in a certain time interval the velocity can be approximated (interpolated) in the following way:

$$\underset{\sim}{v} = \begin{bmatrix} \frac{ay+bz}{1+\alpha t} \\ 0 \\ \frac{cx}{1+\alpha t} \end{bmatrix}.$$

Determine, based on this approximation of the velocity field as a function of time, the associated acceleration field as a function of time, thus: $\underset{\sim}{a}(\underset{\sim}{x}, t)$.

9.3 Consider a (two-dimensional) velocity field for a stationary flowing continuum:

$$v_x = \frac{x}{y^2}$$

$$v_y = \frac{1}{y}$$

with x and y spatial coordinates (expressed in [m]), while v_x and v_y are the velocity components in the x- and y-direction (expressed in [ms^{-1}]). The velocity field holds for the shaded domain in the figure given below.

Consider a material particle, that at time $t = 0$ enters the domain at the position with coordinates $x = 1$ [m], $y = 1$ [m].

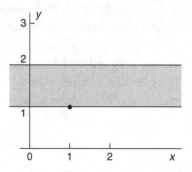

Calculate the time at which the considered particle leaves the shaded domain.

9.4 In a Cartesian xyz-coordinate system a rigid body is rotating around the z-axis with constant angular velocity ω. For the velocity field, in an Eulerian description, the following expression holds:

$$\vec{v}(\vec{x}) = \boldsymbol{\Omega} \cdot \vec{x} \quad \text{with} \quad \boldsymbol{\Omega} = \omega(-\vec{e}_x\vec{e}_y + \vec{e}_y\vec{e}_x).$$

Consider, the associated acceleration field $\vec{a}(\vec{x}) = \dot{\vec{v}}(\vec{x})$ and show that the result can be written as: $\vec{a}(\vec{x}) = \boldsymbol{H} \cdot \vec{x}$, with $\boldsymbol{H}$ a constant tensor.

Give an expression for $\boldsymbol{H}$ formulated as

$$\boldsymbol{H} = H_{xx}\vec{e}_x\vec{e}_x + H_{xy}\vec{e}_x\vec{e}_y + \cdots + H_{zz}\vec{e}_z\vec{e}_z.$$

9.5 A method that is sometimes used to study the mechanical behaviour of cells is based on the so-called cross flow experiment. An example of such an experiment is given in the right figure below (image courtesy of Mr Patrick Anderson). In this case a fluorescent fibroblast is positioned almost in the centre of the cross flow. There are several ways to create such a flow. The set-up, that is shown in the left figure, consists of a reservoir with four cylinders which rotate with the same angular velocity. The figure gives a top view of the set-up. A cell can be trapped in the centre of the cross flow and thus be stretched by the flow.

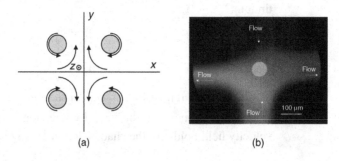

(a)   (b)

The reservoir is assumed to be filled with an incompressible fluid. Close to the origin of the *xyz*-coordinate system the (stationary) two dimensional velocity field is given by

$$\vec{v} = c(-x\vec{e}_x + y\vec{e}_y) \quad \text{with } c \text{ a constant,}$$

where $\vec{e}_x$ and $\vec{e}_y$ are unit vectors along the *x*- and *y*-axis.

Determine the associated acceleration field: $\vec{a}(\vec{x}) = \dot{\vec{v}}(\vec{x})$.

9.6   In the reference configuration (at time $t = 0$) the edges (with length ℓ) of a cubic material specimen are parallel to the axes of a Cartesian *xyz*-coordinate system. See the figure left. The specimen is loaded in shear, cyclically in the time t.

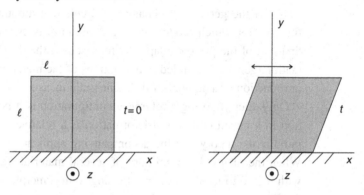

The time-dependent deformation is, in the Lagrangian approach, described by

$$x = x_0 + \frac{y_0}{2} \sin\left(2\pi\frac{t}{T}\right),$$

with T the (constant) time of one cycle.

$$y = y_0$$

$$z = z_0,$$

with x_0, y_0, z_0 the coordinates of the material points at time $t = 0$ and with x, y, z the associated coordinates at time t. Attention is focussed on the material point P that at time $t = T/4$ is located at the position $x = 2\ell/3, y = \ell/3, z = 0$.

Determine the position vector $\underline{x}_P$ of the point P as a function of the time t.

10 Deformation and rotation, deformation rate and spin

10.1 Introduction

Consider the geometrical change of a coherent amount of material or material fraction, for which modelling as a continuum is assumed to be permitted. The first part of the present chapter is focussed on the description of the local deformation (generally coupled with rotations of the material) and along with that, the introduction of a number of different strain measures.

Only after choosing a reference configuration is it possible to define deformation in a meaningful way (deformation is a relative concept). This implies that initially the theory and the accompanying application area are related to solids. When the material is a mixture of several material fractions, each fraction can, with regard to local geometrical changes, in principle, be isolated from the other fractions.

The second part of this chapter discusses geometrical changes in time. Central concepts in this part are deformation rate and rotation velocity (spin). The derivations in this part are not only relevant for solids, but even more important for applications including fluids.

10.2 A material line segment in the reference and current configuration

Consider a coherent amount of material in a fully defined state (the reference configuration). In a material point P, with position vector $\vec{x}_0$ in the reference configuration, we focus our attention on an arbitrary infinitesimally small material line segment $d\vec{x}_0$, see Fig. 10.1.

With respect to the Cartesian xyz-coordinate system the vector $d\vec{x}_0$ can be written as

$$d\vec{x}_0 = dx_0\vec{e}_x + dy_0\vec{e}_y + dz_0\vec{e}_z \quad \text{and also} \quad d\underset{\sim}{x}_0 = \begin{bmatrix} dx_0 \\ dy_0 \\ dz_0 \end{bmatrix}. \qquad (10.1)$$

Figure 10.1

A material line segment in reference and current configuration.

The orientation of the line segment $d\vec{x}_0$ is defined by the unit vector $\vec{e}_0$ with components in the column $\underset{\sim}{e}_0$. In that case it can be written:

$$d\vec{x}_0 = \vec{e}_0 \, d\ell_0 \quad \text{and also} \quad d\underset{\sim}{x}_0 = \underset{\sim}{e}_0 \, d\ell_0$$

$$\text{with} \quad d\ell_0 = \sqrt{d\vec{x}_0 \cdot d\vec{x}_0} = \sqrt{d\underset{\sim}{x}_0^T d\underset{\sim}{x}_0}, \tag{10.2}$$

where $d\ell_0$ specifies the length of vector $d\vec{x}_0$.

The same material line segment, but now considered in the current configuration at time t, is indicated with $d\vec{x}$. It should be emphasized, that the line segment $d\vec{x}$ in the current configuration is composed of the same material points as the line segment $d\vec{x}_0$ in the reference configuration.

With respect to the Cartesian xyz-coordinate system we can write for the vector $d\vec{x}$:

$$d\vec{x} = dx\vec{e}_x + dy\vec{e}_y + dz\vec{e}_z \quad \text{and also} \quad d\underset{\sim}{x} = \begin{bmatrix} dx \\ dy \\ dz \end{bmatrix}. \tag{10.3}$$

The orientation of the line segment $d\vec{x}$ is defined by the unit vector $\vec{e}$ with components in the column $\underset{\sim}{e}$. In that case it can be written:

$$d\vec{x} = \vec{e}d\ell \quad \text{and also} \quad d\underset{\sim}{x} = \underset{\sim}{e}d\ell$$

$$\text{with} \quad d\ell = \sqrt{d\vec{x} \cdot d\vec{x}} = \sqrt{d\underset{\sim}{x}^T d\underset{\sim}{x}}, \tag{10.4}$$

where $d\ell$ specifies the length of vector $d\vec{x}$.

Thus, the directional change (rotation) of the considered line segment, of the current state with respect to the reference state, is described by the difference of the unit vectors $\vec{e}$ and $\vec{e}_0$. For the relation between the components of $d\vec{x}$ at the (fixed) current time t and the components of the accompanying $d\vec{x}_0$ it can be written, using the chain rule for differentiation for (total) differentials:

$$dx = \frac{\partial x}{\partial x_0}dx_0 + \frac{\partial x}{\partial y_0}dy_0 + \frac{\partial x}{\partial z_0}dz_0$$

$$dy = \frac{\partial y}{\partial x_0}dx_0 + \frac{\partial y}{\partial y_0}dy_0 + \frac{\partial y}{\partial z_0}dz_0$$

$$dz = \frac{\partial z}{\partial x_0}dx_0 + \frac{\partial z}{\partial y_0}dy_0 + \frac{\partial z}{\partial z_0}dz_0, \tag{10.5}$$

for which a Lagrangian description has been taken as the point of departure according to: $\vec{x} = \vec{x}(\vec{x}_0, t)$. In a more compact form it can be formulated as

$$d\underset{\sim}{x} = \underline{F}\, d\underset{\sim}{x}_0 \quad \text{with} \quad \underline{F} = \begin{bmatrix} \frac{\partial x}{\partial x_0} & \frac{\partial x}{\partial y_0} & \frac{\partial x}{\partial z_0} \\ \frac{\partial y}{\partial x_0} & \frac{\partial y}{\partial y_0} & \frac{\partial y}{\partial z_0} \\ \frac{\partial z}{\partial x_0} & \frac{\partial z}{\partial y_0} & \frac{\partial z}{\partial z_0} \end{bmatrix} = \left(\nabla_0\, \underset{\sim}{x}^{\mathrm{T}} \right)^{\mathrm{T}} \tag{10.6}$$

and in tensor notation:

$$d\vec{x} = \boldsymbol{F} \cdot d\vec{x}_0 \quad \text{with} \quad \boldsymbol{F} = \left(\vec{\nabla}_0\, \vec{x} \right)^{\mathrm{T}}. \tag{10.7}$$

The tensor $\boldsymbol{F}$, the deformation tensor (or deformation gradient tensor), with matrix representation $\underline{F}$, was already introduced in Section 9.6. This tensor completely describes the (local) geometry change (deformation and rotation). After all, when $\boldsymbol{F}$ is known, it is possible for every line segment (and therefore also for a three-dimensional element) in the reference configuration, to calculate the accompanying line segment (or three-dimensional element) in the current configuration. The tensor $\boldsymbol{F}$ describes for every material line segment the length and orientation change: $\boldsymbol{F}$ determines the transition from $d\ell_0$ to $d\ell$ and the transition from $\vec{e}_0$ to $\vec{e}$.

Fig. 10.2 visualizes a uniaxially loaded bar. It is assumed, that the deformation is homogeneous: for every material point of the bar the same deformation tensor $\boldsymbol{F}$ is applicable. It can simply be verified, that the depicted transition from the reference configuration to the current configuration is defined by

$$\boldsymbol{F} = \lambda_x \vec{e}_x \vec{e}_x + \lambda_y \vec{e}_y \vec{e}_y + \lambda_z \vec{e}_z \vec{e}_z, \tag{10.8}$$

with the stretch ratios:

$$\lambda_x = \frac{\ell}{\ell_0} \quad \text{and} \quad \lambda_y = \lambda_z = \sqrt{\frac{A}{A_0}}. \tag{10.9}$$

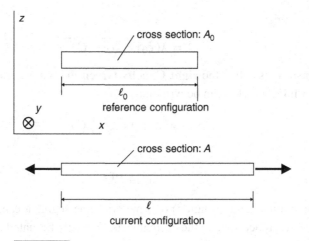

Figure 10.2

A uniaxially loaded bar.

10.3 The stretch ratio and rotation

Consider an infinitesimally small line segment $d\vec{x}_0$ in the reference configuration, directed along the unit vector $\vec{e}_0$ and with length $d\ell_0$, so $d\vec{x}_0 = \vec{e}_0 d\ell_0$. To this line segment belongs, in the current configuration, the line segment $d\vec{x} = \vec{e}d\ell$ directed along the unit vector $\vec{e}$ and with length $d\ell$. The mutual relation satisfies

$$d\vec{x} = \boldsymbol{F} \cdot d\vec{x}_0, \tag{10.10}$$

so

$$\vec{e}\, d\ell = \boldsymbol{F} \cdot \vec{e}_0\, d\ell_0. \tag{10.11}$$

The stretch ratio λ is defined as the ratio between $d\ell$ and $d\ell_0$ (and therefore it will always hold that: $\lambda > 0$). Using Eq. (10.11) it can be written:

$$\vec{e} \cdot \vec{e}\, d\ell^2 = \vec{e}_0 \cdot \boldsymbol{F}^{\mathrm{T}} \cdot \boldsymbol{F} \cdot \vec{e}_0\, d\ell_0^2, \tag{10.12}$$

and consequently

$$\lambda = \lambda(\vec{e}_0) = \sqrt{\vec{e}_0 \cdot \boldsymbol{F}^{\mathrm{T}} \cdot \boldsymbol{F} \cdot \vec{e}_0}. \tag{10.13}$$

This equation can be used to determine the stretch ratio λ for a material line segment with direction $\vec{e}_0$ in the reference configuration (Lagrangian approach). So, for that purpose the tensor (or tensor product) $\boldsymbol{F}^{\mathrm{T}} \cdot \boldsymbol{F}$ has to be known. The tensor $\boldsymbol{C}$ is defined as

$$\boldsymbol{C} = \boldsymbol{F}^{\mathrm{T}} \cdot \boldsymbol{F}, \tag{10.14}$$

and using this:

$$\lambda = \lambda(\vec{e}_0) = \sqrt{\vec{e}_0 \cdot C \cdot \vec{e}_0}. \tag{10.15}$$

The tensor C is called the **right Cauchy Green deformation tensor**. In component form Eq. (10.15) can be written as

$$\lambda = \lambda(\underline{e}_0) = \sqrt{\underline{e}_0^T \underline{C} \, \underline{e}_0}, \tag{10.16}$$

with

$$\underline{C} = \underline{F}^T \underline{F}. \tag{10.17}$$

The direction change (rotation) of a material line segment can, for the transition from the reference state to the current state, formally be stated as

$$\vec{e} = F \cdot \vec{e}_0 \frac{d\ell_0}{d\ell} = F \cdot \vec{e}_0 \frac{1}{\lambda} = \frac{F \cdot \vec{e}_0}{\sqrt{\vec{e}_0 \cdot C \cdot \vec{e}_0}}. \tag{10.18}$$

In component form this equation can be written as

$$\underline{e} = \underline{F} \, \underline{e}_0 \frac{d\ell_0}{d\ell} = \underline{F} \, \underline{e}_0 \frac{1}{\lambda} = \frac{\underline{F} \, \underline{e}_0}{\sqrt{\underline{e}_0^T \underline{C} \, \underline{e}_0}}. \tag{10.19}$$

Above, the current state is considered as a 'function' of the reference state: for a direction $\vec{e}_0$ in the reference configuration, the associated direction $\vec{e}$ and the stretch ratio λ were determined. In the following the 'inverse' procedure is shown.

Based on

$$F^{-1} \cdot d\vec{x} = d\vec{x}_0, \tag{10.20}$$

so

$$F^{-1} \cdot \vec{e} \, d\ell = \vec{e}_0 \, d\ell_0, \tag{10.21}$$

and subsequently

$$\vec{e} \cdot F^{-T} \cdot F^{-1} \cdot \vec{e} \, d\ell^2 = \vec{e}_0 \cdot \vec{e}_0 \, d\ell_0^2, \tag{10.22}$$

it follows for the stretch ratio $\lambda = d\ell/d\ell_0$ that

$$\lambda = \lambda(\vec{e}) = \frac{1}{\sqrt{\vec{e} \cdot F^{-T} \cdot F^{-1} \cdot \vec{e}}}. \tag{10.23}$$

This equation can be used to determine the stretch ratio λ for a material line element with direction $\vec{e}$ in the current configuration (Eulerian description). For this, the tensor (tensor product) $F^{-T} \cdot F^{-1}$ has to be known. The tensor B is defined according to

$$B = F \cdot F^T, \tag{10.24}$$

and using this:

$$\lambda = \lambda(\vec{e}) = \frac{1}{\sqrt{\vec{e} \cdot \boldsymbol{B}^{-1} \cdot \vec{e}}} . \tag{10.25}$$

The tensor $\boldsymbol{B}$ is called the **left Cauchy Green deformation tensor**. In component form Eq. (10.25) can be written as

$$\lambda = \lambda(\underset{\sim}{e}) = \frac{1}{\sqrt{\underset{\sim}{e}^{\mathrm{T}} \underline{B}^{-1} \underset{\sim}{e}}} , \tag{10.26}$$

with

$$\underline{B} = \underline{F}\,\underline{F}^{\mathrm{T}} . \tag{10.27}$$

The direction change (rotation) of a material line segment with direction $\vec{e}$ in the current configuration with respect to the reference configuration can formally be calculated with

$$\vec{e}_0 = \boldsymbol{F}^{-1} \cdot \vec{e}\, \frac{d\ell}{d\ell_0} = \boldsymbol{F}^{-1} \cdot \vec{e}\, \lambda = \frac{\boldsymbol{F}^{-1} \cdot \vec{e}}{\sqrt{\vec{e} \cdot \boldsymbol{B}^{-1} \cdot \vec{e}}} , \tag{10.28}$$

and alternatively, using components:

$$\underset{\sim}{e}_0 = \underline{F}^{-1} \underset{\sim}{e}\, \frac{d\ell}{d\ell_0} = \underline{F}^{-1} \underset{\sim}{e}\, \lambda = \frac{\underline{F}^{-1} \underset{\sim}{e}}{\sqrt{\underset{\sim}{e}^{\mathrm{T}} \underline{B}^{-1} \underset{\sim}{e}}} . \tag{10.29}$$

At the end of the present section we will investigate the influence of an extra displacement as a rigid body of the current configuration for the tensors that were introduced above, see Section 9.7. Properties with respect to the extra rotated (via the rotation tensor $\boldsymbol{P}$) and translated (via the translation vector $\vec{\lambda}$) virtual state are denoted by the superscript *. Because

$$\boldsymbol{F}^* = \boldsymbol{P} \cdot \boldsymbol{F}, \tag{10.30}$$

it can immediately be verified that

$$\boldsymbol{C}^* = \boldsymbol{F}^{*T} \cdot \boldsymbol{F}^* = \boldsymbol{F}^{\mathrm{T}} \cdot \boldsymbol{P}^{\mathrm{T}} \cdot \boldsymbol{P} \cdot \boldsymbol{F} = \boldsymbol{F}^{\mathrm{T}} \cdot \boldsymbol{P}^{-1} \cdot \boldsymbol{P} \cdot \boldsymbol{F}$$
$$= \boldsymbol{F}^{\mathrm{T}} \cdot \boldsymbol{I} \cdot \boldsymbol{F} = \boldsymbol{F}^{\mathrm{T}} \cdot \boldsymbol{F} = \boldsymbol{C}, \tag{10.31}$$

and

$$\boldsymbol{B}^* = \boldsymbol{F}^* \cdot \boldsymbol{F}^{*T} = \boldsymbol{P} \cdot \boldsymbol{F} \cdot \boldsymbol{F}^{\mathrm{T}} \cdot \boldsymbol{P}^{\mathrm{T}} = \boldsymbol{P} \cdot \boldsymbol{B} \cdot \boldsymbol{P}^{\mathrm{T}} . \tag{10.32}$$

Based on Eq. (10.31) the right Cauchy Green tensor $\boldsymbol{C}$ is called **invariant** for extra displacements of the current state as a rigid body. This invariance is completely trivial if the Lagrange description $\lambda = \lambda(\vec{e}_0)$ for the stretch ratio is taken into consideration.

The left Cauchy Green tensor B is, based on the transformation relation Eq. (10.32), called **objective**. In calculating the stretch ratio λ with the earlier derived Eulerian description, the result in the direction $\vec{e}$ in the current state (with associated tensor B) has to be identical to the result in the direction $\vec{e}^* = P \cdot \vec{e}$ in the virtual (extra rotated and translated) current state (using tensor B^*). That this demand is satisfied can be verified easily.

Because the Cauchy stress tensor σ transforms in a similar way as tensor B (see Section 9.7) the tensor σ is also objective. The deformation tensor F is neither invariant, nor objective.

10.4 Strain measures and strain tensors and matrices

In the preceding sections the stretch ratio λ is considered to be a measure for the relative length change of a material line segment in the transition from the reference configuration to the current configuration. If there is no deformation $\lambda = 1$. Often it is more convenient to introduce a variable equal to zero when there is no deformation: the strain. In the present section several different, generally accepted strain measures are treated.

In the previous section it was found, using the Lagrangian description:

$$\lambda^2 = \lambda^2(\vec{e}_0) = \vec{e}_0 \cdot F^{\mathrm{T}} \cdot F \cdot \vec{e}_0 = \vec{e}_0 \cdot C \cdot \vec{e}_0. \tag{10.33}$$

Coupled to this relation, the Green Lagrange strain $\varepsilon_{\mathrm{GL}}$ is defined by:

$$\varepsilon_{\mathrm{GL}} = \varepsilon_{\mathrm{GL}}(\vec{e}_0) = \frac{\lambda^2 - 1}{2} = \frac{1}{2}\,\vec{e}_0 \cdot \left(F^{\mathrm{T}} \cdot F - I\right) \cdot \vec{e}_0$$

$$= \frac{1}{2}\,\vec{e}_0 \cdot (C - I) \cdot \vec{e}_0 . \tag{10.34}$$

This result invites us to introduce the symmetrical **Green Lagrange strain tensor** E according to:

$$E = \frac{1}{2}\left(F^{\mathrm{T}} \cdot F - I\right) = \frac{1}{2}\,(C - I), \tag{10.35}$$

which implies

$$\varepsilon_{\mathrm{GL}} = \varepsilon_{\mathrm{GL}}(\vec{e}_0) = \vec{e}_0 \cdot E \cdot \vec{e}_0 . \tag{10.36}$$

When using matrix notation, Eq. (10.35) can be formulated as

$$\underline{E} = \frac{1}{2}\left(\underline{F}^{\mathrm{T}}\,\underline{F} - \underline{I}\right) = \frac{1}{2}\,(\underline{C} - \underline{I}), \tag{10.37}$$

implying

$$\varepsilon_{\mathrm{GL}} = \varepsilon_{\mathrm{GL}}(\underline{e}_0) = \underline{e}_0^{\mathrm{T}}\,\underline{E}\,\underline{e}_0. \tag{10.38}$$

From a mathematical perspective these are well manageable relations. The Green Lagrange strain tensor $\boldsymbol{E}$ (with matrix representation $\underline{E}$) is invariant for extra rigid body motions of the current state. The components of the (symmetrical, 3×3) Green Lagrange strain matrix $\underline{E}$ can be interpreted as follows:

- The terms on the diagonal are the Green Lagrange strains of material line segments of the reference configuration in the x-, y- and z-directions respectively (the component on the first row in the first column is the Green Lagrange strain of a line segment that is oriented in the x-direction in the reference configuration).
- The off-diagonal terms determine the shear of the material (the component on the first row of the second column is a measure for the change of the angle enclosed by material line segments that are oriented in the x- and y-direction in the undeformed configuration).

For the deformation tensor $\boldsymbol{F}$ and the displacement vector $\vec{u}$, both applying to the current configuration and related to the reference configuration, the following relation was derived in Section 9.6:

$$\boldsymbol{F} = \boldsymbol{I} + \left(\vec{\nabla}_0 \vec{u}\right)^{\mathrm{T}}. \tag{10.39}$$

Substitution into Eq. (10.35) yields

$$\boldsymbol{E} = \frac{1}{2}\left(\left(\vec{\nabla}_0 \vec{u}\right) + \left(\vec{\nabla}_0 \vec{u}\right)^{\mathrm{T}} + \left(\vec{\nabla}_0 \vec{u}\right) \cdot \left(\vec{\nabla}_0 \vec{u}\right)^{\mathrm{T}}\right). \tag{10.40}$$

It can be observed that the first two terms on the right-hand side of this equation are linear in the displacements, while the third term is non-linear (quadratic).

The linear strain $\varepsilon_{\mathrm{lin}}$ is defined, according to:

$$\varepsilon_{\mathrm{lin}} = \varepsilon_{\mathrm{lin}}(\vec{e}_0) = \lambda - 1 = \sqrt{\vec{e}_0 \cdot \boldsymbol{F}^{\mathrm{T}} \cdot \boldsymbol{F} \cdot \vec{e}_0} - 1. \tag{10.41}$$

This expression is not easily manageable for mathematical elaborations.

At small deformations and small rotations, for which: $\boldsymbol{F} \approx \boldsymbol{I}$ (and therefore the components of the tensor $\left(\vec{\nabla}_0 \vec{u}\right)^{\mathrm{T}}$ are much smaller than 1), it can be written:

$$
\begin{aligned}
\varepsilon_{\mathrm{lin}} &= \varepsilon_{\mathrm{lin}}(\vec{e}_0) \\
&= \sqrt{1 + \vec{e}_0 \cdot \left(\boldsymbol{F}^{\mathrm{T}} \cdot \boldsymbol{F} - \boldsymbol{I}\right) \cdot \vec{e}_0} - 1 \\
&\approx \frac{1}{2} \vec{e}_0 \cdot \left(\boldsymbol{F}^{\mathrm{T}} \cdot \boldsymbol{F} - \boldsymbol{I}\right) \cdot \vec{e}_0 \\
&\approx \frac{1}{2} \vec{e}_0 \cdot \left(\boldsymbol{F}^{\mathrm{T}} + \boldsymbol{F} - 2\boldsymbol{I}\right) \cdot \vec{e}_0.
\end{aligned} \tag{10.42}
$$

The last found approximation for the linear strain is denoted by the symbol ε. This strain definition is used on a broad scale. Therefore, the assumption $\boldsymbol{F} \approx \boldsymbol{I}$ leads to the following, mathematically well manageable relation:

$$\varepsilon = \varepsilon(\vec{e}_0) = \vec{e}_0 \cdot \boldsymbol{\varepsilon} \cdot \vec{e}_0 \quad \text{with} \quad \boldsymbol{\varepsilon} = \frac{1}{2}\left(\boldsymbol{F}^{\mathrm{T}} + \boldsymbol{F} - 2\boldsymbol{I}\right), \tag{10.43}$$

where the symmetrical tensor $\boldsymbol{\varepsilon}$ is called the **linear strain tensor**. In displacements this tensor can also be expressed as

$$\boldsymbol{\varepsilon} = \frac{1}{2}\left(\left(\vec{\nabla}_0\vec{u}\right) + \left(\vec{\nabla}_0\vec{u}\right)^{\mathrm{T}}\right). \tag{10.44}$$

The strain tensor $\boldsymbol{\varepsilon}$ is linear in the displacements and can be considered (with respect to the displacements) as a linearized form of the Green Lagrange strain tensor $\boldsymbol{E}$. In component form this results in the well-known and often-used formulation:

$$\underline{\varepsilon} = \begin{bmatrix} \frac{\partial u_x}{\partial x_0} & \frac{1}{2}\left(\frac{\partial u_x}{\partial y_0} + \frac{\partial u_y}{\partial x_0}\right) & \frac{1}{2}\left(\frac{\partial u_x}{\partial z_0} + \frac{\partial u_z}{\partial x_0}\right) \\[2mm] \frac{1}{2}\left(\frac{\partial u_y}{\partial x_0} + \frac{\partial u_x}{\partial y_0}\right) & \frac{\partial u_y}{\partial y_0} & \frac{1}{2}\left(\frac{\partial u_y}{\partial z_0} + \frac{\partial u_z}{\partial y_0}\right) \\[2mm] \frac{1}{2}\left(\frac{\partial u_z}{\partial x_0} + \frac{\partial u_x}{\partial z_0}\right) & \frac{1}{2}\left(\frac{\partial u_z}{\partial y_0} + \frac{\partial u_y}{\partial z_0}\right) & \frac{\partial u_z}{\partial z_0} \end{bmatrix}. \tag{10.45}$$

The components of the (symmetrical, 3×3) linear strain matrix $\underline{\varepsilon}$ can be interpreted as follows:

- The terms on the diagonal are the linear strains of material line segments of the reference configuration in the x-, y- and z-directions respectively (the component on the first row in the first column is the linear strain of a line segment that is oriented in the x-direction in the reference configuration).
- The off-diagonal terms determine the shear of the material (the component on the first row of the second column is a measure for the change of the angle enclosed by material line segments that are oriented in the x- and y-direction in the undeformed configuration. See Fig. 10.3.

Figure 10.3

Interpretation of linear strain components.

In the previous section the Eulerian description for the stretch ratio was derived:

$$\lambda^{-2} = \lambda^{-2}(\vec{e}) = \vec{e} \cdot \boldsymbol{F}^{-T} \cdot \boldsymbol{F}^{-1} \cdot \vec{e} = \vec{e} \cdot \boldsymbol{B}^{-1} \cdot \vec{e}. \tag{10.46}$$

Coupled to this relation, the Almansi Euler strain ε_{AE} is defined by

$$\varepsilon_{AE} = \varepsilon_{AE}(\vec{e}) = \frac{1 - \lambda^{-2}}{2} = \frac{1}{2} \vec{e} \cdot \left(\boldsymbol{I} - \boldsymbol{F}^{-T} \cdot \boldsymbol{F}^{-1} \right) \cdot \vec{e}$$

$$= \frac{1}{2} \vec{e} \cdot \left(\boldsymbol{I} - \boldsymbol{B}^{-1} \right) \cdot \vec{e}. \tag{10.47}$$

This relation gives rise to introduce the symmetrical **Almansi Euler strain tensor** $\boldsymbol{A}$ according to

$$\boldsymbol{A} = \frac{1}{2} \left(\boldsymbol{I} - \boldsymbol{F}^{-T} \cdot \boldsymbol{F}^{-1} \right) = \frac{1}{2} \left(\boldsymbol{I} - \boldsymbol{B}^{-1} \right), \tag{10.48}$$

which implies

$$\varepsilon_{AE} = \varepsilon_{AE}(\vec{e}) = \vec{e} \cdot \boldsymbol{A} \cdot \vec{e}. \tag{10.49}$$

When using matrix notation this may be formulated as

$$\underline{A} = \frac{1}{2} \left(\underline{I} - \underline{F}^{-T} \underline{F}^{-1} \right) = \frac{1}{2} \left(\underline{I} - \underline{B}^{-1} \right), \tag{10.50}$$

implying

$$\varepsilon_{AE} = \varepsilon_{AE}(\underline{e}) = \underline{e}^{T} \underline{A} \, \underline{e}. \tag{10.51}$$

Again from a mathematical perspective this represents well manageable relations. The Almansi Euler strain is used only sporadically, in contrast to the related symmetrical and objective strain tensor $\boldsymbol{\varepsilon}_F$ (Finger) defined by

$$\boldsymbol{\varepsilon}_F = \frac{1}{2} \left(\boldsymbol{F} \cdot \boldsymbol{F}^T - \boldsymbol{I} \right) = \frac{1}{2}(\boldsymbol{B} - \boldsymbol{I}) = \boldsymbol{B} \cdot \boldsymbol{A} = \boldsymbol{A} \cdot \boldsymbol{B}. \tag{10.52}$$

Note the similarity and the difference with the definition of the Green Lagrange strain tensor $\boldsymbol{E}$. The tensor $\boldsymbol{\varepsilon}_F$ can be considered as an objective version of the invariant tensor $\boldsymbol{E}$. The interpretation of the Finger strain tensor reveals some problems which are beyond the scope of the present context.

In this section a number of different strain tensors have been reviewed. It is an interesting exercise to compare these different tensors for a few elementary homogeneous deformations, for example for the case of uniaxial stress (see Fig. 10.2) and for pure shear. It should be noted that for (infinitesimally) small deformations and (infinitesimally) small rotations (so the limiting case that $\boldsymbol{F} \rightarrow \boldsymbol{I}$) the difference between all treated strain tensors vanishes.

10.5   The volume change factor

Consider an arbitrary, infinitesimally small material element (parallellepipedum), in the reference configuration spanned by three linearly independent vectors $d\vec{x}_0^a$, $d\vec{x}_0^b$ and $d\vec{x}_0^c$. The volume of the element is specified by dV_0 and can in principle be calculated when the vectors $d\vec{x}_0^i$ (with $i = a, b, c$) are known. The vectors $d\vec{x}^i$ in the current configuration, associated with the vectors $d\vec{x}_0^i$ in the reference configuration, can be determined using the deformation tensor F via

$$d\vec{x}^i = F \cdot d\vec{x}_0^i \quad \text{for} \quad i = a, b, c. \tag{10.53}$$

Based on the vectors $d\vec{x}^i$ the volume dV of the material element in the current configuration can be calculated. For the volumetric change ratio J it can be shown:

$$J = \frac{dV}{dV_0} = \det(F). \tag{10.54}$$

The result is independent of the originally chosen shape and orientation of the element dV_0.

For **isochoric** deformation (this is a deformation such that locally the volume of the material does not change) the volume change ratio satisfies

$$J = \det(F) = 1. \tag{10.55}$$

For materials that are incompressible, a property which is often attributed to biological materials (related to the high water content of the materials), the deformation is locally always isochoric. For compressible materials it is possible to prescribe an isochoric deformation by means of a special choice of the external mechanical load.

10.6   Deformation rate and rotation velocity

In the preceding sections the current configuration or current state is considered at a (fixed) time t and compared with the reference configuration. Based on that, concepts like deformation and rotation were defined. In this section the attention is focussed on (infinitesimally) small changes of the current state, as seen in the time domain.

A material line segment $d\vec{x}$ (solid or fluid) in the current state at time t converts in the line segment $d\vec{x} + d\dot{\vec{x}}\, dt$ at time $t + dt$, see Fig. 10.4.

It can be written (also see Chapter 7):

$$d\dot{\vec{x}} = L \cdot d\vec{x} \quad \text{with} \quad L = \left(\vec{\nabla}\vec{v}\right)^{\mathrm{T}}, \tag{10.56}$$

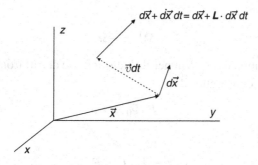

Figure 10.4

Change of a material line segment $d\vec{x}$ after a time increment dt.

where $\vec{v} = \dot{\vec{x}}$ specifies the velocity of the material and L is the velocity gradient tensor. The tensor L is purely a current variable, not in any way related to the reference configuration.

Using the deformation tensor it can be written:

$$d\vec{x} = F \cdot d\vec{x}_0 \tag{10.57}$$

and therefore

$$\dot{d\vec{x}} = \dot{F} \cdot d\vec{x}_0 = \dot{F} \cdot F^{-1} \cdot d\vec{x}, \tag{10.58}$$

accordingly resulting in the relation between the tensors L and F:

$$L = \dot{F} \cdot F^{-1}. \tag{10.59}$$

It is common practice to decompose the velocity gradient tensor L in a symmetrical part D and a skew symmetrical part Ω. The tensor D is called the **rate of deformation tensor** and the tensor Ω the **rotation velocity tensor** or **spin tensor**. The definitions are:

$$D = \frac{1}{2}\left(L + L^{\mathrm{T}}\right) = \frac{1}{2}\left(\left(\vec{\nabla}\vec{v}\right)^{\mathrm{T}} + \left(\vec{\nabla}\vec{v}\right)\right) = \frac{1}{2}\left(\dot{F} \cdot F^{-1} + F^{-\mathrm{T}} \cdot \dot{F}^{\mathrm{T}}\right)$$

$$\Omega = \frac{1}{2}\left(L - L^{\mathrm{T}}\right) = \frac{1}{2}\left(\left(\vec{\nabla}\vec{v}\right)^{\mathrm{T}} - \left(\vec{\nabla}\vec{v}\right)\right) = \frac{1}{2}\left(\dot{F} \cdot F^{-1} - F^{-\mathrm{T}} \cdot \dot{F}^{\mathrm{T}}\right)$$

$$\tag{10.60}$$

and so

$$\dot{d\vec{x}} = (D + \Omega) \cdot d\vec{x}, \tag{10.61}$$

with

$$D^{\mathrm{T}} = D \tag{10.62}$$

and

$$\mathbf{\Omega}^\mathrm{T} = -\mathbf{\Omega}. \tag{10.63}$$

For an interpretation of the symmetrical tensor $\mathbf{D}$ we depart from the relations that have been derived in Section 10.3:

$$\mathbf{F} \cdot \vec{e}_0 = \lambda\, \vec{e} \tag{10.64}$$

and

$$\lambda^2 = \vec{e}_0 \cdot \mathbf{F}^T \cdot \mathbf{F} \cdot \vec{e}_0. \tag{10.65}$$

The material time derivative of the equation for λ^2 can be elaborated as follows:

$$
\begin{aligned}
2\lambda\dot{\lambda} &= \vec{e}_0 \cdot \left(\dot{\mathbf{F}}^\mathrm{T} \cdot \mathbf{F} + \mathbf{F}^\mathrm{T} \cdot \dot{\mathbf{F}} \right) \cdot \vec{e}_0 \\
&= \vec{e}_0 \cdot \left(\mathbf{I} \cdot \dot{\mathbf{F}}^\mathrm{T} \cdot \mathbf{F} + \mathbf{F}^\mathrm{T} \cdot \dot{\mathbf{F}} \cdot \mathbf{I} \right) \cdot \vec{e}_0 \\
&= \vec{e}_0 \cdot \left(\mathbf{F}^\mathrm{T} \cdot \mathbf{F}^{-\mathrm{T}} \cdot \dot{\mathbf{F}}^\mathrm{T} \cdot \mathbf{F} + \mathbf{F}^\mathrm{T} \cdot \dot{\mathbf{F}} \cdot \mathbf{F}^{-1} \cdot \mathbf{F} \right) \cdot \vec{e}_0 \\
&= \vec{e}_0 \cdot \mathbf{F}^\mathrm{T} \cdot \left(\mathbf{F}^{-\mathrm{T}} \cdot \dot{\mathbf{F}}^\mathrm{T} + \dot{\mathbf{F}} \cdot \mathbf{F}^{-1} \right) \cdot \mathbf{F} \cdot \vec{e}_0 \\
&= \lambda^2 \vec{e} \cdot (2\mathbf{D}) \cdot \vec{e},
\end{aligned}
\tag{10.66}
$$

eventually resulting in the simple relation:

$$\frac{\dot{\lambda}}{\lambda} = \vec{e} \cdot \mathbf{D} \cdot \vec{e} \quad (= \ln(\lambda)). \tag{10.67}$$

This equation shows that the deformation velocity tensor $\mathbf{D}$ completely determines the current rate of (logarithmic) strain for an arbitrary line segment in the current state with a direction specified by $\vec{e}$. The analogous equation in component form is written as

$$\frac{\dot{\lambda}}{\lambda} = \underline{e}^\mathrm{T} \underline{D}\, \underline{e}. \tag{10.68}$$

The terms on the diagonal of the matrix $\underline{D}$ represent the rate of strain in the directions of the x-, y- and z-coordinates. The off-diagonal terms represent the rate of shear.

For the interpretation of the skew symmetric spin tensor Eq. (10.61) can be used directly. After all, it is clear that the contribution $\mathbf{\Omega} \cdot d\vec{x}$ to $d\dot{\vec{x}}$ is always perpendicular to $d\vec{x}$, because for all $d\vec{x}$:

$$d\vec{x} \cdot \mathbf{\Omega} \cdot d\vec{x} = 0 \quad \text{because} \quad \mathbf{\Omega}^\mathrm{T} = -\mathbf{\Omega}, \tag{10.69}$$

meaning that the contribution $\mathbf{\Omega} \cdot d\vec{x}$ has to be considered as the effect of a rotation.

For the material time derivative of the volume change factor $J = \det(\mathbf{F})$ it can be derived (without proof):

$$\dot{J} = J\, \mathrm{tr}(\dot{\mathbf{F}} \cdot \mathbf{F}^{-1}) = J\, \mathrm{tr}(\mathbf{L}) = J\, \vec{\nabla} \cdot \vec{v} = J\, \mathrm{tr}(\mathbf{D}). \tag{10.70}$$

Like in previous sections we will examine the influence of an extra rigid body movement of the current configuration, also see Sections 9.7 and 10.3. Variables, associated with the extra rotating (via the rotation tensor $P = P(t)$ that depends explicitly on the time) and extra translating (with $\vec{\lambda} = \vec{\lambda}(t)$ the translation vector) virtual current state, are again denoted by the superscript *. It can be written:

$$\vec{x}^* = P \cdot \vec{x} + \vec{\lambda}$$
$$\vec{v}^* = \dot{\vec{x}}^* = \dot{P} \cdot \vec{x} + P \cdot \dot{\vec{x}} + \dot{\vec{\lambda}} = P \cdot \vec{v} + \dot{P} \cdot \vec{x} + \dot{\vec{\lambda}}. \qquad (10.71)$$

For the velocity gradient tensor L^* in the extra moving configuration it is found:

$$L^* = \left(\vec{\nabla}\vec{v}^*\right)^{*T} = \left(P \cdot \vec{\nabla}\left(\vec{v} \cdot P^T + \vec{x} \cdot \dot{P}^T + \dot{\vec{\lambda}}\right)\right)^T$$
$$= \left(P \cdot L^T \cdot P^T + P \cdot I \cdot \dot{P}^T\right)^T$$
$$= P \cdot L \cdot P^T + \dot{P} \cdot P^T. \qquad (10.72)$$

An identical result is obtained via a different procedure:

$$L^* = \dot{F}^* \cdot (F^*)^{-1}$$
$$= \left(\dot{P} \cdot F + P \cdot \dot{F}\right) \cdot F^{-1} \cdot P^{-1}$$
$$= P \cdot L \cdot P^T + \dot{P} \cdot P^T. \qquad (10.73)$$

It can be observed that the velocity gradient tensor L is neither objective, nor invariant. The contribution $\dot{P} \cdot P^T$ in Eq. (10.73) however is skew symmetric, after all:

$$P \cdot P^T = I \rightarrow \dot{P} \cdot P^T + P \cdot \dot{P}^T = 0$$
$$\rightarrow \dot{P} \cdot P^T = -P \cdot \dot{P}^T = -\left(\dot{P} \cdot P^T\right)^T, \qquad (10.74)$$

meaning that the symmetrical part of the tensor L (which is the rate of deformation tensor D) is objective:

$$D^* = P \cdot D \cdot P^T. \qquad (10.75)$$

The spin tensor $\mathbf{\Omega}$ is not objective.

Exercises

10.1 In a subvolume of a material continuum the deformation of the current state, with respect to the reference state, is homogeneous. In a Cartesian xyz-coordinate system the associated deformation tensor is given as

$$F = I + 3\vec{e}_y\vec{e}_y - 7\vec{e}_y\vec{e}_z - \vec{e}_z\vec{e}_y + \vec{e}_z\vec{e}_z,$$

with I the unit tensor.

For a material point P within the subvolume the position vectors in the reference state as well as the current state are given, respectively:

$$\vec{x}_{0P} = \vec{e}_x + \vec{e}_y + \vec{e}_z, \quad \vec{x}_P = 2\vec{e}_x + 3\vec{e}_y - 2\vec{e}_z.$$

Another point Q within the subvolume appears to be in the origin in the current configuration: $\vec{x}_Q = \vec{0}$.

Calculate the position vector $\vec{x}_{0Q}$ of the point Q in the reference state.

10.2   Within a subvolume of a material continuum the deformation tensor in the deformed current state, with respect to the reference state, is constant. Consider a vector of material points, denoted with $\vec{a}_0$ in the reference state and with $\vec{a}$ in the current state. The angle between $\vec{a}_0$ and $\vec{a}$ is given by ϕ.

Prove that we can write for ϕ:

$$\cos(\phi) = \frac{\vec{a}_0 \cdot \boldsymbol{F} \cdot \vec{a}_0}{\sqrt{(\vec{a}_0 \cdot \vec{a}_0)(\vec{a}_0 \cdot \boldsymbol{F}^{\mathrm{T}} \cdot \boldsymbol{F} \cdot \vec{a}_0)}}.$$

10.3   Within a subvolume of a material continuum the deformation tensor in the deformed current state, with respect to the reference state, is constant. In a Cartesian xyz-coordinate system the following deformation tensor is given:

$$\boldsymbol{F} = \boldsymbol{I} + 4\,\vec{e}_x\vec{e}_x + 2\,\vec{e}_x\vec{e}_y + 2\,\vec{e}_y\vec{e}_x,$$

with $\boldsymbol{I}$ the unit tensor. Within the subvolume two material points P and Q are considered. The position vectors in the reference state are given:

$$\vec{x}_{0P} = \vec{e}_x + \vec{e}_y + \vec{e}_z, \quad \vec{x}_{0Q} = 2\vec{e}_x + 3\vec{e}_y + 2\vec{e}_z.$$

In addition, the position vector for the point P in the current state is given:

$$\vec{x}_P = 2\vec{e}_x + 2\vec{e}_y + 2\vec{e}_z.$$

Calculate the position vector $\vec{x}_Q$ of the point Q in the current state.

10.4   The deformation of a material particle, in the current state with respect to the reference state, is fully described by the Green Lagrange strain matrix $\underline{E}$, with respect to a Cartesian xyz-coordinate system, with

$$\underline{E} = \frac{1}{2} \begin{bmatrix} 4 & 3 & 0 \\ 3 & 1 & 0 \\ 0 & 0 & 3 \end{bmatrix}.$$

Calculate the volume change factor J for this particle.

10.5   A tendon is stretched in a uniaxial stress test. The tendon behaves like an incompressible material. The length axis of the tendon coincides with the x-axis. For this test the following (time-independent) deformation rate matrix $\underline{D}$ is applicable (expressed in $[\mathrm{s}^{-1}]$):

$$\underline{D} = \begin{bmatrix} 0.02 & 0 & 0 \\ 0 & -0.01 & 0 \\ 0 & 0 & -0.01 \end{bmatrix}.$$

At time $t = 0$ [s] the tendon has a length ℓ_0 equal to 3 [cm]. From this time on the above given matrix $\underline{D}$ can be applied. Calculate the length ℓ of the tendon as a function of the time t.

10.6 At some material point the local deformation process is described by means of the deformation tensor as a function of time: $\boldsymbol{F}(t)$. Based on this deformation tensor, the left Cauchy Green tensor $\boldsymbol{B} = \boldsymbol{F} \cdot \boldsymbol{F}^{\mathrm{T}}$ can be derived and subsequently the Finger tensor $\boldsymbol{\varepsilon}_{\mathrm{F}} = \frac{1}{2}(\boldsymbol{B} - \boldsymbol{I})$.

Prove that $\dot{\boldsymbol{\varepsilon}}_{\mathrm{F}} = \frac{1}{2}(\boldsymbol{L} \cdot \boldsymbol{B} + \boldsymbol{B} \cdot \boldsymbol{L}^{\mathrm{T}})$, with: $\boldsymbol{L} = \dot{\boldsymbol{F}} \cdot \boldsymbol{F}^{-1}$ the velocity gradient tensor.

11 Local balance of mass, momentum and energy

11.1 Introduction

A coherent amount of material (a material body or possibly a distinguishable material fraction) is considered to be a continuum with current volume V in three-dimensional space. In Chapter 8 attention was focussed on the local stress state (the internal interaction between neighbouring volume elements), in Chapters 9 and 10 on the local kinematics (shape and volume changes of material particles). To determine the stresses and kinematic variables as a function of the position in the three-dimensional space, we need a description of the material behaviour, which will be the subject of subsequent chapters, and we need local balance laws. In the present chapter the balance of mass (leading to the continuity equation) and the balance of momentum (leading to the equations of motion) for a continuum will be formulated. In addition the balance of mechanical power will be derived based on the balance of momentum.

11.2 The local balance of mass

Let us focus our attention on an infinitesimally small rectangular material element $dV = dxdydz$ in the current state, see Fig. 11.1.

The mass in the current volume element dV equals the mass in the reference configuration of the associated volume element dV_0, while the volumes are related by

$$dV = JdV_0 \text{ with } J = \det(\boldsymbol{F}).\tag{11.1}$$

It is implicitly assumed that during the transformation from the reference state to the current deformed state no material of the considered type is created or lost. So there is no mass exchange with certain other fractions, for example in the form of a chemical reaction. Based on mass conservation, it can be stated:

$$\rho_0 dV_0 = \rho dV = \rho JdV_0,\tag{11.2}$$

Figure 11.1

Infinitesimally small element $dV = dxdydz$ in the current state.

with ρ_0 and ρ the (mass) density in the reference and current configuration, respectively. So, balance of mass leads to the statement, that the product ρJ is time independent and that the material time derivative of the product equals zero:

$$\dot{\rho} J + \rho \dot{J} = 0. \tag{11.3}$$

Using Eq. (10.70), with D the deformation rate tensor we obtain:

$$\dot{\rho} = -\rho \, \text{tr}(D). \tag{11.4}$$

In this final result the reference configuration is no longer represented: all variables in this equation are related to the current configuration.

11.3 The local balance of momentum

In Chapter 8 the equilibrium equations have been derived by considering the force equilibrium of a material cube. The equilibrium equations are a special form of the balance of momentum equations, which are the subject of the present section. This means that some overlap between both derivations can be recognized.

Again, the material volume element dV from Fig. 11.1 is considered. Using the balance of mass allows us to write for the current momentum of the element:

$$\vec{v}\rho dV = \vec{v}\rho J dV_0, \tag{11.5}$$

with $\vec{v}$ the velocity of the material. Using that $\rho J dV_0 = \rho_0 dV_0$ is constant, the change per unit time of the momentum of the volume element equals

$$\dot{\vec{v}}\rho J dV_0 = \dot{\vec{v}}\rho dV = \vec{a}\rho dV, \tag{11.6}$$

with $\vec{a}$ the acceleration vector (see Section 9.2). In the following, the above given momentum change will be postulated, according to Newton's law, to be equal to the total 'external' force acting on the volume element.

For the element in Fig. 11.1 the resulting force in the x-direction is considered. This force is the sum of the forces working in the x-direction on the outer surfaces of the element and the distributed load (force per unit mass) acting on the element in the x-direction, so successively (also see Chapter 8):

$$-\sigma_{xx}dydz \qquad \text{back plane}$$

$$\left(\sigma_{xx} + \frac{\partial \sigma_{xx}}{\partial x}dx\right)dydz \qquad \text{frontal plane}$$

$$-\sigma_{xy}dxdz \qquad \text{left plane}$$

$$\left(\sigma_{xy} + \frac{\partial \sigma_{xy}}{\partial y}dy\right)dxdz \qquad \text{right plane}$$

$$-\sigma_{xz}dxdy \qquad \text{bottom plane}$$

$$\left(\sigma_{xz} + \frac{\partial \sigma_{xz}}{\partial z}dz\right)dxdy \qquad \text{top plane}$$

$$q_x\rho dxdydz \qquad \text{volume,}$$

with resultant (in x-direction):

$$\left(\frac{\partial \sigma_{xx}}{\partial x} + \frac{\partial \sigma_{xy}}{\partial y} + \frac{\partial \sigma_{xz}}{\partial z} + q_x\rho\right)dxdydz = \left(\frac{\partial \sigma_{xx}}{\partial x} + \frac{\partial \sigma_{xy}}{\partial y} + \frac{\partial \sigma_{xz}}{\partial z} + q_x\rho\right)dV.$$

$$(11.7)$$

Similarly, the resulting forces in y- and z-direction can be determined. All external forces applied to the volume element are stored in a column. Using the gradient operator $\underset{\sim}{\nabla}$ introduced earlier and the symmetrical Cauchy stress matrix $\underline{\sigma}$ the column with external forces can be written as

$$\begin{bmatrix} \frac{\partial \sigma_{xx}}{\partial x} + \frac{\partial \sigma_{xy}}{\partial y} + \frac{\partial \sigma_{xz}}{\partial z} + q_x\rho \\\\ \frac{\partial \sigma_{xy}}{\partial x} + \frac{\partial \sigma_{yy}}{\partial y} + \frac{\partial \sigma_{yz}}{\partial z} + q_y\rho \\\\ \frac{\partial \sigma_{xz}}{\partial x} + \frac{\partial \sigma_{yz}}{\partial y} + \frac{\partial \sigma_{zz}}{\partial z} + q_z\rho \end{bmatrix} dV = \left(\left(\underset{\sim}{\nabla}^T \underline{\sigma}\right)^T + \rho\underset{\sim}{q}\right)dV \qquad (11.8)$$

with the column $\underset{\sim}{q}$ for the distributed load defined according to

$$\underset{\sim}{q} = \begin{bmatrix} q_x \\ q_y \\ q_z \end{bmatrix}.$$

In vector/tensor notation the following expression can be given for the resulting force on the element dV:

$$\left(\vec{\nabla}\cdot\sigma + \rho\vec{q}\right)dV.$$

For the considered element the time derivative of the momentum equals the external load, leading to

$$\vec{\nabla}\cdot\sigma + \rho\vec{q} = \rho\vec{a}, \qquad (11.9)$$

and also

$$(\nabla^{\mathrm{T}}\underline{\sigma})^{\mathrm{T}} + \rho\,\underline{q} = \rho\,\underline{a}. \tag{11.10}$$

This equation is called the local equation of motion. Again, all variables in this equation refer to the current configuration; the defined reference configuration (as usual for a solid) is not relevant for this.

11.4 The local balance of mechanical power

It has to be stated explicitly that no new balance law is introduced here; only use will be made of relations that were already introduced before. Again the element is considered, that was introduced in Fig. 11.1 (volume in reference state dV_0, current volume $dV = JdV_0$). The dot product of the local equation of motion Eq. (11.9), with the velocity vector $\vec{v}$ is taken. The result is multiplied by the current volume of the element, yielding

$$\vec{v}\cdot\left(\vec{\nabla}\cdot\sigma\right)dV + \vec{v}\cdot\vec{q}\,\rho\,dV = \vec{v}\cdot\vec{a}\,\rho\,dV. \tag{11.11}$$

The terms in this equation will be interpreted separately.

- For the term on the right-hand side it can be written:

$$\vec{v}\cdot\vec{a}\,\rho\,dV = d\dot{U}_{\mathrm{kin}}, \tag{11.12}$$

  with

$$dU_{\mathrm{kin}} = \frac{1}{2}\vec{v}\cdot\vec{v}\,\rho\,dV = \frac{1}{2}\vec{v}\cdot\vec{v}\,\rho\,JdV_0 = \frac{1}{2}\vec{v}\cdot\vec{v}\,\rho_0 dV_0, \tag{11.13}$$

  where dU_{kin} is the current kinetic energy of the considered volume element. With respect to the material time derivative that is used in Eq. (11.12) it should be realized that $\rho JdV_0 = \rho_0 dV_0$ is constant.
  The term on the right-hand side of Eq. (11.11) can be interpreted as the change of the kinetic energy per unit time.
- The second term on the left-hand side of Eq. (11.11) can be directly interpreted as the mechanical power, externally applied to the volume element by the distributed load $\vec{q}$. It can be noted that

$$dP_{ext_{\vec{q}}} = \vec{v}\cdot\vec{q}\rho dV. \tag{11.14}$$

- For the first term on the left-hand side of Eq. (11.11) some careful mathematical elaboration, using the symmetry of the stress tensor σ, leads to

$$\vec{v}\cdot\left(\vec{\nabla}\cdot\sigma\right)dV = \vec{\nabla}\cdot(\sigma\cdot\vec{v})\,dV - \mathrm{tr}\left(\sigma\cdot(\vec{\nabla}\vec{v})\right)dV = \vec{\nabla}\cdot(\sigma\cdot\vec{v})\,dV - \mathrm{tr}(\sigma\cdot\boldsymbol{D})\,dV. \tag{11.15}$$

For this, the definition of the deformation rate tensor D of Section 10.6 is used.

The first term on the right-hand side of Eq. (11.15) represents the resulting, externally applied power to the volume element by the forces (originating from neighbouring elements) acting on the outer surfaces of the element:

$$dP_{ext_\sigma} = \vec{\nabla} \cdot (\vec{\sigma} \cdot \vec{v}) \ dV. \tag{11.16}$$

This can be proven with a similar strategy as was used in the previous section, to derive the resultant of the forces acting on the outer surfaces of the volume element. Interpretation of the second term on the right-hand side,

$$-\text{tr}(\sigma \cdot D) \ dV,$$

will be given below.

Summarizing, after reconsidering Eq. (11.11), and using the results of the above, it can be stated:

$$dP_{ext_\sigma} - \text{tr}(\sigma \cdot D) \, dV + dP_{ext_{\vec{q}}} = d\dot{U}_{kin}, \tag{11.17}$$

and after some re-ordering:

$$\text{tr}(\sigma \cdot D) \, dV + d\dot{U}_{kin} = dP_{ext}, \tag{11.18}$$

with

$$dP_{ext} = dP_{ext_\sigma} + dP_{ext_{\vec{q}}}. \tag{11.19}$$

So, it can be observed that the total externally applied mechanical power is partly used for a change of the kinetic energy. The remaining part is stored internally, so:

$$dP_{int} = \text{tr}(\sigma \cdot D) \, dV. \tag{11.20}$$

This internally stored mechanical power (increase of the internal mechanical energy per unit of time) can be completely reversible (for elastic behaviour), partly reversible and partly irreversible (for visco-elastic behaviour) or fully irreversible (for viscous behaviour). In the latter case all externally applied mechanical energy to the material is dissipated and converted into other forms of energy (in general a large part is converted into heat).

11.5   Lagrangian and Eulerian description of the balance equations

In summary, the balance equations of mass and momentum as derived in the sections 11.2 and 11.3 can be written as

$$\text{balance of mass} \quad \dot{\rho} = -\rho \text{tr}(D) = -\rho \vec{\nabla} \cdot \vec{v},$$

$$\text{balance of momentum} \quad \vec{\nabla} \cdot \sigma + \rho \vec{q} = \rho \vec{a} = \rho \dot{\vec{v}},$$

and in column/matrix notation:

$$\text{balance of mass}\quad \dot{\rho} = -\rho\,\text{tr}(\underline{D}) = -\rho\underline{\nabla}^{\mathrm{T}}\underline{v},$$

$$\text{balance of momentum}\quad \left(\underline{\nabla}^{\mathrm{T}}\underline{\sigma}\right)^{\mathrm{T}} + \rho\underline{q} = \rho\underline{a} = \rho\dot{\underline{v}}.$$

In a typical Lagrangian description, the field variables are considered to be a function of the material coordinates $\vec{x}_0$, being defined in the reference configuration, and time t. It can be stated that the balance laws have to be satisfied for all $\vec{x}_0$ within the domain V_0 and for all points in time.

Because of the physical relevance (for solids) of the deformation tensor F the balance of mass will usually not be used in the differential form as given above, but rather as

$$\rho = \frac{\rho_0}{\det(F)} \quad \text{and also} \quad \rho = \frac{\rho_0}{\det(\underline{F})}. \tag{11.21}$$

The gradient operator $\vec{\nabla}$ (and also $\underline{\nabla}$) in the balance of momentum equation, built up from derivatives with respect to the spatial coordinates, can be transformed into the gradient operator $\vec{\nabla}_0$ (and $\underline{\nabla}_0$) with respect to the material coordinates, see Section 9.6. The balance of momentum can then be formulated according to

$$\left(F^{-\mathrm{T}} \cdot \vec{\nabla}_0\right) \cdot \sigma + \rho\vec{q} = \rho\vec{a} = \rho\dot{\vec{v}}, \tag{11.22}$$

and also

$$\left(\left(\underline{F}^{-\mathrm{T}}\underline{\nabla}_0\right)^{\mathrm{T}} \underline{\sigma}\right)^{\mathrm{T}} + \rho\underline{q} = \rho\underline{a} = \rho\dot{\underline{v}}. \tag{11.23}$$

In a typical Eulerian description, the field properties are considered to be a function of the spatial coordinates $\vec{x}$, indicating locations in the current configuration, and time t. It can be stated that the balance laws have to be satisfied for all $\vec{x}$ within the domain V and for all points in time.

To reformulate the balance laws the material time derivative is 'replaced' by the spatial time derivative, see Section 9.4. For the mass balance this yields

$$\frac{\delta\rho}{\delta t} + \vec{v} \cdot \vec{\nabla}\rho = -\rho\,\text{tr}(D) = -\rho\vec{\nabla} \cdot \vec{v}, \tag{11.24}$$

and so

$$\frac{\delta\rho}{\delta t} + \vec{\nabla} \cdot (\rho\vec{v}) = 0. \tag{11.25}$$

In column/matrix notation:

$$\frac{\delta\rho}{\delta t} + \underline{v}^{T}\underline{\nabla}\,\rho = -\rho\,\text{tr}(\underline{D}) = -\rho\underline{\nabla}^{T}\,\underline{v}, \tag{11.26}$$

and so

$$\frac{\delta\rho}{\delta t} + \underset{\sim}{\nabla}^{\mathrm{T}}(\rho\underset{\sim}{v}) = 0. \tag{11.27}$$

For the balance of momentum this yields

$$\vec{\nabla}\cdot\boldsymbol{\sigma} + \rho\vec{q} = \rho\left(\left(\vec{\nabla}\vec{v}\right)^{\mathrm{T}}\cdot\vec{v} + \frac{\delta\vec{v}}{\delta t}\right), \tag{11.28}$$

and in column/matrix notation:

$$\left(\underset{\sim}{\nabla}^{\mathrm{T}}\underset{\sim}{\sigma}\right)^{\mathrm{T}} + \rho\underset{\sim}{q} = \rho\left(\left(\underset{\sim}{\nabla}\,\underset{\sim}{v}^{\mathrm{T}}\right)^{\mathrm{T}}\underset{\sim}{v} + \frac{\delta\underset{\sim}{v}}{\delta t}\right). \tag{11.29}$$

For the special case of a stationary flow of a material, the following balance equation results for the mass balance:

$$\vec{\nabla}\cdot(\rho\vec{v}) = 0, \tag{11.30}$$

and also

$$\underset{\sim}{\nabla}^{\mathrm{T}}(\rho\underset{\sim}{v}) = 0. \tag{11.31}$$

The momentum equation reduces to:

$$\vec{\nabla}\cdot\boldsymbol{\sigma} + \rho\,\vec{q} = \rho\left(\vec{\nabla}\vec{v}\right)^{\mathrm{T}}\cdot\vec{v}, \tag{11.32}$$

and also

$$\left(\underset{\sim}{\nabla}^{\mathrm{T}}\underset{\sim}{\sigma}\right)^{\mathrm{T}} + \rho\,\underset{\sim}{q} = \rho\left(\underset{\sim}{\nabla}\,\underset{\sim}{v}^{\mathrm{T}}\right)^{\mathrm{T}}\underset{\sim}{v}. \tag{11.33}$$

Exercises

11.1   Compressible air is flowing through the bronchi. A bronchus is modelled as a straight cylindrical tube. We consider a stationary flow. For each cross section of the tube the velocity of the air V and the density ρ is constant over the cross section. In the direction of the flow, the velocity of the air and the density vary, because of temperature differences. Two cross sections A and B at a distance L are considered. See the figure below.

Which relation can be derived for the variables indicated in the figure?

11.2 For a solid element the deformation matrix $\underline{F}$ is given as a function of time t, with reference to the undeformed configuration at time $t = 0$:

$$\underline{F} = \begin{bmatrix} 1 + \alpha t & 0 & \alpha t \\ 0 & 1 & 0 \\ \alpha t & 0 & 1 + \alpha t \end{bmatrix} \quad \text{with} \quad \alpha = 0.01 \ [\text{s}^{-1}].$$

In the reference configuration the density ρ_0 equals 1500 [kg m^{-3}]. As a result of the deformation process the density ρ of the material will change. Determine the density as a function of the time.

12 Constitutive modelling of solids and fluids

12.1 Introduction

In the first section of this chapter the (biological) material under consideration can be regarded as a solid. This implies that it is possible to define a reference configuration and local deformations can be related to this reference configuration (see Chapter 10). It is assumed that the deformations (or more precisely, the path along which the actual deformations are reached: the deformation history) fully determine the current stress state (see Chapter 8), except for the special case when we deal with incompressible material behaviour. A number of non-mechanical phenomena are not included, such as the influence of temperature variations.

The **constitutive equations** discussed in the present chapter address a relation that can be formally written as

$$\sigma(t) = \mathcal{F}\{F(\tau)\,;\ \tau \leq t\}, \tag{12.1}$$

with $\sigma(t)$ the current Cauchy stress tensor at time t and with $F(\tau)$ the deformation tensor at the relevant (previous) times τ up to the time t, assuming compressible material behaviour. A specification of this relation (the material behaviour) can only be obtained by means of experimental studies. In the current chapter we restrict our attention to **elastic behaviour**. In that case the deformation history is not relevant and we can formally write

$$\sigma = \sigma(F). \tag{12.2}$$

Further elaboration of this relation will initially be done for the case of small deformations and rotations, so under the condition $F \approx I$. After that, the consequences of including large deformations (and large rotations) will be discussed. In that case it is important to account for the fact that large additional rigid body motions are not allowed to induce extra stresses in the material.

In the second part of this chapter we focus on fluid behaviour. Because in a fluid the reference configuration is usually not defined, an Eulerian approach will be chosen and the velocity gradient tensor L and the deformation rate tensor D play a central role.

12.2 Elastic behaviour at small deformations and rotations

To describe the current deformation state of the material a reference configuration has to be specified. Although, in principle, the choice for the reference configuration is free, in this case a fixed, stress free state is chosen. This is not as trivial as it may seem, because for many biological materials a zero-stress state does not exist in vivo, however, this discussion is not within the scope of the current book. Interested readers are referred to e.g. [7]. So

$$\sigma(F = I) = 0. \tag{12.3}$$

It should be noted that this statement only applies to purely elastic material behaviour. If the deformation history is important for the current stress state, the relation given above is generally certainly not valid.

The current deformation tensor F fully describes the local deformations with respect to the reference configuration. This also holds for the linear strain tensor ε that was introduced in Section 10.4 under the strict condition that $F \approx I$. Because rotations can be neglected under these conditions, the Cauchy stress tensor σ can be coupled directly to the linear strain tensor ε via an expression according to the format:

$$\sigma = \sigma(\varepsilon) \quad \text{with} \quad \sigma(\varepsilon = 0) = 0. \tag{12.4}$$

The exact specification of Eq. (12.4) has to be derived from experimental work and is a major issue in biomechanics. In the current section we will explore the commonly used **Hooke's law**, which is often adopted as the first approximation to describe the material behaviour. Hooke's law supplies a linear relation between the components of the stress tensor σ, stored in the (3×3) matrix $\underline{\sigma}$ and the components of the linear strain tensor ε, stored in the (3×3) matrix $\underline{\varepsilon}$. In addition, it is assumed that the material behaviour is **isotropic** (the behaviour is identical in all directions).

Hooke's law will be specified, using the decomposition of the stress tensor in a hydrostatic and a deviatoric part, as described in Section 8.7:

$$\sigma = \sigma^{\mathrm{h}} + \sigma^{\mathrm{d}} = -pI + \sigma^{\mathrm{d}}, \tag{12.5}$$

with p the hydrostatic pressure, defined as

$$p = -\frac{\mathrm{tr}(\sigma)}{3}. \tag{12.6}$$

The linear strain tensor ε can be split in a similar way:

$$\varepsilon = \varepsilon^{\mathrm{h}} + \varepsilon^{\mathrm{d}} = \frac{\varepsilon^{\mathrm{v}}}{3}I + \varepsilon^{\mathrm{d}}, \tag{12.7}$$

with ε^v the (relative) volume change:

$$\varepsilon^v = \mathrm{tr}(\boldsymbol{\varepsilon})\,. \qquad (12.8)$$

This requires some extra elucidation. In general, the relative volume change is defined as (see Sections 10.4 and 10.5)

$$\varepsilon^v = \frac{dV - dV_0}{dV_0} = J - 1 = \det(\boldsymbol{F}) - 1 = \det\left(\boldsymbol{I} + \left(\vec{\nabla}_0 \vec{u}\right)^{\mathrm{T}}\right) - 1\,. \qquad (12.9)$$

However, because the components of $\vec{\nabla}_0 \vec{u}$ can be neglected with respect to 1 it can be written (after linearization):

$$\varepsilon^v = 1 + \mathrm{tr}\left(\left(\vec{\nabla}_0 \vec{u}\right)^{\mathrm{T}}\right) - 1$$

$$= \mathrm{tr}\left(\frac{1}{2}\left(\left(\vec{\nabla}_0 \vec{u}\right) + \left(\vec{\nabla}_0 \vec{u}\right)^{\mathrm{T}}\right)\right) = \mathrm{tr}(\boldsymbol{\varepsilon})\,. \qquad (12.10)$$

Hooke's law for linearly elastic isotropic behaviour can now be written as

$$p = -K\varepsilon^v, \quad \boldsymbol{\sigma}^{\mathrm{d}} = 2G\,\boldsymbol{\varepsilon}^{\mathrm{d}}, \qquad (12.11)$$

and in matrix notation:

$$p = -K\varepsilon^v, \quad \underline{\sigma}^{\mathrm{d}} = 2G\,\underline{\varepsilon}^{\mathrm{d}}, \qquad (12.12)$$

with K the **compression modulus** or **bulk modulus** of the material and G the **shear modulus** (for Hooke's law the relevant material parameters; both positive). In the present section it is assumed, that the material is compressible, meaning that the compression modulus K has a finite value. Based on Hooke's law the above equations can be written as

$$\boldsymbol{\sigma} = \left(K - \frac{2G}{3}\right)\mathrm{tr}(\boldsymbol{\varepsilon})\boldsymbol{I} + 2G\boldsymbol{\varepsilon}, \qquad (12.13)$$

and also

$$\underline{\sigma} = \left(K - \frac{2G}{3}\right)\mathrm{tr}(\underline{\varepsilon})\underline{I} + 2G\underline{\varepsilon}, \qquad (12.14)$$

thus fully establishing the desired tensor relation $\boldsymbol{\sigma} = \boldsymbol{\sigma}(\boldsymbol{\varepsilon})$, and $\underline{\sigma} = \underline{\sigma}(\underline{\varepsilon})$ in matrix notation. For the inverse relationship $\boldsymbol{\varepsilon} = \boldsymbol{\varepsilon}(\boldsymbol{\sigma})$, and $\underline{\varepsilon} = \underline{\varepsilon}(\underline{\sigma})$, it can easily be derived:

$$\boldsymbol{\varepsilon} = \left(\frac{1}{9K} - \frac{1}{6G}\right)\mathrm{tr}(\boldsymbol{\sigma})\boldsymbol{I} + \frac{1}{2G}\boldsymbol{\sigma}, \qquad (12.15)$$

and also

$$\underline{\varepsilon} = \left(\frac{1}{9K} - \frac{1}{6G}\right)\mathrm{tr}(\underline{\sigma})\underline{I} + \frac{1}{2G}\underline{\sigma}. \qquad (12.16)$$

In the following an interpretation of Hooke's law will be given. For this purpose we focus on the matrix formulations $\underline{\sigma} = \underline{\sigma}(\underline{\varepsilon})$ and $\underline{\varepsilon} = \underline{\varepsilon}(\underline{\sigma})$. The symmetrical matrices $\underline{\sigma}$ and $\underline{\varepsilon}$ are composed according to:

$$
\underline{\sigma} = \begin{bmatrix} \sigma_{xx} & \sigma_{xy} & \sigma_{xz} \\ \sigma_{yx} & \sigma_{yy} & \sigma_{yz} \\ \sigma_{zx} & \sigma_{zy} & \sigma_{zz} \end{bmatrix}, \ \underline{\varepsilon} = \begin{bmatrix} \varepsilon_{xx} & \varepsilon_{xy} & \varepsilon_{xz} \\ \varepsilon_{yx} & \varepsilon_{yy} & \varepsilon_{yz} \\ \varepsilon_{zx} & \varepsilon_{zy} & \varepsilon_{zz} \end{bmatrix}. \tag{12.17}
$$

Using $\underline{\varepsilon} = \underline{\varepsilon}(\underline{\sigma})$ it follows from Eq. (12.16) for the diagonal components (the strains in the x-, y- and z-directions):

$$
\varepsilon_{xx} = \left(\frac{1}{9K} - \frac{1}{6G} \right) (\sigma_{xx} + \sigma_{yy} + \sigma_{zz}) + \frac{1}{2G}\sigma_{xx}
$$

$$
\varepsilon_{yy} = \left(\frac{1}{9K} - \frac{1}{6G} \right) (\sigma_{xx} + \sigma_{yy} + \sigma_{zz}) + \frac{1}{2G}\sigma_{yy}
$$

$$
\varepsilon_{zz} = \left(\frac{1}{9K} - \frac{1}{6G} \right) (\sigma_{xx} + \sigma_{yy} + \sigma_{zz}) + \frac{1}{2G}\sigma_{zz}. \tag{12.18}
$$

Eq. (12.18) shows that the strains in the x-, y- and z-directions are determined solely by the normal stresses in the x-, y- and z-directions, which is a consequence of assuming isotropy. It is common practice to use an alternative set of material parameters, namely the Young's modulus E and the Poisson's ratio v. The Young's modulus follows from

$$
\frac{1}{E} = \left(\frac{1}{9K} - \frac{1}{6G} \right) + \frac{1}{2G} = \frac{3K + G}{9KG}, \tag{12.19}
$$

and therefore

$$
E = \frac{9KG}{3K + G}. \tag{12.20}
$$

The Poisson's ratio is defined by

$$
\frac{v}{E} = -\left(\frac{1}{9K} - \frac{1}{6G} \right) = \frac{3K - 2G}{18KG}, \tag{12.21}
$$

so

$$
v = \frac{3K - 2G}{6K + 2G}. \tag{12.22}
$$

From $K > 0$ and $G > 0$ it can easily be derived that $E > 0$ and $-1 < v < 0.5$. Using E and v leads to the commonly used formulation for Hooke's equations:

$$\varepsilon_{xx} = \frac{1}{E} (\quad \sigma_{xx} - \nu\,\sigma_{yy} - \nu\,\sigma_{zz})$$

$$\varepsilon_{yy} = \frac{1}{E} (-\nu\,\sigma_{xx} + \quad \sigma_{yy} - \nu\,\sigma_{zz})$$

$$\varepsilon_{zz} = \frac{1}{E} (-\nu\,\sigma_{xx} - \nu\,\sigma_{yy} + \quad \sigma_{zz}). \tag{12.23}$$

The strain in a certain direction is directly coupled to the stress in that direction via Young's modulus. The stresses in the other directions cause a transverse strain. The reverse equations can also be derived:

$$\sigma_{xx} = \frac{E}{(1+\nu)(1-2\nu)} \left((1-\nu)\,\varepsilon_{xx} + \quad \nu\,\varepsilon_{yy} + \quad \nu\,\varepsilon_{zz}\right)$$

$$\sigma_{yy} = \frac{E}{(1+\nu)(1-2\nu)} \left(\quad \nu\,\varepsilon_{xx} + (1-\nu)\,\varepsilon_{yy} + \quad \nu\,\varepsilon_{zz}\right)$$

$$\sigma_{zz} = \frac{E}{(1+\nu)(1-2\nu)} \left(\quad \nu\,\varepsilon_{xx} + \quad \nu\,\varepsilon_{yy} + (1-\nu)\,\varepsilon_{zz}\right). \tag{12.24}$$

For the shear strains, the off-diagonal components of the matrix $\underline{\varepsilon}$, it follows from Eq. (12.16):

$$\varepsilon_{xy} = \varepsilon_{yx} = \frac{1}{2G}\,\sigma_{xy} = \frac{1}{2G}\,\sigma_{yx}$$

$$\varepsilon_{xz} = \varepsilon_{zx} = \frac{1}{2G}\,\sigma_{xz} = \frac{1}{2G}\,\sigma_{zx}$$

$$\varepsilon_{zy} = \varepsilon_{yz} = \frac{1}{2G}\,\sigma_{yz} = \frac{1}{2G}\,\sigma_{zy}. \tag{12.25}$$

It is clear that shear strains are coupled directly to shear stresses (due to the assumption of isotropy). The inverse relations are trivial. If required, the shear modulus can be written as a function of the Young's modulus E and the Poisson's ratio ν by means of the following equation (follows from Eqs. (12.20) and (12.21) by eliminating K):

$$G = \frac{E}{2(1+\nu)}. \tag{12.26}$$

12.3 The stored internal energy

It is interesting to study the stored internal energy during deformation of a material that is described by means of Hooke's law for linearly elastic behaviour. In Section 11.4 the balance of power was derived for an infinitesimally small element with reference volume dV_0 and current volume $dV = J\,dV_0$. By integrating the internally stored power dP_{int} over the relevant time domain $0 \leq \tau \leq t$, the process time between the reference state and the current state, the total internal energy dE_{int} that is stored by the element is

$$dE_{\text{int}}(t) = \int_{\tau=0}^{t} \text{tr}(\boldsymbol{\sigma} \cdot \boldsymbol{D}) \, dV \, d\tau. \tag{12.27}$$

For this relation, it is assumed that initially (so for $\tau = 0$) the internal mechanical energy was zero. For further elaboration we have to account for dV being a function of time. That is why dV is transformed to dV_0 resulting in

$$dE_{\text{int}}(t) = \int_{\tau=0}^{t} \text{tr}(\boldsymbol{\sigma} \cdot \boldsymbol{D}) \, J dV_0 \, d\tau = \int_{\tau=0}^{t} \text{tr}(\boldsymbol{\sigma} \cdot \boldsymbol{D}) \, J d\tau \, dV_0. \tag{12.28}$$

Subsequently, Φ_{int_0} is introduced as the internal mechanical energy per unit of reference volume, also called the internal energy density (the subscript 0 refers to the fact that the density is defined with respect to the volume in the reference state):

$$\Phi_{\text{int}_0}(t) = \int_{\tau=0}^{t} \text{tr}(\boldsymbol{\sigma} \cdot \boldsymbol{D}) \, J d\tau. \tag{12.29}$$

For a deformation history, given by means of specifying the deformation tensor as a function of time: $\boldsymbol{F}(\tau)$; $0 \leq \tau \leq t$ (with $\boldsymbol{F}(\tau = 0) = \boldsymbol{I}$) and for known material behaviour by specifying the constitutive equations, $\Phi_{\text{int}_0}(t)$ can be calculated. In the following this will be performed using Hooke's law.

Assuming small deformations ($\boldsymbol{F} \approx \boldsymbol{I}$) the rate of deformation tensor $\boldsymbol{D}$ can be written as

$$\boldsymbol{D} = \frac{1}{2}\left(\dot{\boldsymbol{F}} \cdot \boldsymbol{F}^{-1} + \boldsymbol{F}^{-T} \cdot \dot{\boldsymbol{F}}^{T}\right) \approx \frac{1}{2}\left(\dot{\boldsymbol{F}} + \dot{\boldsymbol{F}}^{T}\right) = \dot{\boldsymbol{\varepsilon}}. \tag{12.30}$$

The volume change factor J can be approximated according to

$$J \approx 1 + \text{tr}(\boldsymbol{\varepsilon}) \approx 1. \tag{12.31}$$

Using the definition of the deviatoric form of the strain tensor $\boldsymbol{\varepsilon}$, as defined in Eq. (7.40), and Eq. (12.11), Hooke's law can be written as

$$\boldsymbol{\sigma} = K \text{tr}(\boldsymbol{\varepsilon})\boldsymbol{I} + 2G\boldsymbol{\varepsilon}^{\text{d}}. \tag{12.32}$$

Substituting these results for $\boldsymbol{\sigma}, \boldsymbol{D}$ and J into the expression for Φ_{int_0} yields

$$\begin{aligned}
\Phi_{\text{int}_0} &= \int_{\tau=0}^{t} \left(K\text{tr}(\boldsymbol{\varepsilon})\text{tr}(\dot{\boldsymbol{\varepsilon}}) + 2G\text{tr}(\boldsymbol{\varepsilon}^{\text{d}} \cdot \dot{\boldsymbol{\varepsilon}})\right) d\tau \\
&= \int_{\tau=0}^{t} \left(K\text{tr}(\boldsymbol{\varepsilon})\text{tr}(\dot{\boldsymbol{\varepsilon}}) + 2G\text{tr}(\boldsymbol{\varepsilon}^{\text{d}} \cdot \dot{\boldsymbol{\varepsilon}}^{\text{d}})\right) d\tau \\
&= \left.\left(\frac{1}{2}K\text{tr}^2(\boldsymbol{\varepsilon}) + G\text{tr}(\boldsymbol{\varepsilon}^{\text{d}} \cdot \boldsymbol{\varepsilon}^{\text{d}})\right)\right|_{\tau=0}^{t}.
\end{aligned} \tag{12.33}$$

Taking into account that $\varepsilon(\tau = 0) = \mathbf{0}$ it appears that the 'internal elastic energy per unit volume' Φ_{int_0} at time t is fully determined by the components of the strain tensor at that specific time t; this means that the indication t in this case is redundant and without any problem it can be written:

$$\Phi_{\mathrm{int}_0}(\varepsilon) = \frac{1}{2}K\mathrm{tr}^2(\varepsilon) + G\mathrm{tr}(\varepsilon^{\mathrm{d}} \cdot \varepsilon^{\mathrm{d}}) = \frac{1}{2}K(\varepsilon^{\mathrm{v}})^2 + G\mathrm{tr}(\varepsilon^{\mathrm{d}} \cdot \varepsilon^{\mathrm{d}}). \qquad (12.34)$$

It can be established that in this energy density the hydrostatic (volumetric) part and the deviatoric part are separately identifiable; just like in Hooke's law there is a decoupling. Both parts deliver an always positive (better: non-negative) contribution to the energy density, for every arbitrary ε.

Using Hooke's law, the energy density $\Phi_{\mathrm{int}_0}(\varepsilon)$, according to the above given equation, can be transformed to $\Phi_{\mathrm{int}_0}(\sigma)$ resulting in

$$\Phi_{\mathrm{int}_0}(\sigma) = \frac{1}{2}\frac{1}{K}p^2 + \frac{1}{4G}\mathrm{tr}(\sigma^{\mathrm{d}} \cdot \sigma^{\mathrm{d}}). \qquad (12.35)$$

Finally, it should be remarked that from the previous it can be concluded that a cyclic process in the deformation or in the stress will always be energetically neutral. This means that the 'postulated' constitutive equation (Hooke's law) indeed gives a correct description of elastic material behaviour. Every cyclic process is reversible. No energy is dissipated or released. If we compare the second term on the right-hand side of Eq. (12.35), representing the 'distortion energy density' with Eq. (8.75), defining the von Mises stress, it is clear that both are related. In other words, if we would like to define some threshold based on the maximum amount of distortional energy that can be stored in a material before it becomes damaged, the von Mises stress can be used for this purpose.

12.4   Elastic behaviour at large deformations and/or large rotations

In this section the attention is focussed on constitutive equations for elastic, compressible, isotropic material behaviour at large deformations. The formulation in Sections 12.2 and 12.3 is no longer valid, because under those circumstances the linear strain tensor ε cannot be used.

The Cauchy stress tensor σ is objective (σ transforms in a very specific way when an extra rigid body rotation is enforced, see Section 9.7). This implies that σ can certainly not be coupled to an invariant measure for the strain, such as the right Cauchy Green deformation tensor C or the related Green Lagrange strain tensor E). But it is allowed to relate σ to the objective left Cauchy Green tensor B, or the associated strain tensors A (Almansi Euler) and ε_{F} (Finger).

Just like in the previous sections, the contribution originating from the volume change to the stress tensor is considered separately from the distortion (shape change). For this reason the deformation tensor F is decomposed according to:

$$F = J^{1/3}\tilde{F} \quad \text{with} \quad J = \det(F).$$ (12.36)

The tensor $\tilde{F}$ is called the isochoric deformation tensor (because of the equality: $\det(\tilde{F}) = 1$). The multiplication factor $J^{1/3}$ represents the volume change. Departing from the isochoric deformation tensor $\tilde{F} = J^{-1/3}F$, the associated objective, isochoric left Cauchy Green tensor $\tilde{B}$ is defined according to:

$$\tilde{B} = \tilde{F} \cdot \tilde{F}^{\mathrm{T}} = J^{-2/3}F \cdot F^{\mathrm{T}} = J^{-2/3}B,$$ (12.37)

and subsequently the isochoric Finger strain tensor $\tilde{\varepsilon}_{\mathrm{F}}$ according to:

$$\tilde{\varepsilon}_{\mathrm{F}} = \frac{1}{2}\left(\tilde{B} - I\right) = \frac{1}{2}\left(J^{-2/3}B - I\right).$$ (12.38)

Analogous to the formulation in Section 12.2 linear relations for the hydrostatic and deviatoric part of the stress tensor can be postulated:

$$\sigma^{\mathrm{h}} = -pI = K(J - 1)I,$$ (12.39)

$$\sigma^{\mathrm{d}} = 2G\tilde{\varepsilon}_F^{\mathrm{d}} = G\tilde{B}^{\mathrm{d}},$$ (12.40)

with K the compression modulus and G the shear modulus of the material. Summation leads to:

$$\sigma = K(J - 1)I + G\tilde{B}^{\mathrm{d}}.$$ (12.41)

This coupling between the stress and deformation state is often indicated as 'compressible Neo-Hookean' material behaviour, thus referring to the linearity.

It can be shown that Eq. (12.41) does not exactly satisfy the requirement that in a cyclic process no energy is dissipated. It can be proven that a small (but not trivial) modification, according to:

$$\sigma = K(J - 1)I + \frac{G}{J}\tilde{B}^{\mathrm{d}}$$ (12.42)

does satisfy this requirement.

Substituting Eq. (12.42) into the definition equation for $\Phi_{\mathrm{int}_0}(t)$, Eq. (12.29) yields

$$\Phi_{\mathrm{int}_0}(t) = \int_{\tau=0}^{t}\left(K(J - 1)\,J\mathrm{tr}(D) + G\mathrm{tr}(\tilde{B}^{\mathrm{d}} \cdot D)\right)d\tau.$$ (12.43)

Based on the relations given in Chapter 10 the following expressions can be derived for the first and second term in the integrand:

$$J\mathrm{tr}(D) = \dot{J},$$ (12.44)

and

$$\mathrm{tr}(\tilde{\boldsymbol{B}}^{\mathrm{d}} \cdot \boldsymbol{D}) = \mathrm{tr}\left(\left(\tilde{\boldsymbol{B}} - \frac{1}{3}\mathrm{tr}(\tilde{\boldsymbol{B}})\boldsymbol{I}\right) \cdot \boldsymbol{D}\right)$$

$$= J^{-2/3}\mathrm{tr}(\boldsymbol{B} \cdot \boldsymbol{D}) - \frac{1}{3}J^{-2/3}\mathrm{tr}(\boldsymbol{B})\,\mathrm{tr}(\boldsymbol{D})$$

$$= \frac{1}{2}J^{-2/3}\mathrm{tr}\left(\boldsymbol{F} \cdot \boldsymbol{F}^{\mathrm{T}} \cdot \dot{\boldsymbol{F}} \cdot \boldsymbol{F}^{-1} + \boldsymbol{F} \cdot \boldsymbol{F}^{\mathrm{T}} \cdot \boldsymbol{F}^{-\mathrm{T}} \cdot \dot{\boldsymbol{F}}^{\mathrm{T}}\right)$$

$$- \frac{1}{3}J^{-5/3}\dot{J}\mathrm{tr}(\boldsymbol{B})$$

$$= \frac{1}{2}J^{-2/3}\mathrm{tr}\left(\dot{\boldsymbol{F}} \cdot \boldsymbol{F}^{\mathrm{T}} + \boldsymbol{F} \cdot \dot{\boldsymbol{F}}^{\mathrm{T}}\right) + \frac{1}{2}(J^{-2/3})\,\mathrm{tr}(\boldsymbol{B})$$

$$= \frac{1}{2}J^{-2/3}\mathrm{tr}\left(\dot{\boldsymbol{B}}\right) + \frac{1}{2}(J^{-2/3})\,\mathrm{tr}(\boldsymbol{B}).$$

$$(12.45)$$

With Eq. (12.45) the integral expression for $\Phi_{\mathrm{int}_0}(t)$ can be elaborated further:

$$\Phi_{\mathrm{int}_0} = \left.\left(\frac{1}{2}K(J-1)^2 + \frac{1}{2}GJ^{-2/3}\mathrm{tr}(\boldsymbol{B})\right)\right|_{\tau=0}^{t}. \qquad (12.46)$$

Using $J = 1$ and $\boldsymbol{B} = \boldsymbol{I}$ for $\tau = 0$ results in the current energy density Φ_{int_0}, which only depends on the current left Cauchy Green tensor $\boldsymbol{B}(t)$, so it can be noted:

$$\Phi_{\mathrm{int}_0}(\boldsymbol{B}) = \frac{1}{2}K(J-1)^2 + \frac{1}{2}G\left(J^{-2/3}\mathrm{tr}(\boldsymbol{B}) - 3\right), \qquad (12.47)$$

with

$$J = (\det(\boldsymbol{B}))^{1/2}. \qquad (12.48)$$

Again, it can be established, that in Φ_{int_0} the volumetric and the deviatoric part are clearly distinguishable and that both parts deliver an always positive contribution to the energy density, for every arbitrary deformation process. Finally, it can again be observed that a cyclic process in the deformation or in the stress will always be energetically neutral.

More general expressions for (non-linearly) elastic behaviour can formally be written as

$$p = p(J), \qquad \sigma^{\mathrm{d}} = \sigma^{\mathrm{d}}(J, \tilde{\boldsymbol{B}}). \qquad (12.49)$$

For a detailed specification many possibilities exist and have been published in the scientific literature. An extensive treatment for biological materials is beyond the scope of the present discussion. For this the reader is referred to more advanced textbooks on Biomechanics.

12.5 Constitutive modelling of viscous fluids

For viscous fluids, as considered in this book, in contrast to solids a reference state is not important. Therefore an Eulerian approach is pursued. Accordingly, the velocity field is written as

$$\vec{v} = \vec{v}(\vec{x}, t) \quad \text{and also} \quad \underset{\sim}{v} = \underset{\sim}{v}(\underset{\sim}{x}, t). \tag{12.50}$$

So the velocity is a function of the coordinates x, y and z, associated with a fixed coordinate system in space, and the time t. The current local velocity does not include any information with respect to deformation changes in the fluid, contrary to the velocity gradient tensor L with matrix representation $\underset{\sim}{L}$, defined in Sections 7.5 and 10.6 according to:

$$L = \left(\vec{\nabla}\vec{v}\right)^{\mathrm{T}} \quad \text{and also} \quad \underset{\sim}{L} = \left(\underset{\sim}{\nabla}\underset{\sim}{v}^{\mathrm{T}}\right)^{\mathrm{T}}. \tag{12.51}$$

In Section 10.6 the velocity gradient tensor L is split into the symmetrical rate of deformation tensor D and the skew symmetric spin tensor $\boldsymbol{\Omega}$. The tensor D is a measure for the deformation changes. After all the tensor D determines the current length changes of all material line segments and if $D = 0$, all those line segments have a (temporarily) constant length, independent of $\boldsymbol{\Omega}$. In components the matrix $\underset{\sim}{D}$ associated with tensor D can be written as

$$\underset{\sim}{D} = \begin{bmatrix} \frac{\partial v_x}{\partial x} & \frac{1}{2}\left(\frac{\partial v_x}{\partial y} + \frac{\partial v_y}{\partial x}\right) & \frac{1}{2}\left(\frac{\partial v_x}{\partial z} + \frac{\partial v_z}{\partial x}\right) \\ \frac{1}{2}\left(\frac{\partial v_y}{\partial x} + \frac{\partial v_x}{\partial y}\right) & \frac{\partial v_y}{\partial y} & \frac{1}{2}\left(\frac{\partial v_y}{\partial z} + \frac{\partial v_z}{\partial y}\right) \\ \frac{1}{2}\left(\frac{\partial v_z}{\partial x} + \frac{\partial v_x}{\partial z}\right) & \frac{1}{2}\left(\frac{\partial v_z}{\partial y} + \frac{\partial v_y}{\partial z}\right) & \frac{\partial v_z}{\partial z} \end{bmatrix}. \tag{12.52}$$

When comparing this elaborated expression for matrix $\underset{\sim}{D}$ with the linear strain matrix $\underset{\sim}{\varepsilon}$, see Section 10.4, it is not surprising that $\underset{\sim}{D}$ is called the rate of deformation matrix. Should the current state be chosen to be the reference state (in that special case in the current state: $F = I$) the following relation holds: $\underset{\sim}{D} = \dot{\underset{\sim}{\varepsilon}}$ (and in tensor notation: $D = \dot{\varepsilon}$).

As a constitutive equation for the behaviour of incompressible viscous fluids, based on the above considerations, the following (general) expression for the Cauchy stress tensor will be used:

$$\sigma = -pI + \sigma^{\mathrm{d}}(D). \tag{12.53}$$

Because of the assumption of incompressibility, the pressure p is undetermined, while the deformation rate tensor D has to satisfy the constraint: $\mathrm{tr}(D) = 0$. It should be noted that both the Cauchy stress tensor and the deformation rate tensor are objective tensors. In Sections 12.6 and 12.7 two types of constitutive behaviour for fluids will be discussed by means of a specification of $\sigma^d(D)$.

12.6 Newtonian fluids

For a Newtonian fluid the relation between the deviatoric stress tensor and the deformation rate tensor is linear, yielding:

$$\sigma = -pI + 2\eta D \quad \text{and also} \quad \underline{\sigma} = -p\underline{I} + 2\eta \underline{D}, \tag{12.54}$$

with η the **viscosity** (a material parameter that is assumed to be constant) of the fluid. The typical behaviour of a Newtonian fluid can be demonstrated by applying Eq. (12.54) to two elementary examples of fluid flow: pure shear and uniaxial extensional flow.

A pure shear flow 'in the xy-plane' can be created with the following velocity field (in column notation):

$$\underline{v} = \begin{bmatrix} v_x \\ v_y \\ v_z \end{bmatrix} = \dot{\gamma} \begin{bmatrix} y \\ 0 \\ 0 \end{bmatrix}, \tag{12.55}$$

with $\dot{\gamma}$ (the shear velocity) constant. For the associated deformation rate matrix $\underline{D}$ it can easily be derived that

$$\underline{D} = \frac{1}{2} \begin{bmatrix} 0 & \dot{\gamma} & 0 \\ \dot{\gamma} & 0 & 0 \\ 0 & 0 & 0 \end{bmatrix} \tag{12.56}$$

and verified that the constraint $\text{tr}(D) = \text{tr}(\underline{D}) = 0$ is satisfied. Applying the constitutive equation it follows for the relevant shear stress $\sigma_{xy} = \sigma_{yx}$ in the fluid:

$$\sigma_{xy} = \eta \dot{\gamma}, \tag{12.57}$$

which appears to be constant. Thus, the viscosity can be interpreted as the 'resistance' of the fluid against 'shear' (shear rate actually).

To a uniaxial extensional flow (incompressible) in the x-direction the following deformation rate matrix is applicable:

$$\underline{D} = \begin{bmatrix} \dot{\epsilon} & 0 & 0 \\ 0 & -\dot{\epsilon}/2 & 0 \\ 0 & 0 & -\dot{\epsilon}/2 \end{bmatrix}, \tag{12.58}$$

with $\dot{\epsilon}$ (the rate of extension) constant. For the stress matrix it is immediately found that

$$\underline{\sigma} = -p\,\underline{I} + 2\eta \begin{bmatrix} \dot{\epsilon} & 0 & 0 \\ 0 & -\dot{\epsilon}/2 & 0 \\ 0 & 0 & -\dot{\epsilon}/2 \end{bmatrix}. \tag{12.59}$$

Assuming, that in the y- and in the z-direction the flow can develop freely: $\sigma_{yy} = \sigma_{zz} = 0$ and with that it follows for the hydrostatic pressure:

$$p = -\eta\dot{\epsilon}. \tag{12.60}$$

This leads to the required uniaxial stress for the extensional flow:

$$\sigma_{xx} = 3\eta\dot{\epsilon}. \tag{12.61}$$

The fact that the effective uniaxial extensional viscosity (3η) is three times as high as the shear viscosity (η) is known as **Trouton's law**.

12.7   Non-Newtonian fluids

For a non-Newtonian incompressible viscous fluid the constitutive equation has the same form as the equation for a Newtonian fluid:

$$\sigma = -pI + 2\eta D \qquad \text{and also} \qquad \underline{\sigma} = -p\underline{I} + 2\eta\underline{D}, \tag{12.62}$$

however, the viscosity η is now a function of the deformation rate tensor: $\eta = \eta(D)$. Specification of the relation for the viscosity has to be based on experimental research. Here we limit ourselves to an example, the three parameter 'power law' model (with the temperature influence T according to Arrhenius):

$$\eta = me^{(A/T)} \left(\sqrt{2\,\mathrm{tr}(D \cdot D)} \right)^{(n-1)}, \tag{12.63}$$

with m, A and n material constants (for $n = 1$ the viscosity is independent of D and the behaviour is 'Newtonian' again). Substitution of the deformation rate tensor for pure shear (see previous section) leads to

$$\eta = m\,e^{(A/T)}\,|\dot{\gamma}|^{(n-1)}, \tag{12.64}$$

making the mathematical format of the equation more transparent.

12.8   Diffusion and filtration

Although somewhat beyond the scope of the present chapter in this last section material flow due to diffusion or filtration is considered, i.e. transport of material through a stationary porous medium (no convection). The flowing material can indeed be considered to be a continuum, but the constitutive equations are completely different from those treated before. Here, the constitutive equations describe the transport of material depending on the driving mechanisms, while above the constitutive equations related the characteristics of the flow to the internal stresses.

Diffusion of a certain material through a porous medium is generated by concentration differences of the material in the medium. The material will in general strive for a homogeneous density distribution (provided that the porous medium is homogeneous) implying that material will flow from regions with a high concentration to regions with a low concentration. The mathematical form for this phenomenon is given by **Fick's law**:

$$\rho \vec{v} = -D\vec{\nabla}\rho, \tag{12.65}$$

with on the left-hand side the mass flux vector (also see Section 7.5) and on the right-hand side the driving mechanism for transport, $\vec{\nabla}\rho$, multiplied with a certain factor D. This factor D is called the diffusion coefficient, and can be considered to be a constitutive parameter, that is determined by the combination of materials (and the temperature).

In case of filtration of material through a stationary porous medium the driving force is formed by pressure differences in that material. The material strives for an equal pressure, so in the presence of pressure differences, a flow will occur from areas with a high pressure to areas with a lower pressure. This is expressed by **Darcy's law**:

$$\rho \vec{v} = -\kappa \vec{\nabla}p \tag{12.66}$$

with the permeability κ as a constitutive parameter.

Exercises

12.1 Consider a cubic material element of which the edges are oriented in the direction of a Cartesian xyz-coordinate system, see the figure. In the figure also the normal and shear stresses are depicted (expressed in [kPa]) acting on the (visible) faces of the element. The mechanical behaviour of the element is described by the linear Hooke's law, with Young's modulus $E = 8$ [MPa] and Poisson's ratio $v = 1/3$.

Determine the volume change V/V_0, with V_0 the volume of the element in the unloaded reference configuration and V the volume in the current loaded configuration, assuming small deformations.

12.2 An element of an incompressible material (in the reference state a cube: $\ell \times \ell \times \ell$), is placed in a Cartesian xyz-coordinate system as given in the figure below. Because of a load in the z-direction the height of the element is reduced to $2\ell/3$. In the x-direction the displacement is suppressed. The element can expand freely in the y-direction. The deformation is assumed to be homogeneous.

The material behaviour is described by a Neo-Hookean relation, according to:

$$\underline{\sigma} = -p\underline{I} + G\underline{B}^d,$$

with $\underline{\sigma}$ the stress matrix, p the hydrostatic pressure (to be determined), $\underline{I}$ the unit matrix, G the shear modulus and $\underline{B}$ the left Cauchy Green deformation matrix.

Determine the compressive force F_v in z-direction that is necessary to realize this deformation.

12.3 A frequently applied test to determine the stiffness properties of biological materials is the 'confined compression test'. A schematic of such a test is given in the figure. A cylindrical specimen, Young's modulus E, (cylindrically shaped with a circular cross section with radius R, height h) of the material is placed in a tight fitting die, that can be regarded as a rigid mould with a smooth inner wall. The height of the specimen is reduced with a value δ (with $\delta \ll h$) by means of an indenter. A vertical force K_v is acting on the indenter in the direction of the arrow. It can be assumed that the deformation of the specimen is homogeneous. As the strains are small, the material behaviour is described by the linear Hooke's law. The Poisson's ratio is known (measured in another experiment): $\nu = 1/4$.

Express the Young's modulus E of the material in the parameters R, h, δ and K_v (i.e. in the parameters that can easily be measured).

12.4   Because confined compression tests have several disadvantages, unconfined compression tests are also performed frequently. Again a cylindrical specimen is used. In this case the specimen has in the reference state a radius $R_0 = 2.000$ [cm] and thickness $h_0 = 0.500$ [cm]. It is compressed to $h = 0.490$ [cm]. Now, the specimen can expand freely in a radial direction. In the deformed state: $R = 2.008$ [cm].

The material behaves linearly elastic according to Hooke's law, characterized by Young's modulus E and Poisson's ratio ν.

Calculate the Poisson's ratio ν based on the above described experiment.

12.5   During a percutaneous angioplasty the vessel wall is expanded by inflating a balloon. At a certain moment during inflation the internal pressure $p_i = 0.2$ [MPa] and the internal radius R_i of the vessel is increased by 10%. Assume that the length of the balloon and the length of the vessel wall in contact with the balloon do not change. Further it is known, that the wall behaviour can be modelled according to Hooke's law, with Young's modulus of the wall $E = 8$ [MPa] and the Poisson's ratio of the wall $\nu = 1/3$.

(a) Calculate the strain in circumferential direction at the inner side of the wall.

(b) Calculate the strain in the $\vec{e}_x$-direction at point A at the inner side of the vessel wall.

(c) Calculate the stress in circumferential direction at point A.

12.6 Consider a bar (length $2L$) with a circular cross section (radius R). The axis of the bar coincides with the z-axis of a xyz-coordinate system. With respect to the mid plane ($z = 0$) the top and bottom plane of the bar are rotated around the z-axis with an angle α (with $\alpha \ll 1$), thus loading the bar with torsion. See the figure.

The position vector $\vec{x}$ of a material point in the deformed configuration is

$$\vec{x} = \vec{x}_0 - \frac{\alpha(\vec{x}_0 \cdot \vec{e}_y)(\vec{x}_0 \cdot \vec{e}_z)}{L}\,\vec{e}_x + \frac{\alpha(\vec{x}_0 \cdot \vec{e}_x)(\vec{x}_0 \cdot \vec{e}_z)}{L}\,\vec{e}_y,$$

with $\vec{x}_0$ the position vector of that point in the undeformed state. The material behaviour is linearly elastic according to Hooke's law, with Young's modulus E and Poisson's ratio ν.

Determine, for the material point defined with $\vec{x}_0 = R\vec{e}_y$, the linear strain tensor $\boldsymbol{\varepsilon}$, the stress tensor $\boldsymbol{\sigma}$ and the equivalent stress, according to von Mises $\bar{\sigma}_M$.

13 Solution strategies for solid and fluid mechanics problems

13.1 Introduction

The goal of the present chapter is to describe a procedure to formally determine solutions for solid mechanics problems, fluid mechanics problems and problems with filtration and diffusion. Mechanical problems in biomechanics can be very diverse and most problems are so complex, that it is impossible to derive analytical solutions and often very complicated to determine numerical solutions. Fortunately, in most cases it is not necessary to describe all phenomena related to the problem in full detail and simplifying assumptions can be made, thus reducing the complexity of the set of equations that have to be solved. The present chapter deals with formulating problems and solution strategies, starting from the most general set of equations and gradually reducing the generality by imposing simplifying assumptions. In Section 13.2 this will be done for solids. Section 13.3 is devoted to solving fluid mechanics problems. The last section of this chapter discusses diffusion and filtration.

13.2 Solution strategies for deforming solids

In this section it is assumed that the material (or material fraction) to be considered can be modelled as a deforming solid continuum. This implies that it is possible and significant to define a reference configuration or reference state. With respect to the reference state, the displacement field as a function of time supplies a full description of the deformation process to which the continuum is subjected (at least under the restrictions given in previous chapters, such as for example a constant temperature). After all, for a displacement field that is known as a function of time, it is possible to directly calculate the local deformation history (applying the kinematics, see Chapter 10) and subsequently, the stress state as a function of time (using the constitutive equations, see Chapter 12). The relevant fields, from a mechanical point of view, that are obtained in this way for the continuum have to satisfy the balance equations (see Chapter 11). In addition, the initial conditions have to be fulfilled: at the beginning of the process the positions of the material points in space have to be prescribed, in general in accordance with the reference configuration, and also the initial velocities of the material points have to be in agreement with the specification of the initial state. In addition, during the entire process the boundary conditions have to be satisfied: along the outer surface of the continuum the displacement field and stress field have to be consistent with the process specification.

The goal of the present section is to outline a procedure to formally determine (as a function of time) the displacement field, such that all requirements are satisfied. For (almost) no realistic problem exact analytical solutions can be found, not even when the mathematical description is drastically simplified via assumptions. A global description will be given of strategies to derive approximate solutions.

In Section 13.2.1 the general (complete) formulation of the problem will be given. Subsequently, in the sections that follow, the generality will gradually be limited. Initially, in Section 13.2.2 the general description will be restricted with respect to the magnitude of the displacements, deformations and rotations (geometrically linear behaviour). In Section 13.2.3 the restriction to linearly elastic behaviour follows (physical linearity), leading to the set of equations, establishing the so-called 'linear elasticity theory'. In Section 13.2.4 processes are considered for which inertia effects can be neglected (quasi-static processes). Then time (and thus also process rate) is no longer relevant. In Section 13.2.5 the attention is concentrated on configurations that, because of geometry and external loading, can be regarded as two-dimensional (plane stress theory). Finally, Section 13.2.6 is focussed on formulating boundary conditions for continuum problems. Sometimes extra constitutive equations are necessary to describe the interaction of the considered continuum with the environment.

13.2.1 General formulation for solid mechanics problems

For general solid mechanics problems addressing the deformation process in a Lagrange description, the following fields have to be determined:

- the displacement field: $\vec{u}(\vec{x}_0, t)$ for all $\vec{x}_0$ in V_0 and for all t and
- the stress field: $\sigma(\vec{x}_0, t)$ for all $\vec{x}_0$ in V_0 and for all t.

These fields have to be connected for all $\vec{x}_0$ in V_0 in accordance with the local constitutive equation (see Chapter 12):

$$\sigma(\vec{x}_0, t) = \mathcal{F}\{F(\vec{x}_0, \tau)\,;\ \tau \leq t\}, \tag{13.1}$$

with

$$F = I + \left(\vec{\nabla}_0 \vec{u}\right)^{\mathrm{T}},$$

(13.2)

reflecting history-dependent material behaviour. In addition the local balance of momentum has to be satisfied (see Chapter 11), as well as the local mass balance:

$$\left(F^{-\mathrm{T}} \cdot \vec{\nabla}_0\right) \cdot \sigma + \rho\vec{q} = \rho\ddot{\vec{u}},$$

(13.3)

with

$$\rho = \frac{\rho_0}{\det(F)}.$$

(13.4)

The equations given above form a set of non-linear, coupled partial differential equations (derivatives with respect to the three material coordinates in $\vec{x}_0$ and the time t are dealt with). Consequently, for a unique solution of the displacement field $\vec{u}(\vec{x}_0, t)$ and the stress field $\sigma(\vec{x}_0, t)$ boundary conditions and initial conditions are required. With respect to the boundary conditions, for all t at every point of the outer surface of V_0 three (scalar) relations have to be specified: either completely formulated in stresses (dynamic or natural boundary conditions), completely formulated in displacements (kinematic or essential boundary conditions) or in mixed formulations. With respect to the initial conditions, at the initial time point ($t = 0$), for all the points in V_0, the displacement and velocity have to be specified. If the initial state is used as the reference configuration, $\vec{u}(\vec{x}_0, t = 0) = \vec{0}$ for all $\vec{x}_0$ in V_0.

13.2.2 Geometrical linearity

Provided that displacements, strains and rotations are small (so $F \approx I$ and consequently $\det(F) \approx 1$) the general set for solid continuum problems as presented in the previous section can be written as

$$\sigma(\vec{x}_0, t) = \mathcal{G}\{\varepsilon(\vec{x}_0, \tau) \; ; \; \tau \le t\},$$

(13.5)

where $\mathcal{F}\{\}$, as used in Eq. (13.1), has been replaced by $\mathcal{G}\{\}$ due to adaptations in the argument, with

$$\varepsilon = \frac{1}{2}\left(\left(\vec{\nabla}_0\vec{u}\right) + \left(\vec{\nabla}_0\vec{u}\right)^{\mathrm{T}}\right),$$

(13.6)

and

$$\vec{\nabla}_0 \cdot \sigma + \rho_0\,\vec{q} = \rho_0\,\ddot{\vec{u}}.$$

(13.7)

Again these equations have to be satisfied for all $\vec{x}_0$ in V_0 and for all times t. The first equation (a formal functional expression) indicates that the current local

Cauchy stress tensor $\boldsymbol{\sigma}$ is fully determined by the (history of the) local linear strain tensor $\boldsymbol{\varepsilon}$. The second equation actually implies that the balance of momentum (equation of motion) is now related to the reference configuration instead of the current configuration. The balance of mass is now redundant through replacing ρ by ρ_0. With respect to the boundary conditions and initial conditions, the same statements apply as in the previous section.

13.2.3 Linear elasticity theory, dynamic

In Section 13.2.2 geometrical linearity has been introduced to simplify the general set of equations given in Section 13.2.1. If we add physical linearity to this (meaning that the relation between stress and strain is described by Hooke's law as formulated in Chapter 12) linear elasticity theory results. The associated (linear) equations, that have to be satisfied for all $\vec{x}_0$ in V_0 and for all times t, read:

$$\boldsymbol{\sigma} = \left(K - \frac{2G}{3} \right) \text{tr}(\boldsymbol{\varepsilon})\boldsymbol{I} + 2G\boldsymbol{\varepsilon}, \tag{13.8}$$

with

$$\boldsymbol{\varepsilon} = \frac{1}{2}\left(\left(\vec{\nabla}_0\vec{u}\right) + \left(\vec{\nabla}_0\vec{u}\right)^{\text{T}} \right),$$

and

$$\vec{\nabla}_0 \cdot \boldsymbol{\sigma} + \rho_0\,\vec{q} = \rho_0\,\ddot{\vec{u}}. \tag{13.9}$$

With respect to boundary conditions and initial conditions the same procedure must be followed as in the previous section.

13.2.4 Linear elasticity theory, static

For slowly evolving (quasi-static) processes inertia effects can be neglected. In that case it is sufficient to consider only the relevant current state and time is no longer of interest. The current displacement field $\vec{u}(\vec{x}_0)$ and the current stress field $\boldsymbol{\sigma}(\vec{x}_0)$ have to be determined on the basis of the equations:

$$\boldsymbol{\sigma} = \left(K - \frac{2G}{3} \right) \text{tr}(\boldsymbol{\varepsilon})\boldsymbol{I} + 2G\boldsymbol{\varepsilon}, \tag{13.10}$$

with

$$\boldsymbol{\varepsilon} = \frac{1}{2}\left(\left(\vec{\nabla}_0\vec{u}\right) + \left(\vec{\nabla}_0\vec{u}\right)^{\text{T}} \right), \tag{13.11}$$

and

$$\vec{\nabla}_0 \cdot \boldsymbol{\sigma} + \rho_0 \, \vec{q} = \vec{0}. \tag{13.12}$$

The equation of motion is reduced to the equilibrium equation. Only boundary conditions are necessary to solve this set. Initial conditions are no longer applicable. In every point of the outer surface of V_0 three (scalar) relations have to be specified. In this context, to find a unique solution for the displacement field, it is necessary to suppress movement of the configuration as a rigid body. This will be further elucidated in the following example.

Example 13.1   Consider a homogeneous body with reference volume V_0 under a given hydrostatic pressure p. For the stress field in V_0 satisfying the (dynamic) boundary conditions and the equilibrium equations, it can be written:

$$\boldsymbol{\sigma}(\vec{x}_0) = -p\boldsymbol{I} \text{ for all } \vec{x}_0 \text{ in } V_0.$$

For the strain field, according to Hooke's law it is found:

$$\boldsymbol{\varepsilon}(\vec{x}_0) = -\frac{p}{3K}\boldsymbol{I} \text{ for all } \vec{x}_0 \text{ in } V_0.$$

A matching displacement field is, for example,

$$\vec{u}(\vec{x}_0) = -\frac{p}{3K}\vec{x}_0.$$

It is easy to verify that the above given solution satisfies all equations. However, because in the given problem description the displacement as a rigid body is not prescribed the solution is not unique. It can be proven that the general solution for the components of the displacement vector reads:

$$u_x = -\frac{p}{3K}x_0 + c_1 - c_6 y_0 + c_5 z_0$$

$$u_y = -\frac{p}{3K}y_0 + c_2 + c_6 x_0 - c_4 z_0$$

$$u_z = -\frac{p}{3K}z_0 + c_3 - c_5 x_0 + c_4 y_0,$$

with c_i ($i = 1, 2, \ldots 6$) yet undetermined constants. The constants c_1, c_2 and c_3 represent translations in the coordinate directions and the constants c_4, c_5 and c_6 (small) rotations around the coordinate axes.

13.2.5 Linear plane stress theory, static

Consider a flat membrane with, in the reference configuration, constant thickness h. The 'midplane' of the membrane coincides with the $x_0 y_0$-plane, while in the

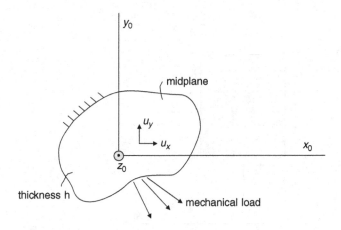

Configuration with plane stress state.

direction perpendicular to that plane the domain for z_0 is given by: $-h/2 \leq z_0 \leq h/2$. The thickness h is supposed to be small with respect to the dimensions 'in the plane'. The loading is parallel to the midplane of the membrane, see Fig. 13.1.

The midplane will continue to be a symmetry plane after deformation. It is assumed that straight material line segments, initially perpendicular to the midplane will remain straight (and perpendicular to the midplane) after deformation. For the displacements this means

$$u_x = u_x(x_0, y_0), \qquad u_y = u_y(x_0, y_0). \tag{13.13}$$

The relevant strain components of the linear strain matrix $\underline{\varepsilon}$ for the membrane are

$$\varepsilon_{xx} = \frac{\partial u_x}{\partial x_0}$$

$$\varepsilon_{yy} = \frac{\partial u_y}{\partial y_0} \tag{13.14}$$

$$\varepsilon_{xy} = \varepsilon_{yx} = \frac{1}{2}\left(\frac{\partial u_x}{\partial y_0} + \frac{\partial u_y}{\partial x_0}\right).$$

In general, from the other components ε_{zz} will certainly not be zero, while $\varepsilon_{xz} = \varepsilon_{zx}$ and $\varepsilon_{yz} = \varepsilon_{zy}$ will vanish. Actually, these other strain components are not important to set-up the theory.

With respect to the stress, it is assumed that only components 'acting in the x_0y_0-plane' play a role and that for those components (see Fig. 13.2) it can be written:

Figure 13.2

Stress components in a plane stress state.

$$\sigma_{xx} = \sigma_{xx}(x_0, y_0)$$
$$\sigma_{yy} = \sigma_{yy}(x_0, y_0) \tag{13.15}$$
$$\sigma_{xy} = \sigma_{yx} = \sigma_{xy}(x_0, y_0).$$

The stress components σ_{zz}, $\sigma_{xy} = \sigma_{yx}$ and $\sigma_{xz} = \sigma_{zx}$ are assumed to be zero (negligible).

Based on Hooke's law (see Section 12.2 for the fully elaborated expression in components), using $\sigma_{zz} = 0$ it is found:

$$\varepsilon_{zz} = -\frac{v}{1-v}(\varepsilon_{xx} + \varepsilon_{yy}), \tag{13.16}$$

and by exploiting this equation, the description of the (linearly elastic) material behaviour for plane stress becomes

$$\sigma_{xx} = \frac{E}{1-v^2}(\varepsilon_{xx} + v\varepsilon_{yy}) \tag{13.17}$$

$$\sigma_{yy} = \frac{E}{1-v^2}(v\varepsilon_{xx} + \varepsilon_{yy}) \tag{13.18}$$

$$\sigma_{xy} = \frac{E}{(1+v)}\varepsilon_{xy}, \tag{13.19}$$

thus coupling the relevant stresses and strains. It should be noted that in this case for the material parameters the Young's modulus E and the Poisson's ratio v have been used, instead of the compression modulus K and the shear modulus G in previous sections.

The stresses have to satisfy the equilibrium equations. For plane stress this means that only equilibrium 'in the plane' (in the case considered therefore in the x_0- and y_0-direction) results in non-trivial equations:

$$\frac{\partial \sigma_{xx}}{\partial x_0} + \frac{\partial \sigma_{xy}}{\partial y_0} + \rho_0 \, q_x = 0 \tag{13.20}$$

$$\frac{\partial \sigma_{xy}}{\partial x_0} + \frac{\partial \sigma_{yy}}{\partial y_0} + \rho_0 \, q_y = 0. \tag{13.21}$$

Summarizing, it can be stated that in case of plane stress eight scalar functions of the coordinates in the midplane have to be calculated: the displacements u_x and u_y, the strains ε_{xx}, ε_{yy} and ε_{xy}, and the stresses σ_{xx}, σ_{yy} and σ_{xy}. For this objective we have eight equations at our disposal: the strain/displacement relations (three), the constitutive equations (three) and the equilibrium equations (two). In addition, for each boundary point of the midplane two scalar boundary conditions have to be specified and for uniqueness of the displacement field rigid body motion has to be suppressed.

Example 13.2 In Fig. 13.3 a simple plane stress problem is defined for a rectangular membrane (length $2l$, width $2b$ and thickness h) with linearly elastic material behaviour (Young's modulus E and Poisson's ratio ν). The mathematical form for the boundary conditions reads:

$$\text{for } x_0 = \pm l \quad \text{and} \quad -b \le y_0 \le b \text{ it holds: } \sigma_{xx} = \alpha + \beta \frac{y_0}{b}$$

$$\sigma_{xy} = 0$$

$$\text{for } y_0 = \pm b \quad \text{and} \quad -l \le x_0 \le l \text{ it holds: } \sigma_{yy} = 0$$

$$\sigma_{xy} = 0.$$

Figure 13.3
A simple plane stress problem.

Herewith, the external load is specified using the constants α and β; the constant α is representative for the 'normal' force in the x_0 direction and the constant β for the 'bending'. With the above given boundary conditions the displacement of the membrane as a rigid body is not suppressed; uniqueness of the displacement solution is obtained, if it is additionally required that the material point coinciding with origin O is fixed in space and if in the deformed state the symmetry with respect to the y_0-axis is maintained. For the given problem an exact analytical solution can be calculated. It is easy to verify that the solution has the following form:

$$u_x(x_0, y_0) = \frac{1}{E}\left(\alpha + \beta\frac{y_0}{b}\right)x_0$$

$$u_y(x_0, y_0) = -\frac{1}{E}\left(\nu\alpha y_0 + \nu\beta\frac{y_0^2}{2b} + \beta\frac{x_0^2}{2b}\right)$$

$$\varepsilon_{xx}(x_0, y_0) = \frac{1}{E}\left(\alpha + \beta\frac{y_0}{b}\right)$$

$$\varepsilon_{yy}(x_0, y_0) = -\frac{\nu}{E}\left(\alpha + \beta\frac{y_0}{b}\right)$$

$$\varepsilon_{xy}(x_0, y_0) = 0$$

$$\sigma_{xx}(x_0, y_0) = \alpha + \beta\frac{y_0}{b}$$

$$\sigma_{yy}(x_0, y_0) = 0$$

$$\sigma_{xy}(x_0, y_0) = 0.$$

It has to be considered as an exception, when for a specified plane stress problem an analytical solution exists. In general, only approximate solutions can be determined. A technique to do this is the Finite Element Method, which is the subject of the Chapters 14 to 18. Chapter 18 is especially devoted to the solution of linear elasticity problems as described in the present chapter.

13.2.6 Boundary conditions

In the previous sections only simply formulated boundary conditions have been considered. In the case of dynamic boundary conditions, the components of the stress vector $\vec{p}$ (see Fig. 8.1) are prescribed at a point along the boundary of the volume of the continuum (in case of plane stress along the boundary of the configuration surface). In case of kinematic boundary conditions the displacement vector $\vec{u}$ is prescribed. Sometimes the boundary conditions are less explicitly defined, however, for example when the considered continuum interacts with its environment. In the following an example of such a situation will be outlined.

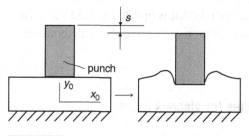

Figure 13.4

Rigid indenter impressing a deforming continuum.

Consider a plane stress continuum of which the midplane (coinciding with the x_0y_0-plane) has a rectangular shape in the reference configuration. The current state arises because the top edge is indented by means of a rigid punch. The punch displacement is specified by s (Fig. 13.4 shows the deformation process, the displacements are magnified). At the location of the contact between indenter and continuum, the interaction is described with a friction model according to Coulomb (which can be considered to be a constitutive description for the contact interaction). For a material point at the top contour of the continuum the following three distinguishable situations may arise:

(i)   There is no contact between the point of the continuum and the indenter. In this case the boundary conditions are

$$\sigma_{yy} = 0, \quad \sigma_{xy} = 0, \tag{13.22}$$

with as an extra constraint that in the current state the vector $\vec{x}_0 + \vec{u}$ does not cross the edge of the indenter.

(ii) There is contact between the point of the continuum and the indenter, and that with 'stick' boundary conditions (no relative tangential displacement between continuum and indenter):

$$u_x = 0, \quad u_y = -s, \tag{13.23}$$

with as extra constraints $\sigma_{yy} \leq 0$ and $|\sigma_{xy}| \leq -\mu\sigma_{yy}$ with μ the friction coefficient.

(iii) There is contact between the continuum and the indenter, and that with 'slip' boundary conditions:

$$u_y = -s, \quad \sigma_{xy} = \mu\frac{u_x}{|u_x|}\sigma_{yy}, \tag{13.24}$$

with as additional constraint $\sigma_{yy} \leq 0$.

The principal problem in accounting for the interaction between the continuum and the indenter is that it is not a priori known to which of the three categories described above the points of the top layer of the continuum belong. In general an estimation is made for this that is updated in case the constraints are violated. In this way an iterative solution can eventually be determined.

13.3   Solution strategies for viscous fluids

Consider a fixed volume V in three-dimensional space, through which (or within which) a certain amount of material flows, while this material can be considered as an incompressible viscous fluid (see Section 12.5). Because, for such a fluid a (possibly defined) reference state is not of interest at all, an Eulerian description is used for relevant fields within the volume V.

The velocity field has to fulfil the incompressibility constraint at each point in time. The current velocity field fully determines the deviatoric part of the stress state (via the constitutive modelling). The hydrostatic part of the stress field cannot be determined on the basis of the velocity field. The stress field (the combination of the hydrostatic and deviatoric part) has to satisfy the momentum balance equation (see Section 11.3). The problem definition is completed by means of initial conditions and boundary conditions. The initial velocity field has to be described in consistency with the incompressibility constraint and along the boundary of V for every point in time velocities and/or stresses have to be in agreement with reality. It should be emphasized explicitly, that considering a fixed volume in space implies a serious limitation for the prospects to apply the theory.

The goal of the present section is to outline a routine to formally determine the velocity field and the (hydrostatic) stress field, both as a function of time, such that all the above mentioned equations are satisfied. However, for (almost all) realistic problems it is not possible to derive an exact analytical solution, not even via assumptions that simplify the mathematical description drastically. A global description will be given of strategies to derive approximate solutions.

In Section 13.3.1 the general (complete) formulation of the problem will be given, including the relevant equations. Thereupon, in the following sections the complexity of the formulation will be gradually reduced. For this, firstly in Section 13.3.2 the material will be modelled as a Newtonian fluid (see Section 12.6). This leads to the so-called Navier–Stokes equation (an equation with the pressure field and the velocity field as unknowns) that has to be solved in combination with the continuity equation (mass balance). In section 13.3.3 the limitation for a stationary flow is dealt with (the time dependency, including the need for initial conditions is no longer relevant). After a section on boundary conditions, a few elementary analytical solutions of the equations for a stationary viscous flow are presented in Section 13.3.5.

13.3.1 General equations for viscous flow

Consider the flow of an incompressible fluid through (or within) a spatially fixed, time-independent volume V. For general viscous flow problems it can be stated that, in an Eulerian description, the following physical fields have to be determined:

- the velocity field: $\vec{v}(\vec{x}, t)$ for all $\vec{x}$ in V and all t and
- the stress field: $\sigma(\vec{x}, t)$ for all $\vec{x}$ in V and all t.

Starting from the incompressibility condition the velocity field has to satisfy the continuity equation (mass balance) for all $\vec{x}$ in V for all t:

$$\text{tr}(\boldsymbol{D}) = 0 \quad \text{with} \quad \boldsymbol{D} = \frac{1}{2}\left(\left(\vec{\nabla}\vec{v}\right)^{\text{T}} + \vec{\nabla}\vec{v}\right), \tag{13.25}$$

while next to that the velocity field and the stress field should be related for all $\vec{x}$ in V and for all times t according to the local constitutive equation (see Section 12.5):

$$\sigma = -p\boldsymbol{I} + \sigma^{\text{d}}(\boldsymbol{D}). \tag{13.26}$$

Also the local balance of momentum (see Section 11.5 for the Eulerian description) has to be satisfied; so for all $\vec{x}$ in V and for all t:

$$\vec{\nabla} \cdot \sigma + \rho\vec{q} = \rho\left(\left(\vec{\nabla}\vec{v}\right)^{\text{T}} \cdot \vec{v} + \frac{\delta\vec{v}}{\delta t}\right), \tag{13.27}$$

with the (mass) density ρ constant.

The equations above form a set of coupled partial differential equations. Consequently, for a unique solution of the velocity field $\vec{v}(\vec{x}, t)$ and the stress field $\sigma(\vec{x}, t)$ boundary conditions and initial conditions are indispensable.

With respect to boundary conditions it can be stated that for each t at every point on the outer surface of V three (scalar) relations have to be specified: either completely formulated in stresses (dynamic boundary conditions), or completely expressed in velocities (kinematic boundary conditions) or in a mixed format.

A detailed description of the interpretation of the initial conditions is not considered.

13.3.2 The equations for a Newtonian fluid

For a Newtonian fluid, see Section 12.5, the stress tensor can be written as

$$\sigma = -p\boldsymbol{I} + 2\eta\boldsymbol{D}, \tag{13.28}$$

where the viscosity η is constant. Substitution of this constitutive equation into the local momentum balance leads to the equation:

$$- \vec{\nabla} p + \eta \vec{\nabla} \cdot \left(\left(\vec{\nabla} \vec{v} \right)^{\mathrm{T}} + \vec{\nabla} \vec{v} \right) + \rho \vec{q} = \rho \left(\left(\vec{\nabla} \vec{v} \right)^{\mathrm{T}} \cdot \vec{v} + \frac{\delta \vec{v}}{\delta t} \right). \qquad (13.29)$$

The left-hand side of this equation can be simplified by using the following identities (to be derived by elaboration in components):

$$\vec{\nabla} \cdot \left(\vec{\nabla} \vec{v} \right)^{\mathrm{T}} = \vec{\nabla} \left(\vec{\nabla} \cdot \vec{v} \right) \qquad (13.30)$$

$$\vec{\nabla} \cdot \left(\vec{\nabla} \vec{v} \right) = \left(\vec{\nabla} \cdot \vec{\nabla} \right) \vec{v}. \qquad (13.31)$$

This leads to:

$$- \vec{\nabla} p + \eta \left(\vec{\nabla} \left(\vec{\nabla} \cdot \vec{v} \right) + \left(\vec{\nabla} \cdot \vec{\nabla} \right) \vec{v} \right) + \rho \vec{q} = \rho \left(\left(\vec{\nabla} \vec{v} \right)^{\mathrm{T}} \cdot \vec{v} + \frac{\delta \vec{v}}{\delta t} \right). \qquad (13.32)$$

Using the expression for incompressibility of the fluid:

$$\vec{\nabla} \cdot \vec{v} = 0, \qquad (13.33)$$

results in the so-called Navier–Stokes equation:

$$- \vec{\nabla} p + \eta \left(\vec{\nabla} \cdot \vec{\nabla} \right) \vec{v} + \rho \vec{q} = \rho \left(\left(\vec{\nabla} \vec{v} \right)^{\mathrm{T}} \cdot \vec{v} + \frac{\delta \vec{v}}{\delta t} \right). \qquad (13.34)$$

The last two relations, the incompressibility condition (continuity equation) and the Navier–Stokes equation, together form a set that allows the determination of the velocity field $\vec{v}(\vec{x}, t)$ and the pressure field $p(\vec{x}, t)$. For the solution boundary conditions and initial conditions have to be supplied to the equations.

13.3.3 Stationary flow of an incompressible Newtonian fluid

For a stationary flow the relevant field variables are only a function of the position vector $\vec{x}$ within the volume V and no longer a function of time. To determine the velocity field $\vec{v}(\vec{x})$ and the pressure field $p(\vec{x})$ the set of equations that has to be solved is reduced to:

$$\vec{\nabla} \cdot \vec{v} = 0 \qquad (13.35)$$

$$- \vec{\nabla} p + \eta \left(\vec{\nabla} \cdot \vec{\nabla} \right) \vec{v} + \rho \vec{q} = \rho \left(\vec{\nabla} \vec{v} \right)^{\mathrm{T}} \cdot \vec{v}. \qquad (13.36)$$

In addition, it is necessary to specify a full set of boundary conditions. Initial conditions do not apply for stationary problems. Note that the equation is non-linear as a consequence of the term on the right-hand side of the last equation; this has a seriously complicating effect on the solution process. Exact analytical solutions can only be found for very simple problems.

13.3.4 Boundary conditions

In the present section the attention is focussed on the formulations of simple boundary conditions with respect to an arbitrary point on the outer surface of the considered (fixed) volume V, with local outward unit normal $\vec{n}$. A number of different possibilities will separately be reviewed.

- Locally prescribed velocity $\vec{v}$ along the boundary, i.e. the component $\vec{v} \cdot \vec{n}$ in normal direction as well as the component $\vec{v} - (\vec{v} \cdot \vec{n})\vec{n}$ in tangential direction. A well-known example of this, is the set of boundary conditions for 'no slip' contact of a fluid with a fixed wall: $\vec{v} = \vec{0}$. In fact, the impermeability of the wall (there is no flux through the outer surface, also see Fig. 7.9) is expressed by $\vec{v} \cdot \vec{n} = 0$, while suppressing of slip is expressed by $\vec{v} - (\vec{v} \cdot \vec{n})\vec{n} = \vec{0}$.

- A locally prescribed stress vector $\sigma \cdot \vec{n}$ along the boundary, i.e. the component $\vec{n} \cdot \sigma \cdot \vec{n}$ in normal direction as well as the components of $\sigma \cdot \vec{n} - (\vec{n} \cdot \sigma \cdot \vec{n})\vec{n}$ in tangential direction. Boundary conditions of this type can be transformed into boundary conditions expressed in $\vec{v}$ and p by means of the constitutive equations. A known example of this is the set of boundary conditions at a free surface: the normal component of the stress vector is related to the atmospheric pressure (equal with opposite sign) and the tangential components are equal to zero.

- For a frictionless flow along a fixed wall it should be required that $\vec{v} \cdot \vec{n} = 0$ combined with $\sigma \cdot \vec{n} - (\vec{n} \cdot \sigma \cdot \vec{n})\vec{n} = \vec{0}$. Again the last condition can, by using the constitutive equation, be expressed in $\vec{v}$ and p.

13.3.5 Elementary analytical solutions

Figure 13.5 visualizes a stationary flow of a fluid between two 'infinitely extended' stationary parallel flat plates (mutual distance h). The flow in the positive x-direction is activated by means of an externally applied pressure gradient. We consider that part between the plates (the specific domain with x- and z-coordinates) where the flow is fully developed. This means that no influence is noticeable anymore from the detailed conditions near the inflow or outflow of the fluid domain: for all relevant values of x, the velocity profile is the same. For $z = 0$ and $z = h$ the fluid adheres to the bottom and the top plate respectively ('no slip' boundary conditions, see previous section). It is simple to verify that the

Figure 13.5

Flow between two stationary parallel plates.

solution for (the components of) the velocity field $\vec{v}(\vec{x})$ and the pressure field $p(\vec{x})$ according to:

$$v_x = -\frac{1}{2\eta}\frac{\partial p}{\partial x}z(h-z), \qquad v_y = 0, \qquad v_z = 0 \qquad (13.37)$$

$$\frac{\partial p}{\partial x} \text{ constant,} \qquad \frac{\partial p}{\partial y} = 0, \qquad \frac{\partial p}{\partial z} = 0, \qquad (13.38)$$

satisfies exactly the continuity equation, the Navier–Stokes equation (with $\vec{q} = \vec{0}$) and the prescribed boundary conditions. It can be observed that the inertia term (the right-hand side of the Navier–Stokes equation) for the defined problem vanishes. This is quite obvious as material particles of the fluid move with constant velocity (no acceleration) in the x-direction.

The situation described above is called the 'plane Poiseuille flow'. The flow is characterized by a parabolic (in the z-coordinate) profile for the velocity in the x-direction, coupled to a constant pressure gradient in the x-direction. Note, that the pressure itself cannot be determined. In order to do that the pressure should be prescribed for a certain value of x, in combination with the pressure gradient in the x-direction or in combination with for example the total mass flux (per unit time) in the x-direction.

In Fig. 13.6 the stationary flow is depicted of a fluid between two parallel flat plates (distance h) in the case where the bottom plate ($z = 0$) is spatially fixed and the top plate ($z = h$) translates in the x-direction with a constant velocity V. There is no pressure gradient. In this example a fully developed flow is the starting point (plates are assumed to be 'infinitely wide' and peripheral phenomena at the inflow and outflow are left out of consideration). Also in this case for $z = 0$ and $z = h$ a perfect adhesion between the fluid and plates occurs. Again, it is simple to verify that for the velocity components the following solution holds:

$$v_x = \frac{z}{h}V, \qquad v_y = 0, \qquad v_z = 0. \qquad (13.39)$$

Figure 13.6

Flow between two mutually translating parallel plates.

Thus, for the velocity in x-direction a linear profile with respect to the z-coordinate is found. The flow in this example is known as the 'Couette flow'.

In this section two very special cases (yet both of practical interest) of fluid flows are treated that permit an analytical solution. For very few practical problems this is possible. The Finite Element Method as discussed in Chapters 14 to 18 enables us to construct approximate solutions for very complex flows, however the specific algorithms to do that for viscous flows are beyond the scope of the contents of this book.

13.4 Diffusion and filtration

In Section 12.8 the constitutive equations for diffusion and filtration to describe transport of material ('fluid') through a stationary porous medium, have been discussed. By adding the relevant balance laws (see Chapter 11) the problem description can be further elaborated. This will be the subject of the present section.

For diffusion Fick's law can be applied:

$$\rho \vec{v} = -D\vec{\nabla}\rho, \tag{13.40}$$

with D the diffusion coefficient.
The mass balance, see Section 11.2, can be written as

$$\frac{\delta\rho}{\delta t} + \vec{\nabla} \cdot (\rho\vec{v}) = 0. \tag{13.41}$$

Elimination of the velocity $\vec{v}$ from the two equations above, leads to a linear partial differential equation for the density $\rho = \rho(\vec{x}, t)$ according to:

$$\frac{\delta\rho}{\delta t} - D\vec{\nabla} \cdot (\vec{\nabla}\rho) = 0. \tag{13.42}$$

This differential equation should be satisfied for all $\vec{x}$ in the considered domain V and for all times t. By specifying the initial conditions (prescribed ρ for all $\vec{x}$ in V at $t = 0$) and with one single boundary condition for each boundary point of V (either the density ρ or the outward mass flux $\rho\vec{v} \cdot \vec{n} = -D(\vec{\nabla}\rho)\cdot\vec{n}$ should be prescribed), in principle a solution for $\rho(\vec{x}, t)$ can be calculated.

To illustrate some of the problems that arise, we confine ourselves to an attempt to solve a simple one-dimensional problem. A domain is given by $0 \leq x \leq L$. Diffusion of a certain material (diffusion coefficient D) in the x-direction can take place. For the density $\rho = \rho(x, t)$ the following partial differential equation holds:

$$\frac{\partial\rho}{\partial t} - D\frac{\partial^2\rho}{\partial x^2} = 0, \qquad (13.43)$$

emphasizing that the spatial derivative $\delta\rho/\delta t$ is written here as the partial derivative of ρ to the time t (with constant x). Misunderstandings because of this will not be introduced, because exclusively an Eulerian description will be used. The differential equation in this example is completed with:

- the initial condition:

$$\rho = 0 \quad \text{for } 0 \leq x \leq L \text{ and } t = 0,$$

- the boundary conditions:

$$\rho = \rho_0 \quad \text{(with } \rho_0 \text{ a constant) for } x = 0 \text{ and } t > 0$$

$$\frac{\partial\rho}{\partial x} = 0 \quad \text{(no outflow of material) for } x = L \text{ and } t > 0.$$

Even for this very simple situation an exact solution is very difficult to determine. A numerical approach (for example by means of the Finite Element Method) can lead to a solution in a simple way. This is the topic of Chapter 14. Here, it can be stated that the solution for $\rho(x, t)$ at $t \rightarrow \infty$ has to satisfy $\rho(x, t) = \rho_0$ for all x. A large number of closed form solutions can be found in [4].

For filtration problems Darcy's law can be applied:

$$\rho\vec{v} = -\kappa\vec{\nabla}p, \qquad (13.44)$$

with κ the permeability. The mass balance, see Section 11.2, can be written as follows:

$$\frac{\delta\rho}{\delta t} + \vec{\nabla} \cdot (\rho\vec{v}) = 0. \qquad (13.45)$$

Further elaboration is limited to stationary filtration (time t does not play a role). In that case

$$\frac{\delta\rho}{\delta t} = 0 \quad \text{and so} \quad \vec{\nabla} \cdot (\rho\vec{v}) = 0.$$

Combination of this equation with Darcy's constitutive law leads to:

$$\vec{\nabla} \cdot \left(\vec{\nabla} p \right) = 0. \tag{13.46}$$

This equation for the pressure p can formally be solved when for every boundary point of the volume V one single condition is specified. This can either be formulated in the pressure p, or in the outward mass flux $\rho \vec{v} \cdot \vec{n} = -\kappa (\vec{\nabla} p) \cdot \vec{n}$. When the solution for p is determined it is easy to calculate directly the mass flux $\rho \vec{v}$ with Darcy's law.

In the one-dimensional case (with x as the only relevant independent variable) the differential equation for p reduces to

$$\frac{d^2 p}{dx^2} = 0. \tag{13.47}$$

In this case p will be a linear function of x.

Exercises

13.1 Consider a material element with the shape of a cube (length of the edges ℓ). The cube is placed in a Cartesian xyz-coordinate system, see figure.

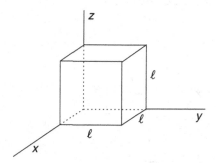

All displacements from the bottom face of the element (coinciding with the xy-plane) are suppressed. The top face has a prescribed displacement in the y-direction, which is small with respect to ℓ. The side faces are unloaded. Assume that a homogeneous stress state occurs with $\sigma_{yz} = \sigma_{zy}$, the only components of the stress matrix $\underline{\sigma}$ unequal to zero.
       Why can this assumption not be correct?

13.2 Consider a thin rectangular piece of material (constant thickness h). The midplane of the material coincides with the xy-plane of a Cartesian xyz-coordinate system. The material behaviour is described by means of Hooke's law (Young's modulus E and Poisson's ratio v). The plate is statically loaded with pure shear (plane stress state). The accompanying boundary conditions are shown in the figure below.

For an arbitrary point of the midplane, with the coordinates (x_0, y_0) in the unloaded reference configuration, the displacements in the x- and y-direction, as a result of the stress τ, are indicated with $u_x(x_0, y_0)$, $u_y(x_0, y_0)$, respectively.

Why does no unique solution exist for the fields $u_x(x_0, y_0)$, $u_y(x_0, y_0)$, based on the information that is given above?

13.3   A thin trapezium shaped plate is clamped on the left side and statically loaded on the right side with an extensional load (local extensional stress p), as given in the figure below.

In the plate plane stress conditions occur ($\sigma_{xz} = \sigma_{yz} = \sigma_{zz} = 0$). With respect to the stress components still relevant for the plane stress conditions, the following assumptions are made:

$$\sigma_{xx} = p\frac{x+\ell}{2\ell}, \qquad \sigma_{yy} = 0, \qquad \sigma_{xy} = 0.$$

Give two reasons why this assumed stress field cannot be correct.

13.4   Consider a rectangular plate of some material ($\ell \times b$). The thickness h is small with respect to ℓ and b, see figure.

The left side ($x = 0$) is clamped. The right side ($x = \ell$) is loaded with a distributed tangential load (the resultant force P is known) in the negative y-direction. The top and bottom surface of the plate are unloaded. A plane stress condition is supposed. With respect to the stress field the following assumption is proposed:

$$\sigma_{xx} = c_1(\ell - x)y$$
$$\sigma_{yy} = 0$$
$$\sigma_{xy} = \left(\frac{b^2}{4} - y^2\right),$$

with c_1 and c_2 constants.

Determine c_2 based on the relationship between σ_{xy} and P and subsequently, determine c_1 by means of the local equilibrium equations.

13.5 In the environment of the origin of a Cartesian xyz-coordinate system it is given that a (two-dimensional) stationary velocity field in an incompressible Newtonian fluid (with density ρ) can be described as

$$\vec{v} = \alpha(-y\vec{e}_x + x\vec{e}_y) \quad \text{with } \alpha \text{ a constant.}$$

Based on this velocity field the deformation rate tensor D can be calculated: $D = 0$ with 0 the zero tensor. With substitution into the constitutive equation:

$$\sigma = -pI + 2\eta D,$$

with η the viscosity and σ the stress tensor (and with $p = 0$ originated by the applied boundary conditions) it follows that $\sigma = 0$.

Then, by means of the Navier–Stokes equation it is found that the distributed load $\vec{q}$ (force per unit mass) necessary to realize the described flow field is given by

$$\vec{q} = -\alpha^2\vec{r} \quad \text{with} \quad \vec{r} = x\vec{e}_x + y\vec{e}_y.$$

Interpret the stated results: why $D = 0$ is obtained for the given flow field and what is the physical meaning of $\vec{q} = -\alpha^2 \vec{r}$?

13.6 An incompressible Newtonian fluid (density ρ, viscosity η) is located in a cavity (rectangular) with a cross section as given in the figure: $-a \leq x \leq a$, $-b \leq y \leq b$. Because the top wall closing the cavity is translating in the

x-direction with velocity V, a stationary (two-dimensional) fluid flow (in the shape of a vortex) will develop in the cavity: $v_x = v_x(x, y)$, $v_y = v_y(x, y)$, $v_z = 0$. No slip conditions hold at all walls of the cavity (moving as well as stationary).

Which inconsistency ('something leading to contradiction') can be observed in the above given problem definition?

13.7 A compressible medium (for example a gas) is forced in the x-direction through a contraction (see figure). With respect to the components of the

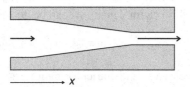

velocity field $\underset{\sim}{v}$ of the medium, defined in the Cartesian xyz-coordinate system, it is assumed that v_x only depends on x, thus $v_x = v_x(x)$, completed with $v_y = v_z = 0$.

Why is it impossible that this assumption is correct?

13.8 In the neighbourhood of the origin of a Cartesian xyz-coordinate system the stationary (two-dimensional) flow field in an incompressible Newtonian fluid (with density ρ) can be described as

$$\vec{v} = \alpha(-y\vec{e}_x + x\vec{e}_y) \quad \text{with } \alpha \text{ a constant.}$$

With respect to the constitutive equation:

$$\boldsymbol{\sigma} = -p\boldsymbol{I} + 2\eta\boldsymbol{D},$$

with η the viscosity, $\boldsymbol{\sigma}$ the stress tensor and $\boldsymbol{D}$ the deformation rate tensor, it can be stated (based on the prescribed boundary conditions) that p satisfies $p = 0$.

Determine by means of the Navier–Stokes equation the distributed load $\vec{q}$ (per unit mass) necessary to realize the given flow pattern.

14 Solution of the one-dimensional diffusion equation by means of the Finite Element Method

In the present and following chapters extensive use will be made of a simple finite element code `mlfem_nac`. This code, including a manual, can be freely downloaded from the website: www.mate.tue.nl/biomechanicsbook.

The code is written in the program environment MATLAB. To be able to use this environment a licence for MATLAB has to be obtained. For information about MATLAB see: www.mathworks.com.

14.1 Introduction

It will be clear from the previous chapters that many problems in biomechanics are described by (sets of) partial differential equations and for most problems it is difficult or impossible to derive closed form (analytical) solutions. However, by means of computers, approximate solutions can be determined for a very large range of complex problems, which is one of the reasons why biomechanics as a discipline has grown so fast in the last three decades. These computer-aided solutions are called numerical solutions, as opposed to analytical or closed form solutions of equations. The present and following chapters are devoted to the numerical solution of partial differential equations, for which several methods exist. The most important ones are the Finite Difference Method and the Finite Element Method. The latter is especially suitable for partial differential equations on domains with complicated geometries, material properties and boundary conditions (which is nearly always the case in biomechanics). That is why the next chapters focus on the Finite Element Method. The basic concepts of the method are explained in the present chapter. The one-dimensional diffusion equation will be used as an example to illustrate the key features of the finite element method. Clearly, the one-dimensional diffusion problem can be solved analytically for a wide variety of parameter choices, but the structure of the differential equation is representative of a much larger class of problems to be discussed later. In addition, the diffusion equation and the more extended, instationary (convection) diffusion equation play an important role in many processes in biomechanics.

14.2   The diffusion equation

The differential equation that describes the one-dimensional diffusion problem is given by

$$\frac{d}{dx}\left(c\frac{du}{dx}\right) + f = 0, \tag{14.1}$$

where $u(x)$ is the unknown function, $c(x) > 0$ a given material characteristic function and $f(x)$ a given source term. This differential equation is defined on a one-dimensional domain Ω that spans the x-axis between $x = a$ and $x = b$ while the boundary, which is formally denoted by Γ, is located at $x = a$ and $x = b$.

Eq. (14.1) is an adapted form of the diffusion equation Eq. (13.43), introduced in the previous chapter. Different symbols for the unknown (u instead of ρ) and coefficient (c instead of D) are used to emphasize the general character of the equation, applicable to different kinds of problems (see below). Furthermore, the coefficient c can be a function of x and a source term $f(x)$ is introduced.

Two types of boundary conditions can be discerned. Firstly, the essential boundary condition, which must be specified in terms of u. For the derivations that follow the boundary at $x = a$ is chosen, to specify this type of boundary condition:

$$u = U \text{ at } \Gamma_u, \tag{14.2}$$

where Γ_u denotes the boundary of the domain Ω at $x = a$. Secondly, a natural boundary condition may be specified. Here the boundary at $x = b$ is chosen to specify the flux $c\,du/dx$:

$$c\frac{du}{dx} = P \text{ at } \Gamma_p, \tag{14.3}$$

where Γ_p denotes the boundary at $x = b$. For diffusion problems, an essential boundary condition must be specified to have a well-posed boundary value problem. This is not necessarily the case for natural boundary conditions; they may be absent.

Example 14.1   The diffusion equation describes a large range of problems in biomechanics and is applicable in many different areas. In the way it was introduced in Section 13.4 the unknown u represents the concentration of some matter, for example: oxygen in blood, proteins or other molecules in an extracellular matrix or inside a cell, medication in blood or tissue. In that case the term f can be either a source term (where matter is produced) or a sink term (where matter is consumed).

Figure 14.1

(a) Schematic view of a bioreactor system designed for growing articular cartilage tissue (b) result of a numerical calculation of the concentration of glucose in the tissue engineered construct. In the analysis it is assumed that the glucose concentration in the medium surrounding the construct is constant (essential boundary condition) and that glucose is consumed by the cells in the construct (sink term). Because of symmetry, only the right half of the construct is modelled. Adapted from [17].

In the human body diffusion is very important, but also in in vitro experimental set-ups in the laboratory, for example in tissue engineering applications (see Fig. 14.1).

Example 14.2   Another completely different process which is also relevant in biomechanics is steady state one-dimensional heat conduction with a source term:

$$\frac{d}{dx}\left(\lambda\frac{dT}{dx}\right) + f = 0,$$

where λ is the Fourier coefficient of heat conduction and f a heat source term. Further, $T(x)$ represents the temperature. The boundary conditions might be formulated as: prescribed temperature at Γ_u and prescribed heat flux at Γ_p. It should be noticed that the mathematical structure of this heat conduction problem is fully equivalent to the structure of the diffusion problem.

Example 14.3   A third example of a completely different diffusion type problem is the uniaxial tension or compression of a bar as introduced in Chapter 6, governed by the equation:

$$\frac{d}{dx}\left(EA\frac{du}{dx}\right) + f = 0,$$

where E is the Young's modulus, A the cross section of the bar and f a distributed force per unit length. The unknown $u(x)$ is the displacement field of the bar (also see Eq. (6.20)). Boundary conditions may be imposed as a prescribed displacement on Γ_u and a prescribed force at Γ_p.

The approximate solution of Eq. (14.1) is found by transforming the differential equation into a discrete set of ordinary equations:

$$\frac{d}{dx}\left(c\frac{du}{dx}\right) + f = 0 \longrightarrow \underline{K}\,\underline{u} = \underline{f}. \tag{14.4}$$

The array $\underline{u}$ contains **approximations** of the continuous solution u of the differential equation at a **finite** number of locations on the x-axis. Increasing the number of points defining $\underline{u}$ should lead to an increased accuracy of the approximation of u. A particularly attractive feature of the Finite Element Method is that the spatial distribution of these points does not need to be equidistant and can be chosen such that accurate solutions can be obtained with a limited number of points, even on complicated geometries (in the multi-dimensional case) or problems with large gradients in the solution.

The finite element method proceeds along three well-defined steps.

(i) Transformation of the original differential equation into an integral equation by means of the principle of **weighted residuals**.

(ii) **Discretization** of the solution u by interpolation. If an approximation of the solution u is known at a finite number of points (nodes) an approximation field may be constructed by interpolation between these point (nodal) values.

(iii) Using the discretization the integral equation is transformed into a **linear set of equations** from which the nodal values $\underline{u}$ can be solved.

14.3 Method of weighted residuals and weak form of the model problem

First of all the differential equation is transformed into an integral equation by means of the **weighted residuals** method. Suppose that a given function $g(x) = 0$ on a certain domain $a \leq x \leq b$, then this formulation is equivalent to requiring

$$\int_a^b w(x)\,g(x)\;dx = 0 \qquad \text{for all } w, \tag{14.5}$$

and to emphasize this important equivalence:

$$g(x) = 0 \quad \text{on} \quad a \leq x \leq b \Leftrightarrow \int_a^b w(x)\,g(x)\;dx = 0 \qquad \text{for all } w. \tag{14.6}$$

The function $w(x)$ is called the **weighting function**, and is assumed to be a continuous function on the integration domain. The equivalence originates from the requirement that the integral equation must hold for **all** possible weighting functions w. It therefore should also hold for $w = g(x)$. For this particular choice of w, the integral expression yields

$$\int_a^b g^2(x)\ dx = 0 \quad \Rightarrow \quad g(x) = 0 \quad \text{on } a \le x \le b. \tag{14.7}$$

This follows immediately from the observation that the square of a function $g(x)$ is always greater than or equal to zero for any value of x, i.e. $g^2(x) \ge 0$, such that the integral of $g^2(x)$ over the domain $a \le x \le b$ must be greater than or equal to zero, i.e.

$$\int_a^b g^2(x)\ dx \ge 0, \tag{14.8}$$

and can only be **equal** to zero if $g(x) = 0$ for all $a \le x \le b$.

Effectively, the method of weighted residuals transforms the requirement that a function, say $g(x)$, must be equal to zero on a given domain at an **infinite** number of points into a **single evaluation** of the integral, that must be equal to zero.

Using the method of weighted residuals, the differential equation, Eq. (14.1), is transformed into an integral equation:

$$\int_a^b w \left[\frac{d}{dx} \left(c \frac{du}{dx} \right) + f \right] dx = 0, \tag{14.9}$$

which should hold for all weighting functions $w(x)$. The term between the square brackets contains second order derivatives $d^2 u/dx^2$ of the function u. As has been outlined in the introduction, approximate solutions of u are sought by defining an interpolation of u on the domain of interest and transforming the integral equation into a discrete set of linear equations. Defining interpolation functions that are both second-order differentiable and still integrable is far from straightforward, in particular in the multi-dimensional case on arbitrarily shaped domains. Fortunately, the second-order derivatives can be removed by means of an integration by parts:

$$w \left(c \frac{du}{dx} \right) \bigg|_a^b - \int_a^b \frac{dw}{dx} c \frac{du}{dx}\ dx + \int_a^b wf\ dx = 0. \tag{14.10}$$

This introduces the boundary terms:

$$w \left(c \frac{du}{dx} \right) \bigg|_a^b = -w(a) \left(c \frac{du}{dx} \right) \bigg|_{x=a} + w(b) \left(c \frac{du}{dx} \right) \bigg|_{x=b}. \tag{14.11}$$

At the boundary either u is prescribed (i.e. the essential boundary condition at $x = a$) or the derivative $c\ du/dx$ is prescribed (i.e. the natural boundary condition at $x = b$). Along the boundary where u is prescribed the corresponding flux, say p_u, with

$$p_u = c \frac{du}{dx} \bigg|_{x=a}, \tag{14.12}$$

is unknown, while along the other boundary the flux $c\frac{du}{dx}\Big|_{x=b} = P$ is known. For the time being the combination of boundary terms in Eq. (14.11) is written as

$$B = w\left(c\frac{du}{dx}\right)\Big|_a^b = -w(a)p_u + w(b)P,\tag{14.13}$$

realizing that p_u is considered as yet unknown. The term B is introduced as an abbreviation for the boundary contribution.

Consequently, the following integral equation, known as the **weak form**, results:

$$\int_a^b \frac{dw}{dx}c\frac{du}{dx}\,dx = \int_a^b wf\,dx + B.\tag{14.14}$$

Eq. (14.14) is the point of departure for the Finite Element Method, and is called the weak form, because the differentiability requirements imposed on u have been reduced: in the original differential equation, Eq. (14.1), the second-order derivative d^2u/dx^2 appears, while in the weak form, Eq. (14.14), only the first-order derivative du/dx has to be dealt with.

14.4 Polynomial interpolation

Suppose that at a finite number of points x_i in the domain Ω, the function values $u_i = u(x_i)$ are known, then a polynomial approximation, denoted by u_h, of $u(x)$ can be constructed. The polynomial approximation u_h of degree $n - 1$ can be constructed by

$$u_h(x) = a_0 + a_1 x + a_2 x^2 + \cdots + a_{n-1}x^{n-1},\tag{14.15}$$

if u is known at n points. The coefficients a_i can be identified uniquely and expressed in terms of u_i, by solving the set of equations:

$$\begin{bmatrix} 1 & x_1 & x_1^2 & \cdots & x_1^{n-1} \\ 1 & x_2 & x_2^2 & \cdots & x_2^{n-1} \\ 1 & x_3 & x_3^2 & \cdots & x_3^{n-1} \\ \vdots & \vdots & \vdots & \vdots & \vdots \\ 1 & x_n & x_n^2 & \cdots & x_n^{n-1} \end{bmatrix} \begin{bmatrix} a_0 \\ a_1 \\ a_2 \\ \vdots \\ a_{n-1} \end{bmatrix} = \begin{bmatrix} u_1 \\ u_2 \\ u_3 \\ \vdots \\ u_n \end{bmatrix}.\tag{14.16}$$

An example is given in Fig. 14.2, where, in the domain $x_{i-1} \le x \le x_{i+2}$, a third-order polynomial (dashed curve) is used to approximate a given function (solid curve) based on the function values u_{i-1}, u_i, u_{i+1} and u_{i+2}.

Clearly, the coefficients a_i are linearly dependent on the values u_i, therefore the polynomial may be rewritten in terms of u_i by

Figure 14.2

Solid line: $u(x)$, dashed line: polynomial approximation of $u(x)$.

$$u_h(x) = \sum_{i=1}^{n} N_i(x)\, u_i, \tag{14.17}$$

where the functions $N_i(x)$ are polynomial expressions of order $n-1$ in terms of the coordinate x. These functions $N_i(x)$ are called **shape functions** because they define the shape of the interpolation of u_h, for instance linear, quadratic etc. To illustrate this, consider a first-order polynomial on the domain $[x_1, x_2]$, with $u(x)$ known at x_1 and x_2. In that case

$$u_h(x) = \left(1 - \frac{x - x_1}{x_2 - x_1}\right) u_1 + \frac{x - x_1}{x_2 - x_1} u_2, \tag{14.18}$$

implying that

$$N_1 = \left(1 - \frac{x - x_1}{x_2 - x_1}\right), \quad N_2 = \frac{x - x_1}{x_2 - x_1}. \tag{14.19}$$

Rather than approximating $u(x)$ with a single polynomial of a certain degree over the entire domain Ω, the domain Ω may also be divided into a number of non-overlapping subdomains, say Ω_e. Within each subdomain Ω_e a **local** polynomial approximation of u may be constructed. A typical example is a piecewise linear approximation within each subdomain Ω_e, see Fig. 14.3. Consider one of the subdomains $\Omega_e = [x_i, x_{i+1}]$. Then within Ω_e the function $u(x)$ is approximated by

$$u_h(x)|_{\Omega_e} = N_1(x)\, u_i + N_2(x)\, u_{i+1}, \tag{14.20}$$

with, in conformity with Eq. (14.19):

$$N_1 = \left(1 - \frac{x - x_i}{x_{i+1} - x_i}\right) \quad N_2 = \frac{x - x_i}{x_{i+1} - x_i}. \tag{14.21}$$

More generally, if within a subdomain Ω_e an n-th order polynomial approximation of u is applied, the subdomain should have $n+1$ points at which the function u is

Figure 14.3

Solid line: $u(x)$, dashed line: piecewise linear approximation of $u(x)$.

known. For instance, to use a second-order (quadratic) polynomial, the subdomain should cover at least three consecutive points, for example $\Omega_e = [x_i, x_{i+2}]$, such that

$$u_h(x)|_{\Omega_e} = N_1(x)\,u_i + N_2(x)\,u_{i+1} + N_3(x)\,u_{i+2}. \qquad (14.22)$$

The subdomain Ω_e within which a certain polynomial approximation is used is referred to as an **element**. The points at which the values of u are defined are called **nodes**.

The shape functions N_i may not be chosen arbitrarily. The most stringent requirement is that u_h must be interpolated continuously over the total domain Ω (the first-order derivative of $u_h(x)$ should exist). Suppose that the nodes x_j within an element are numbered $j = 1, \ldots, n$ and that the associated shape functions N_i are numbered $i = 1, \ldots, n$. In that case, for consistency, the shape functions must be chosen such that

$$N_i(x_j) = \delta_{ij}, \qquad (14.23)$$

with

$$\delta_{ij} = 0 \ \text{if} \ i \neq j$$
$$= 1 \ \text{if} \ i = j. \qquad (14.24)$$

14.5 Galerkin approximation

To transform the weak form into a linear set of equations in order to derive an approximate solution the following steps are taken.

Step 1. Element division As shown in the previous section the domain Ω may be split into a number of subdomains Ω_e, elements, and within each element a polynomial interpolation can be made of the function u. An example of such an element division is given in Fig. 14.4. This distribution of elements is called a **mesh**. Then, the integration over the domain Ω can be performed by summing up the integrals over each element. Consequently, Eq. (14.14) yields:

$$\sum_{e=1}^{N_{el}} \int_{\Omega_e} \frac{dw}{dx} c \frac{du}{dx} \, dx = \sum_{e=1}^{N_{el}} \int_{\Omega_e} wf \, dx + B, \qquad (14.25)$$

where N_{el} denotes the number of elements.

Step 2. Interpolation Suppose that the domain Ω has been divided into three linear elements, as depicted in Fig. 14.4. Then the nodal values u_i may be collected in an array $\underset{\sim}{u}$:

$$\underset{\sim}{u} = \begin{bmatrix} u_1 \\ u_2 \\ u_3 \\ u_4 \end{bmatrix}. \qquad (14.26)$$

The unknowns associated with each of the elements Ω_e are collected in the arrays $\underset{\sim}{u}_e$, such that

$$\underset{\sim}{u}_1 = \begin{bmatrix} u_1 \\ u_2 \end{bmatrix}, \quad \underset{\sim}{u}_2 = \begin{bmatrix} u_2 \\ u_3 \end{bmatrix}, \quad \underset{\sim}{u}_3 = \begin{bmatrix} u_3 \\ u_4 \end{bmatrix}. \qquad (14.27)$$

So, it is important to realize that each particular element array $\underset{\sim}{u}_e$ contains a subset of the total, or global, array $\underset{\sim}{u}$. Within each element array $\underset{\sim}{u}_e$ a local numbering may be used, such that for this particular example with linear elements:

Figure 14.4
Element distribution and unknowns at local and global levels.

$$\underset{\sim}{u}_e = \begin{bmatrix} u_1^e \\ u_2^e \end{bmatrix}. \tag{14.28}$$

For instance in case of the second element the element array $\underset{\sim}{u}_e$ contains

$$\underset{\sim}{u}_2 = \begin{bmatrix} u_1^2 \\ u_2^2 \end{bmatrix} = \begin{bmatrix} u_2 \\ u_3 \end{bmatrix}. \tag{14.29}$$

Clearly, in the case of quadratic elements the element array $\underset{\sim}{u}_e$ contains three unknowns, while for cubic elements $\underset{\sim}{u}_e$ contains four unknowns, and so on.

Within each element a polynomial approximation for both the unknown function $u(x)$ and the weighting function $w(x)$ is introduced. Use of the shape function approach as outlined in the previous section yields

$$u_h|_{\Omega_e} = \sum_{i=1}^{n} N_i(x)\, u_i^e = \underset{\sim}{N}^T \underset{\sim}{u}_e, \tag{14.30}$$

$$w_h|_{\Omega_e} = \sum_{i=1}^{n} N_i(x)\, w_i^e = \underset{\sim}{N}^T \underset{\sim}{w}_e, \tag{14.31}$$

where N_i are the **shape** functions

$$\underset{\sim}{N}^T = [N_1\ N_2 \cdots N_n], \tag{14.32}$$

and $\underset{\sim}{u}_e$ and $\underset{\sim}{w}_e$ contain the unknowns u_i^e and the weighting values w_i^e, respectively, associated with element Ω_e. The fact that the same shape functions (and hence polynomial interpolation order) are chosen for both the unknown function u_h and the weighting function w_h means that the so-called **(Bubnov) Galerkin** method is used.

As a consequence of the interpolation of u_h governed by the shape functions N_i, the differentiation of u_h is straightforward since the nodal values u_i are independent of the coordinate x, while the shape functions N_i are simple, known, functions of x:

$$\begin{aligned}
\frac{du_h}{dx}\bigg|_{\Omega_e} &= \sum_{i=1}^{n} \frac{d}{dx}\left(N_i(x)\, u_i^e\right) \\
&= \sum_{i=1}^{n} \frac{dN_i(x)}{dx} u_i^e \\
&= \frac{d\underset{\sim}{N}^T}{dx} \underset{\sim}{u}_e.
\end{aligned} \tag{14.33}$$

Clearly a similar expression holds for the weighting function:

$$\left.\frac{dw_h}{dx}\right|_{\Omega_e} = \frac{d\underset{\sim}{N}^{\mathrm{T}}}{dx}\underset{\sim}{w}_e. \tag{14.34}$$

Substitution of this result into the left-hand side of Eq. (14.25) and considering one element only yields:

$$\int_{\Omega_e} \frac{dw_h}{dx}c\frac{du_h}{dx}\,dx = \int_{\Omega_e} \frac{d\underset{\sim}{N}^{\mathrm{T}}}{dx}\underset{\sim}{w}_e\,c\,\frac{d\underset{\sim}{N}^{\mathrm{T}}}{dx}\underset{\sim}{u}_e\,dx$$

$$= \int_{\Omega_e} \underset{\sim}{w}_e^{\mathrm{T}}\frac{d\underset{\sim}{N}}{dx}\,c\,\frac{d\underset{\sim}{N}^{\mathrm{T}}}{dx}\underset{\sim}{u}_e\,dx$$

$$= \underset{\sim}{w}_e^{\mathrm{T}}\int_{\Omega_e} \frac{d\underset{\sim}{N}}{dx}\,c\,\frac{d\underset{\sim}{N}^{\mathrm{T}}}{dx}\,dx\,\underset{\sim}{u}_e. \tag{14.35}$$

In the last step use has been made of the fact that $\underset{\sim}{w}_e$ and $\underset{\sim}{u}_e$ are both independent of the coordinate x. Likewise, the integral expression on the right-hand side of Eq. (14.25) yields for a single element:

$$\int_{\Omega_e} w_h f\,dx = \int_{\Omega_e} \underset{\sim}{N}^{\mathrm{T}}\underset{\sim}{w}_e f\,dx = \underset{\sim}{w}_e^{\mathrm{T}}\int_{\Omega_e} \underset{\sim}{N} f\,dx. \tag{14.36}$$

Notice that the integral appearing on the right-hand side of Eq. (14.35) is in fact a matrix, called the element coefficient or (in mechanical terms) stiffness matrix:

$$\underline{K}_e = \int_{\Omega_e} \frac{d\underset{\sim}{N}}{dx}\,c\,\frac{d\underset{\sim}{N}^{\mathrm{T}}}{dx}\,dx. \tag{14.37}$$

Similarly, the integral on the right-hand side of Eq. (14.36) is the element array corresponding to the internal source or distributed load:

$$\underset{\sim}{f}_e = \int_{\Omega_e} \underset{\sim}{N} f\,dx. \tag{14.38}$$

Substitution of the expression for the element coefficient matrix and element column in Eq. (14.25) yields

$$\sum_{e=1}^{n_{el}} \underset{\sim}{w}_e^{\mathrm{T}}\underline{K}_e\underset{\sim}{u}_e = \sum_{e=1}^{n_{el}} \underset{\sim}{w}_e^{\mathrm{T}}\underset{\sim}{f}_e + B. \tag{14.39}$$

Step 3. Assembling the global set of equations The individual element contributions

$$\underset{\sim}{w}_e^{\mathrm{T}}\underline{K}_e\underset{\sim}{u}_e, \tag{14.40}$$

using the local unknowns and weighting values ($\underset{\sim}{u}_e$ and $\underset{\sim}{w}_e$, respectively) only, may also be rewritten in terms of the global unknowns $\underset{\sim}{u}$ and the associated weighting function values w. This can be done by introducing an auxiliary matrix $\underline{\hat{K}}_e$ on element level such that

$$w_e^{\mathrm{T}} \underline{K}_e \underline{u}_e = \underline{w}^T \underline{\hat{K}}_e \underline{u}. \tag{14.41}$$

To illustrate this, consider once more the element distribution of Fig. 14.4. For the second element it holds that

$$w_2^{\mathrm{T}} \underline{K}_2 \underline{u}_2 = \begin{bmatrix} w_1^2 & w_2^2 \end{bmatrix} \begin{bmatrix} K_{11}^2 & K_{12}^2 \\ K_{21}^2 & K_{22}^2 \end{bmatrix} \begin{bmatrix} u_1^2 \\ u_2^2 \end{bmatrix} \tag{14.42}$$

$$= \begin{bmatrix} w_2 & w_3 \end{bmatrix} \begin{bmatrix} K_{11}^2 & K_{12}^2 \\ K_{21}^2 & K_{22}^2 \end{bmatrix} \begin{bmatrix} u_2 \\ u_3 \end{bmatrix}. \tag{14.43}$$

Notice that in Eq. (14.42) the local nodal values have been used, while in Eq. (14.43) the global values have been used. Eq. (14.43) may also be rewritten as

$$w_2^{\mathrm{T}} \underline{K}_2 \underline{u}_2 = \begin{bmatrix} w_2 & w_3 \end{bmatrix} \begin{bmatrix} K_{11}^2 & K_{12}^2 \\ K_{21}^2 & K_{22}^2 \end{bmatrix} \begin{bmatrix} u_2 \\ u_3 \end{bmatrix}$$

$$= \begin{bmatrix} w_1 & w_2 & w_3 & w_4 \end{bmatrix} \underbrace{\begin{bmatrix} 0 & 0 & 0 & 0 \\ 0 & K_{11}^2 & K_{12}^2 & 0 \\ 0 & K_{21}^2 & K_{22}^2 & 0 \\ 0 & 0 & 0 & 0 \end{bmatrix}}_{\underline{\hat{K}}_2} \begin{bmatrix} u_1 \\ u_2 \\ u_3 \\ u_4 \end{bmatrix}$$

$$= \underline{w}^{\mathrm{T}} \underline{\hat{K}}_2 \, \underline{u}. \tag{14.44}$$

Consequently,

$$\sum_{e=1}^{N_{el}} \underline{w}_e^{\mathrm{T}} \underline{K}_e \, \underline{u}_e = \sum_{e=1}^{N_{el}} \underline{w}^{\mathrm{T}} \underline{\hat{K}}_e \, \underline{u}, \tag{14.45}$$

and by summing the individual $\underline{\hat{K}}_e$ matrices the result:

$$\sum_{e=1}^{N_{el}} \underline{w}_e^{\mathrm{T}} \underline{K}_e \underline{u}_e = \underline{w}^{\mathrm{T}} \underline{K} \, \underline{u}, \tag{14.46}$$

is obtained, with

$$\underline{K} = \sum_{e=1}^{N_{el}} \underline{\hat{K}}_e. \tag{14.47}$$

In the case of the element distribution of Fig. 14.4, it holds that

$$\hat{\underline{K}}_1 = \begin{bmatrix} K_{11}^1 & K_{12}^1 & 0 & 0 \\ K_{21}^1 & K_{22}^1 & 0 & 0 \\ 0 & 0 & 0 & 0 \\ 0 & 0 & 0 & 0 \end{bmatrix} \tag{14.48}$$

$$\hat{\underline{K}}_2 = \begin{bmatrix} 0 & 0 & 0 & 0 \\ 0 & K_{11}^2 & K_{12}^2 & 0 \\ 0 & K_{21}^2 & K_{22}^2 & 0 \\ 0 & 0 & 0 & 0 \end{bmatrix} \tag{14.49}$$

$$\hat{\underline{K}}_3 = \begin{bmatrix} 0 & 0 & 0 & 0 \\ 0 & 0 & 0 & 0 \\ 0 & 0 & K_{11}^3 & K_{12}^3 \\ 0 & 0 & K_{21}^3 & K_{22}^3 \end{bmatrix}, \tag{14.50}$$

such that

$$\underline{K} = \hat{\underline{K}}_1 + \hat{\underline{K}}_2 + \hat{\underline{K}}_3, \tag{14.51}$$

leading to

$$\underline{K} = \begin{bmatrix} K_{11}^1 & K_{12}^1 & 0 & 0 \\ K_{21}^1 & K_{22}^1 + K_{11}^2 & K_{12}^2 & 0 \\ 0 & K_{21}^2 & K_{22}^2 + K_{11}^3 & K_{12}^3 \\ 0 & 0 & K_{21}^3 & K_{22}^3 \end{bmatrix}. \tag{14.52}$$

In computer codes, however, the individual matrices $\hat{\underline{K}}_i$ are never formed explicitly. The non-trivial components of $\hat{\underline{K}}_i$ are supplied directly to the appropriate position within $\underline{K}$. This process of assembling the global matrix $\underline{K}$ based on the contributions of the individual element matrices $\underline{K}_e$ is called the **assembly** process.

For the right-hand side the same procedure is followed. For element 2 it can be written analogously:

$$\underline{w}_2^T \underline{f}_2 = \begin{bmatrix} w_1^2 & w_2^2 \end{bmatrix} \begin{bmatrix} f_1^2 \\ f_2^2 \end{bmatrix} = \begin{bmatrix} w_2 & w_3 \end{bmatrix} \begin{bmatrix} f_1^2 \\ f_2^2 \end{bmatrix}$$

$$= \begin{bmatrix} w_1 & w_2 & w_3 & w_4 \end{bmatrix} \underbrace{\begin{bmatrix} 0 \\ f_1^2 \\ f_2^2 \\ 0 \end{bmatrix}}_{\hat{\underline{f}}_2} = \underline{w}^T \hat{\underline{f}}_2. \tag{14.53}$$

Following the same procedure for the other elements and adding up the contribution for the internal source term for all (three) elements gives

$$\underset{\sim}{f} = \hat{\underset{\sim}{f}}_1 + \hat{\underset{\sim}{f}}_2 + \hat{\underset{\sim}{f}}_3. \tag{14.54}$$

This leads to

$$\underset{\sim}{w}^T \underset{\sim}{f}_{\text{int}} = \begin{bmatrix} w_1 & w_2 & w_3 & w_4 \end{bmatrix} \begin{bmatrix} f_1^1 \\ f_2^1 + f_1^2 \\ f_2^2 + f_1^3 \\ f_2^3 \end{bmatrix}. \tag{14.55}$$

What remains is the term B in Eq. (14.25). The effect of this boundary term B may be included via

$$B = -w(a)p_a + w(b)p_b = \underset{\sim}{w}^T \underset{\sim}{f}_{\text{ext}}, \tag{14.56}$$

where $\underset{\sim}{f}_{\text{ext}}$ contains p_a and p_b at the appropriate positions, according to

$$B = \begin{bmatrix} w_1 & w_2 & w_3 & w_4 \end{bmatrix} \begin{bmatrix} -p_a \\ 0 \\ 0 \\ p_b \end{bmatrix}. \tag{14.57}$$

This finally leads to an equation of the form:

$$\underset{\sim}{w}^T \underline{K} \underline{u} = \underset{\sim}{w}^T (\underset{\sim}{f}_{\text{int}} + \underset{\sim}{f}_{\text{ext}}). \tag{14.58}$$

Using the fact that Eq. (14.58) must hold for 'all' $\underset{\sim}{w}$, this results in the so-called **discrete set of equations**:

$$\begin{bmatrix} K_{11}^1 & K_{12}^1 & 0 & 0 \\ K_{21}^1 & K_{22}^1 + K_{11}^2 & K_{12}^2 & 0 \\ 0 & K_{21}^2 & K_{22}^2 + K_{11}^3 & K_{12}^3 \\ 0 & 0 & K_{21}^3 & K_{22}^3 \end{bmatrix} \begin{bmatrix} u_1 \\ u_2 \\ u_3 \\ u_4 \end{bmatrix} = \begin{bmatrix} f_1^1 - p_a \\ f_2^1 + f_1^2 \\ f_2^2 + f_1^3 \\ f_2^3 + p_b \end{bmatrix}. \tag{14.59}$$

Assume that in the example of Fig. 14.4 at $x = x_1$, an essential boundary condition is prescribed $u(x_1) = U$. This means that $p_a = p_u$ is unknown beforehand. At $x = x_4$ a natural boundary condition is prescribed, so $p_b = P$ is known beforehand.

Then

$$\begin{bmatrix} K_{11}^1 & K_{12}^1 & 0 & 0 \\ K_{21}^1 & K_{22}^1 + K_{11}^2 & K_{12}^2 & 0 \\ 0 & K_{21}^2 & K_{22}^2 + K_{11}^3 & K_{12}^3 \\ 0 & 0 & K_{21}^3 & K_{22}^3 \end{bmatrix} \begin{bmatrix} U \\ u_2 \\ u_3 \\ u_4 \end{bmatrix} = \begin{bmatrix} f_1^1 - p_u \\ f_2^1 + f_1^2 \\ f_2^2 + f_1^3 \\ f_2^3 + P \end{bmatrix}. \tag{14.60}$$

It is clear that in Eq. (14.60) the unknowns are u_2, u_3, u_4 on the left-hand side of the equation, and p_u on the right-hand side of the equation. So both columns can be divided in a known part and an unknown part, depending on the essential and natural boundary conditions that have been prescribed. The next section will outline how this equation is partitioned to facilitate the solution process.

14.6 Solution of the discrete set of equations

Let Eq. (14.60) be written as

$$\underline{K}\, \underset{\sim}{u} = \underset{\sim}{f} \qquad (14.61)$$

where $\underset{\sim}{f} = \underset{\sim}{f}_{\text{int}} + \underset{\sim}{f}_{\text{ext}}$.

The unknowns can be partitioned into two groups. First, some of the components of the column $\underset{\sim}{u}$ will be prescribed. This subset of $\underset{\sim}{u}$ is labeled $\underset{\sim}{u}_p$. The remaining components of $\underset{\sim}{u}$ are the actual unknowns, labeled $\underset{\sim}{u}_u$. In a similar manner $\underline{K}$ and $\underset{\sim}{f}$ can be partitioned. Consequently, Eq. (14.60) can be rewritten as

$$\begin{bmatrix} \underline{K}_{uu} & \underline{K}_{up} \\ \underline{K}_{pu} & \underline{K}_{pp} \end{bmatrix} \begin{bmatrix} \underset{\sim}{u}_u \\ \underset{\sim}{u}_p \end{bmatrix} = \begin{bmatrix} \underset{\sim}{f}_u \\ \underset{\sim}{f}_p \end{bmatrix}. \qquad (14.62)$$

It is emphasized that the right-hand side partition $\underset{\sim}{f}_u$ associated with the unknowns $\underset{\sim}{u}_u$ will be known, and reversely, $\underset{\sim}{f}_p$ will be unknown as $\underset{\sim}{u}_p$ is known. Since $\underset{\sim}{u}_p$ is known, the actual unknowns $\underset{\sim}{u}_u$ can be solved from

$$\underline{K}_{uu}\, \underset{\sim}{u}_u = \underset{\sim}{f}_u - \underline{K}_{up}\, \underset{\sim}{u}_p. \qquad (14.63)$$

Notice that at the part of the boundary where u is prescribed, the associated external load f_p is unknown.

The components of $\underset{\sim}{f}_p$ can be obtained by simple multiplication, after having solved $\underset{\sim}{u}_u$ from Eq. (14.63):

$$\underset{\sim}{f}_p = \underline{K}_{pu}\, \underset{\sim}{u}_u + \underline{K}_{pp}\, \underset{\sim}{u}_p. \qquad (14.64)$$

14.7 Isoparametric elements and numerical integration

In Section 14.4 the concept of shape functions has been introduced. Within each element u_h has been written as

$$u_h|_{\Omega_e} = \underline{N}^T(x)\, \underset{\sim}{u}_e. \qquad (14.65)$$

The shape functions are simple polynomial expressions in terms of the coordinate x. For instance for a linear interpolation the shape functions are linear polynomials, according to

$$N_1 = 1 - \frac{x - x_1}{x_2 - x_1}, \quad N_2 = \frac{x - x_1}{x_2 - x_1}, \qquad (14.66)$$

where x_1 and x_2 denote the position of the nodes of the element. In this case the shape functions are linear functions of the global coordinate x. It is appropriate in the context of a generalization to more-dimensional problems to introduce a local coordinate $-1 \leq \xi \leq 1$ within each element such that $\xi = -1$ and $\xi = 1$ correspond to the edges of the element. With respect to this local coordinate system, the shape functions may be written as

$$N_1 = -\frac{1}{2}(\xi - 1), \quad N_2 = \frac{1}{2}(\xi + 1). \qquad (14.67)$$

This is visualized in Fig. 14.5.

Computation of components of the element coefficient matrix and the element load array requires the evaluation of integrals of the form

$$\int_{\Omega_e} f(x)\, dx . \qquad (14.68)$$

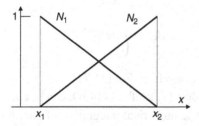

Figure 14.5

Shape functions with respect to the global x-coordinate.

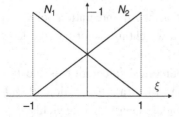

Figure 14.6

Shape functions with respect to the local ξ-coordinate.

These integrals may be transformed into integrals using the local coordinate system, according to

$$\int_{\Omega_e} f(x) \, dx = \int_{-1}^{1} f(x(\xi)) \, \frac{dx}{d\xi} \, d\xi. \tag{14.69}$$

This requires the computation of the derivative $dx/d\xi$. For this purpose the concept of **isoparametric elements** is introduced. The shape functions N_i introduced for the interpolation of the unknown function u_h are also used for the relation between the coordinates x and the coordinates ξ within an element:

$$x(\xi) = \sum_{i=1}^{n} N_i(\xi) x_i^e = \underset{\sim}{N}^{T} \underset{\sim}{x_e}, \tag{14.70}$$

where x_i^e are the coordinates of the nodes of the element. As a result the derivative $dx/d\xi$ is obtained easily:

$$\frac{dx}{d\xi} = \frac{d\underset{\sim}{N}^{T}}{d\xi} \underset{\sim}{x_e}. \tag{14.71}$$

The element coefficient matrix requires the derivatives of the shape functions with respect to the coordinate x. For this purpose

$$\frac{dN_i}{dx} = \frac{dN_i}{d\xi} \frac{d\xi}{dx}, \tag{14.72}$$

where

$$\frac{d\xi}{dx} = \left(\frac{dx}{d\xi} \right)^{-1}, \tag{14.73}$$

is used, which is easily obtained from Eq. (14.71).

The integral on the right-hand side of Eq. (14.69) can be approximated by means of numerical integration. The numerical integration of an arbitrary function $g(\xi)$ over the domain $-1 \leq \xi \leq 1$ is approximated by

$$\int_{-1}^{1} g(\xi) \, d\xi = \sum_{i=1}^{n_{int}} g(\xi_i) \, W_i, \tag{14.74}$$

where ξ_i denotes the location of the i-th integration point, n_{int} is the total number of integration points and W_i a weighting factor, i.e. the length of the ξ-domain, associated with this integration point.

A simple example of a numerical integration is the trapezoidal integration scheme, as illustrated in Fig. 14.7 (a). Integration of $g(\xi)$ over the domain $-1 \leq \xi \leq 1$ using the trapezoidal integration rule yields:

$$\int_{-1}^{1} g(\xi) \, d\xi \approx g(\xi = -1) + g(\xi = 1), \tag{14.75}$$

Table 14.1   Gaussian quadrature points up to $n_{\text{int}} = 3$.

n_{int}	ξ_i	W_i
1	$\xi_1 = 0$	$W_1 = 2$
2	$\xi_1 = \frac{-1}{\sqrt{3}}, \xi_2 = \frac{1}{\sqrt{3}}$	$W_1 = W_2 = 1$
3	$\xi_1 = -\sqrt{\frac{3}{5}}, \xi_2 = 0, \xi_3 = \sqrt{\frac{3}{5}}$	$W_1 = W_3 = \frac{5}{9}, W_2 = \frac{8}{9}$

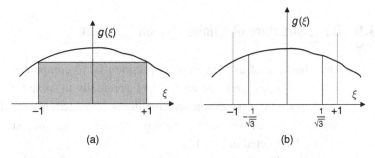

(a)   (b)

Figure 14.7

(a) Trapezoidal integration (b) 2-point Gauss integration.

which corresponds to the shaded area in Fig. 14.7(a). For trapezoidal integration the integration point positions ξ_i are given by

$$\xi_1 = -1, \qquad \xi_2 = 1, \tag{14.76}$$

while the associated weighting factors are

$$W_1 = 1, \qquad W_2 = 1. \tag{14.77}$$

The trapezoidal integration rule integrates a linear function exactly. A 2-point Gaussian integration rule, as depicted in Fig. 14.7(b) may yield a more accurate result since this integration rule integrates up to a third order function exactly using two integration points only. In this case the integral is approximated by

$$\int_{-1}^{1} g(\xi) \, d\xi \approx g\left(\xi = \frac{-1}{\sqrt{3}}\right) + g\left(\xi = \frac{1}{\sqrt{3}}\right). \tag{14.78}$$

The location of the Gaussian integration (quadrature) points and the associated weighting factors are summarized in Table 14.1.

Application to element coefficient matrix Use of the local coordinate system, with isoparametric formulation and numerical integration to the element coefficient matrix yields

$$\underline{K}_e = \int_{\Omega_e} \frac{d\underline{N}}{dx} c \frac{d\underline{N}^{\mathrm{T}}}{dx} \, dx$$

$$= \int_{\xi=-1}^{1} \frac{d\underline{N}}{d\xi} \frac{d\xi}{dx} c \frac{d\underline{N}^{\mathrm{T}}}{d\xi} \frac{d\xi}{dx} \left(\frac{d\xi}{dx} \right)^{-1} d\xi$$

$$\approx \sum_{i=1}^{n_{\mathrm{int}}} \left(\frac{d\underline{N}}{d\xi} c \frac{d\underline{N}^{\mathrm{T}}}{d\xi} \frac{d\xi}{dx} \right)_{\xi=\xi_i} W_i. \qquad (14.79)$$

14.8 Basic structure of a finite element program

The objective of a finite element program is, to compute the coefficient matrix $\underline{K}$ and the right-hand side array $\underset{\sim}{f}$ and eventually to solve the resulting system of equations taking the boundary conditions into account. To illustrate the typical data structure and the layout of a finite element program, consider, as an example, the mesh depicted in Fig. 14.8.

The MATLAB programming language is used for explanation purposes. The following data input is needed:

- **Element topology** First of all the domain is divided into a number of elements and each node is given a unique global number. In this example two elements have been used, the first element Ω_1 is a quadratic element connecting nodes 3, 4 and 2 (in that order) and the second element is a linear element having nodes 1 and 3 (again, in that order). The node numbers of each element are stored in the topology array top, such that the i-th row of this array corresponds to the i-th element. In the current example the topology array would be:

$$\mathtt{top} = \begin{bmatrix} 3 & 4 & 2 \\ 1 & 3 & 0 \end{bmatrix}.$$

Besides the node numbers of the element, a number of identifiers may be included for each element, for instance to refer to different material parameters c or different

Figure 14.8

Mesh for a one-dimensional problem, consisting of a linear and a quadratic element.

element types, e.g. linear versus quadratic elements. In fact, the MATLAB code provided to experiment with, has two identifiers per element. Please consult the manual of the code: `mlfem_nac`.

- **Nodal coordinates** The nodal coordinates $\underset{\sim}{x}$ are stored in the array `coord`, hence in this example:

$$\underset{\sim}{x} = \texttt{coord} = \begin{bmatrix} x_1 \\ x_2 \\ x_3 \\ x_4 \end{bmatrix}.$$

The nodal coordinates of the nodes associated with the element e can be retrieved from `coord` using `top`. The node numbers of element e can be extracted from the array `top` by:

$$\texttt{ii = nonzeros(top(e,:))},$$

such that the nodal coordinates of the e-th element are obtained via

$$\underset{\sim}{x}_e = \texttt{nodcoord = coord(ii,:)}.$$

- **Solution array** The nodal unknowns, also called degrees of freedom, $\underset{\sim}{u}$ are stored in the array `sol`:

$$\underset{\sim}{u} = \texttt{sol} = \begin{bmatrix} u_1 \\ u_2 \\ u_3 \\ u_4 \end{bmatrix}.$$

It is not necessary to store the degrees of freedom in a sequential manner, in fact, any ordering may be chosen, as long as each array component corresponds to a unique degree of freedom. To extract the nodal degrees of freedom for element e, $\underset{\sim}{u}_e$ from `sol`, a separate index array is needed: `pos`. The e-th row of the array `pos` contains the location of the nodal degrees of freedom of element e in the array `sol`. For this example

$$\underset{\sim}{u}_1 = \begin{bmatrix} u_3 \\ u_4 \\ u_2 \end{bmatrix}, \quad \underset{\sim}{u}_2 = \begin{bmatrix} u_1 \\ u_3 \end{bmatrix},$$

hence the index array `pos` should contain

$$\texttt{pos} = \begin{bmatrix} 3 & 4 & 2 \\ 1 & 3 & 0 \end{bmatrix}.$$

Using this array, the nodal degrees of freedom of element e can be extracted from `sol` via

$$\texttt{ii = nonzeros(pos(e,:))}$$
$$\underset{\sim}{u}_e = \texttt{nodu = sol(ii)}$$

- **Shape functions** The shape functions and their derivatives are needed at the integration points $\underset{\sim}{N}(\xi_i)$ and $d\underset{\sim}{N}/d\xi$. The shape function values are stored in the array n such that at the i-th integration point

$$\underset{\sim}{N}^T(\xi_i) = \mathtt{n(i,:)}.$$

Using these shape functions the coordinate $x(\xi_i)$ within element e can be computed at the i-th integration point:

$$x(\xi_i) = \underset{\sim}{N}^T(\xi_i)\underset{\sim}{x}_e = \mathtt{n(i,:)*nodcoord}.$$

Similarly, the value of the solution at the i-th integration point of element e is obtained via

$$u(\xi_i) = \underset{\sim}{N}^T(\xi_i)\underset{\sim}{u}_e = \mathtt{n(i,:)*nodu}.$$

In a similar fashion the shape function derivatives with respect to the local coordinate ξ are stored in an array, called dndxi.

Structure of the Finite Element code Typically the structure of a finite element programme is as follows.

(i) Pre-processing: mesh generation, boundary condition specification and parameter declaration. This should provide the topology array top, the coordinate array coord and a number of auxiliary arrays containing boundary conditions and material parameters.

(ii) Based on the mesh and element types used, the index array pos can be computed.

(iii) Assembly of the coefficient matrix $\mathtt{q} = \underset{\sim}{K}$ and the element array $\mathtt{rhs} = \underset{\sim}{f}$. Let $\mathtt{qe} = \underset{\sim}{K}_e$ and $\mathtt{rhse} = \underset{\sim}{f}_e$, then the assembly process in a MATLAB environment would look like

```
% nelem: the number of elements

for ielem=1:nelem

  % compute qe and rhse

  [qe,rhse]=<elementfunction>(ielem,coord,top,.....)

  % get the location of the degrees of freedom
  % in the solution array

  ii = nonzeros(pos(ielem,:));

  % add the element coefficient matrix
  % and the element right-hand side array
```

```
       % to the total coefficient matrix q
       % and the load array rhs
         q(ii,ii) = q(ii,ii) + qe;
         rhs(ii)  = rhs(ii)  + rhse;

    end
```

(iv) Solution of the set of equations taking into account the boundary conditions.
(v)   Post-processing based on the solution, for instance by computing associated quanti-
      ties such as heat fluxes or stresses.

14.9 Example

As an example consider the diffusion problem with the following parameter setting. We consider the domain $\Omega : 0 \leq x \leq 1$, with prescribed essential boundary
conditions at $x = 0$ and $x = 1$. These conditions are: $u(0) = 0$ and $u(1) = 0$. There
are no natural boundary conditions. The material constant satisfies: $c = 1$ and the
source term: $f = 1$.

The domain Ω is divided into five elements of equal length. Fig. 14.9 shows
the solution. The left part displays the computed solution u (solid line) as well as
the exact solution (dashed line). Remarkably, in this one-dimensional case with

Figure 14.9

Five element solution. Left: (solid line) approximate solution $u_h(x)$, (dashed line) exact solution $u(x)$.
Right: (solid line) approximate flux $p_h(x)$, (dashed line) exact flux $p(x)$.

the current choice of parameters, the nodal solutions are exact. The right part of the figure shows the computed flux, say flux $p = c\, du/dx$. Again, the solid line denotes the computed flux p, which is clearly discontinuous from one element to the next, and the dashed line denotes the exact solution. The discontinuity of the computed flux field is obvious: the field u is piecewise linear, therefore the derivative du/dx is piecewise constant. The flux p does not necessarily have to be piecewise constant: if the parameter c is a function of x, the flux p will be varying within an element.

Mesh refinement leads to an improved approximate solution. For instance, using ten rather than five elements, yields the results depicted in Fig. 14.10. The impact of a varying c, say $c = 1 + x$ is depicted in Fig. 14.11.

Changing the interpolation order of the shape functions N_i from linear to quadratic, also has a significant impact on the results, in particular on the quality of the flux prediction. For the constant c case, the solution becomes exact. Also for $c = 1 + x$ a significant improvement can be observed, as depicted in Fig. 14.13.

 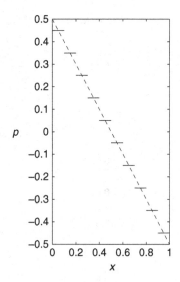

Figure 14.10

Ten element solution. Left: (solid line) approximate solution $u_h(x)$, (dashed line) exact solution $u(x)$. Right: (solid line) approximate flux $p_h(x)$, (dashed line) exact flux $p(x)$.

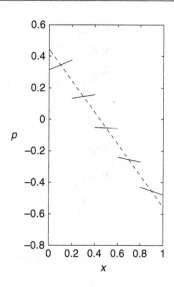

Figure 14.11

Five element solution for $c = 1 + x$. Left: (solid line) approximate solution $u_h(x)$, (dashed line) exact solution $u(x)$. Right: (solid line) approximate flux $p_h(x)$, (dashed line) exact flux $p(x)$.

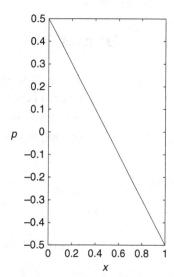

Figure 14.12

Solution for $c = 1$ using five quadratic elements. Left: (solid line) approximate solution $u_h(x)$, (dashed line) exact solution $u(x)$. Right: (solid line) approximate flux $p_h(x)$, (dashed line) exact flux $p(x)$.

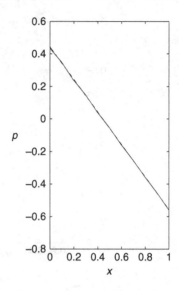

Figure 14.13

Solution for $c = 1 + x$ using five quadratic elements. Left: (solid line) approximate solution $u_h(x)$, (dashed line) exact solution $u(x)$. Right: (solid line) approximate flux $p_h(x)$, (dashed line) exact flux $p(x)$.

Exercises

14.1   The method of weighted residuals can be used to find approximations of a given function. Let $f(x)$ be a function that one would like to approximate with a polynomial of the order n in a certain domain, say $0 \leq x \leq 1$. Let the polynomial be given by

$$h(x) = a_0 + a_1 x + \cdots + a_n x^n.$$

Ideally

$$g(x) = h(x) - f(x) = 0 \text{ for all } x \text{ with } 0 \leq x \leq 1.$$

In terms of the weighted residuals equation this may also be expressed as

$$\int_0^1 w(x) \left[\left(a_0 + a_1 x + \cdots + a_n x^n \right) - f(x) \right] \, dx = 0 \text{ for all } w.$$

Now, suppose that w is also a polynomial of the order n:

$$w(x) = b_0 + b_1 x + \cdots + b_n x^n.$$

This may also be written in an alternative format:

$$w(x) = \underset{\sim}{b}^{\mathrm{T}} \underset{\sim}{p},$$

with

$$
b = \begin{bmatrix} b_0 \\ b_1 \\ \vdots \\ b_n \end{bmatrix}, \quad p = \begin{bmatrix} 1 \\ x \\ \vdots \\ x^n \end{bmatrix}.
$$

Likewise $h(x)$ can be written as

$$
h(x) = a^{\mathrm{T}} p,
$$

with

$$
a = \begin{bmatrix} a_0 \\ a_1 \\ \vdots \\ a_n \end{bmatrix}.
$$

Use of these expressions yields

$$
\int_0^1 w(x)\,[h(x) - f(x)]\ dx
$$

$$
= \int_0^1 b^{\mathrm{T}} p p^{\mathrm{T}} a\ dx - \int_0^1 b^{\mathrm{T}} p f(x)\ dx = 0.
$$

Notice that both a and b are arrays with polynomial coefficients independent of x, while p is an array with known functions of the coordinate x. Therefore, this equation may also be written as

$$
b^{\mathrm{T}} \int_0^1 p p^{\mathrm{T}}\ dx\ a = b^{\mathrm{T}} \int_0^1 p f(x)\ dx.
$$

The integral expression on the left-hand side is a matrix:

$$
\underline{K} = \int_0^1 p p^{\mathrm{T}}\ dx,
$$

while the integral on the right-hand side is a column:

$$
f = \int_0^1 p f(x)\ dx.
$$

The equation must be satisfied for all b hence, with the use of the above matrix-array notation it follows that:

$$
\underline{K}\, a = f.
$$

(a) Suppose that $w(x)$ and $h(x)$ are both first-order polynomials.

- Show that in that case

$$\underset{\sim}{K} = \begin{bmatrix} 1 & \frac{1}{2} \\ \frac{1}{2} & \frac{1}{3} \end{bmatrix}.$$

- If $f(x) = 3$, show that

$$\underset{\sim}{f} = \begin{bmatrix} 3 \\ \frac{3}{2} \end{bmatrix}.$$

- Compute the coefficients of the polynomial $h(x)$ collected in $\underset{\sim}{a}$. Explain the results. (Hint: it is recommended to use MATLAB for this purpose.)

(b) If $w(x)$ and $h(x)$ are both polynomials of the order n, show that

$$\underset{\sim}{K} = \begin{bmatrix} 1 & \frac{1}{2} & \cdots & \frac{1}{n+1} \\ \frac{1}{2} & \frac{1}{3} & \cdots & \frac{1}{n+2} \\ \vdots & \vdots & \vdots & \vdots \\ \frac{1}{n+1} & \frac{1}{n+2} & \cdots & \frac{1}{2n+1} \end{bmatrix}.$$

(c) Let $f(x)$ be such that

$$f(x) = 1 \text{ for } 0 \leq x \leq 0.5,$$
$$f(x) = 0 \text{ for } 0.5 < x \leq 1.$$

- Show that in this case

$$\underset{\sim}{f} = \int_0^1 \underset{\sim}{p} f(x) \, dx = \begin{bmatrix} \frac{1}{2} \\ \frac{1}{2}(\frac{1}{2})^2 \\ \vdots \\ \frac{1}{n+1}(\frac{1}{2})^{n+1} \end{bmatrix}.$$

- Use MATLAB to find the polynomial approximation $h(x)$ of $f(x)$ for $n = 2$, $n = 3$, etc. up to $n = 10$. Plot the original function $f(x)$ as well as the polynomial approximation $h(x)$. Hint: use the function polyval. If the MATLAB array a represents $\underset{\sim}{a}$ and n denotes the order of the polynomial, then to plot the function $h(x)$ you may use

    `x=0:0.01:1; plot(x,polyval(a(n+1:-1:1),x))`

- Investigate the condition number of the matrix $\underset{\sim}{K}$ with increasing n. Hint: use the MATLAB function cond. What does this condition number mean and what does this imply with respect to the coefficients of $h(x)$, collected in $\underset{\sim}{a}$?

14.2 Consider the differential equation

$$u + \frac{du}{dx} + \frac{d}{dx}\left(c\frac{du}{dx}\right) + f = 0,$$

on the domain $a \le x \le b$.

Derive the weak form of this differential equation, and explain what steps are taken.

14.3 Let $f(x)$ be a function on the domain $0 \le x \le 1$. Let $f(x)$ be known at n points, denoted by x_i, homogeneously distributed on the above domain. Hence the distance Δx between two subsequent points equals

$$\Delta x = \frac{1}{n-1}.$$

A polynomial $f_h(x)$ of order $n-1$ can be constructed through these points, which generally will form an approximation of $f(x)$:

$$f_h(x) = a_0 + a_1 x + \cdots + a_{n-1} x^{n-1}.$$

(a) Show that the coefficients of a_i can be found by solving

$$\begin{bmatrix} 1 & x_1 & x_1^2 & \cdots & x_1^{n-1} \\ 1 & x_2 & x_2^2 & \cdots & x_2^{n-1} \\ 1 & x_3 & x_3^2 & \cdots & x_3^{n-1} \\ \vdots & \vdots & \vdots & \vdots & \vdots \\ 1 & x_n & x_n^2 & \cdots & x_n^{n-1} \end{bmatrix} \begin{bmatrix} a_0 \\ a_1 \\ a_2 \\ \vdots \\ a_{n-1} \end{bmatrix} = \begin{bmatrix} f_1 \\ f_2 \\ f_3 \\ \vdots \\ f_n \end{bmatrix},$$

where $f_i = f(x_i)$.

(b) Use this to find a polynomial approximation for different values of n to the function:

$$f(x) = 1 \text{ for } 0 \le x \le 0.5,$$
$$f(x) = 0 \text{ for } 0.5 < x \le 1.$$

Compare the results to those obtained using the weighted residuals formulation in Exercise 14.1(c). Explain the differences.

14.4 Consider the domain $-1 \le x \le 1$. Assume that the function u is known at $x_1 = -1$, $x_2 = 0$ and $x_3 = 1$, say u_1, u_2 and u_3 respectively. The polynomial approximation of u, denoted by u_h is written as

$$u_h = a_0 + a_1 x + a_2 x^2.$$

(a) Determine the coefficients a_0, a_1 and a_2 to be expressed as a function of u_1, u_2 and u_3.

(b) If the polynomial u_h is written in terms of the shape functions N_i:

$$u_h = \sum_{i=1}^{n} N_i(x) \, u_i,$$

then determine N_i, $i = 1, 2, 3$ as a function of x.

(c) Sketch the shape functions N_i.

(d) Is it possible that the shape function $N_i \neq 1$ at $x = x_i$? Explain.

(e) Is it possible that the shape function $N_i \neq 0$ at $x = x_j$ with $j \neq i$? Explain.

14.5 Consider the differential equation

$$\frac{d}{dx}\left(c\frac{du}{dx}\right) = 1$$

on the domain $0 \leq x \leq h_1 + h_2$. At both ends of this domain u is set to zero. Consider the element distribution as depicted in the figure below.

(a) Derive the weak formulation of this problem.

(b) The elements employed have linear shape functions stored in array $\underset{\sim}{N}(x)$. Express these shape functions in terms of x and the element lengths h_1 and h_2.

(c) Show that the coefficient matrix of an element is given by

$$\underline{K}_e = \int_{\Omega_e} \frac{d\underset{\sim}{N}}{dx} c \frac{d\underset{\sim}{N}^T}{dx} \, dx.$$

(d) Demonstrate that the coefficient matrix of element e is given by

$$\underline{K}_e = \frac{c}{h_e}\begin{bmatrix} 1 & -1 \\ -1 & 1 \end{bmatrix},$$

if c is a constant.

(e) Show that the source term leads to a right-hand side column on element level given by

$$\underset{\sim}{f}_e = -\frac{h_e}{2}\begin{bmatrix} 1 \\ 1 \end{bmatrix}.$$

(f) After discretization the resulting set of equations is expressed as

$$\underline{K}\underset{\sim}{u} = \underset{\sim}{f}.$$

Define the solution array $\underset{\sim}{u}$ and derive the coefficient matrix $\underline{K}$ and the array f for the two element mesh depicted in the figure.

(g) Determine the solution array $\underset{\sim}{u}$.

14.6 Using an isoparametric formulation, the shape functions are defined with respect to a local coordinate system $-1 \leq \xi \leq 1$. Within an element the unknown u_h is written as

$$u_h = \sum_{i=1}^{n} N_i(\xi)\, u_i.$$

(a) What are the above shape functions with respect to the local coordinate system if a linear or quadratic interpolation is used?

(b) How is the derivative of the shape functions

$$\frac{dN_i}{dx}$$

obtained?

(c) Assume a quadratic shape function, and let $x_1 = 0$, $x_2 = 1$, and $x_3 = 3$. Compute

$$\frac{dN_i}{dx} \text{ for } i = 1, 2, 3.$$

(d) Compute from array $\underset{\sim}{h}$ given by

$$\underset{\sim}{h} = \int_0^3 \frac{d\underset{\sim}{N}}{dx} x\, dx,$$

the first component using the same quadratic shape functions as above.

14.7 Consider in the code `mlfem_nac` the directory `oneD`. The one-dimensional finite element program `fem1d` solves the diffusion problem:

$$\frac{d}{dx}\left(c\frac{du}{dx}\right) + f = 0,$$

on a domain $a \leq x \leq b$ subject to given boundary conditions for a certain problem. The input data for this program are specified in the m-file `demo_fem1d`, along with the post-processing statements.

(a) Modify the m-file `demo_fem1d` such that five linear elements are used to solve the above differential equation, using the boundary conditions

$$u = 0 \text{ at } x = 0,$$

and

$$p = c\frac{du}{dx} = 1 \text{ at } x = 1,$$

with $c = 1$ and $f = 0$. Compute the FEM solution (array `sol`) and compare the nodal values with the exact solution.

(b) Use the array `pos` to extract the nodal solutions within the second element from the array `sol`.

(c) Determine the solution for the third node using the array `dest`.

14.8 One of the major problems in the tissue engineering of cartilage is to make thick constructs. Nutrients coming from the surrounding medium have to reach the cells in the middle of the construct by diffusion, but cells close to the edge of the construct may consume so much nutrients that there is nothing left for cells in the middle. Consider the experimental set-up as given in the figure below, representing a schematic drawing of a bioreactor to culture articular cartilage.

The construct is fixed between two highly permeable membranes allowing free contact with the culture medium. The thickness t is a trade-off between the diffusion coefficient c, the consumption rate of the cells f (both cannot be influenced) and the amount of the molecules that can be supplied via the medium, which is usually bound to a maximum.

The diffusion problem can be considered as a one-dimensional problem. The current analysis is meant to determine the glucose concentration $u(x)$, which is an essential nutrient for the cells. The following properties are given: $c = 9.2 \times 10^{-6} [\text{cm}^2 \text{ s}^{-1}]$ and $f = 56 \times 10^{-7} [\text{Mol hour}^{-1} \text{cm}^{-3}]$. Low glucose medium is used, concentration $= 5 \times 10^{-3} [\text{Mol litre}^{-1}]$.

(a) Assuming the consumption rate f is constant and the medium is refreshed continuously, a stable glucose concentration as a function of the location in the construct will be reached after a while. Give the differential equation and boundary conditions describing this process.

(b) Give the analytical solution of this problem by integrating the differential equation.

(c) Adjust `demo_fem1d` to solve the problem with the finite element method. Solve the problem for $t = 1, 2, 3, 4$ and 5 [mm].

(d) At which thickness do the cells appear to die in the middle of the construct?

14.9 To determine the material properties of a skeletal muscle a uniaxial tensile test is performed as shown in the figure. The muscle is clamped on one side and a force $F = 10$ [N] is applied on the other side. The muscle has a total length of $\ell = 12$ [cm]. The muscle has a circular cross section. The radius of the cross section can be approximated by:

$$r = a_1 \sin^3(a_2 x + a_3),$$

with $a_1 = 1.6$ [cm], $a_2 = 0.15$ [cm^{-1}] and $a_3 = 0.8$ [-]. The estimated Young's modulus is $E = 10^5$ [N m^{-2}].

(a) Give the differential equation for the axial displacement $u(x)$ and boundary conditions that describe the current problem.

(b) Determine the displacement field by adjusting the file demo_fem1d.

15 Solution of the one-dimensional convection-diffusion equation by means of the Finite Element Method

15.1 Introduction

This chapter extends the formulation of the previous chapter for the one-dimensional diffusion equation to the time-dependent convection-diffusion equation. Although a good functioning of the human body relies on maintaining a homeostasis or equilibrium in the physiological state of the tissues and organs, it is a dynamic equilibrium. This means that all processes have to respond to changing inputs, which are caused by changes of the environment. The diffusion processes taking place in the body are not constant, but instationary, so time has to be included as an independent variable in the diffusion equation. Thus, the instationary diffusion equation becomes a **partial** differential equation.

Convection is the process whereby heat or particles are transported by air or fluid moving from one point to another point. Diffusion could be seen as a process of transport through immobilized fluid or air. When the fluid itself moves, particles in that fluid are dragged along. This is called convection and also plays a major role in biomechanics. An example is the loss of heat because moving air is passing the body. The air next to the body is heated by conduction, moves away and carries off the heat just taken from the body. Another example is a drug that is released at some spot in the circulation and is transported away from that spot by means of the blood flow. In larger blood vessels the prime mechanism of transportation is convection.

15.2 The convection-diffusion equation

Assuming that the source term $f = 0$, the unsteady one-dimensional convection-diffusion equation can be written as

$$\frac{\partial u}{\partial t} + v\frac{\partial u}{\partial x} = \frac{\partial}{\partial x}\left(c\frac{\partial u}{\partial x}\right), \tag{15.1}$$

with u a function of both position x and time t:

$$u = u(x, t). \tag{15.2}$$

The convective velocity is denoted by v and c is the diffusion coefficient. Both v and c are assumed to be constant in the present chapter. Compared to the diffusion equation of the previous chapter, two terms have been added: the time dependency term $\partial u/\partial t$ (inertia term) and the convective term $v\partial u/\partial x$. The convection-diffusion equation holds within a given spatial domain $\Omega = [a, b]$, i.e. with boundaries at $x = a$ and $x = b$, as well as within a time domain, say $S = [0, T]$, and is assumed to be subject to the boundary conditions:

$$u = U \text{ at } \Gamma_u \text{ (located at } x = a),\qquad(15.3)$$

and

$$c\frac{\partial u}{\partial x} = P \text{ at } \Gamma_p \text{ (located at } x = b).\qquad(15.4)$$

Furthermore, one initial condition on u must be specified, say at $t = 0$:

$$u(x, t = 0) = u_{ini}(x) \text{ in } \Omega.\qquad(15.5)$$

Under certain conditions the transient character of the solution of the unsteady convection-diffusion equation vanishes. In that case we deal with a steady convection-diffusion problem, described by the differential equation

$$v\frac{du}{dx} = \frac{d}{dx}\left(c\frac{du}{dx}\right).\qquad(15.6)$$

Example 15.1   A typical example of a solution of the **steady** convection-diffusion problem is represented in Fig. 15.1. In this example the domain spans $0 \le x \le 1$, while at $x = 0$ the solution is set to 0 and at $x = 1$ the value $u = 1$ is imposed. That means that in this example two essential boundary conditions are used and that there is no natural boundary condition prescribed. The analytical solution for $v \ne 0$ in this case is

$$u = \frac{1}{1 - e^{\frac{v}{c}}}(1 - e^{\frac{v}{c}x}).$$

Clearly, without any convection, i.e. $v = 0$, a spatially linear distribution of u results, while with increasing convective velocity v a boundary layer develops at $x = 1$.

The convection-diffusion equation may be written in a dimensionless form by introducing an appropriate length scale L, a time scale Θ and a reference solution U, for example the reference temperature or concentration. Then, the dimensionless solution u^*, the dimensionless coordinate x^* and the dimensionless time t^* are defined as

$$u^* = \frac{u}{U} \qquad x^* = \frac{x}{L} \qquad t^* = \frac{t}{\Theta}.\qquad(15.7)$$

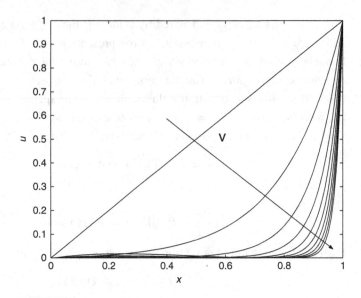

Figure 15.1

Solution to steady 1-D convection-diffusion problem.

Assuming v and c constant, the dimensionless form may be written as

$$\frac{1}{\text{Fo}}\frac{\partial u^*}{\partial t^*} + \text{Pe}\frac{\partial u^*}{\partial x^*} = \frac{\partial}{\partial x^*}\left(\frac{\partial u^*}{\partial x^*}\right), \tag{15.8}$$

with the Fourier number given by

$$\text{Fo} = \frac{\Theta c}{L^2}, \tag{15.9}$$

and the Peclet number given by

$$\text{Pe} = \frac{vL}{c}. \tag{15.10}$$

The Peclet number reflects the relative importance of convection compared to diffusion. It will be demonstrated that with increasing Peclet number the numerical solution of the convection-diffusion problem becomes more difficult.

In the next section we will start by studying the time discretization of an instationary equation. After that in Section 15.4 the spatial discretization of the convection-diffusion equation will be discussed.

15.3 Temporal discretization

Many algorithms have been developed for the temporal discretization of the convection-diffusion equation. Only one method is discussed here: the so-called

θ-scheme. Getting ahead of the spatial discretization in the next section, which will lead to a set of linear differential equations, the time discretization is best illustrated with a set of first-order linear differential equations:

$$\frac{d\underset{\sim}{u}}{dt} + \underline{A}\,\underset{\sim}{u} = \underset{\sim}{f}(t), \tag{15.11}$$

where $\underline{A}$ is a constant matrix and $\underset{\sim}{f}(t)$ a given column. To solve this time-dependent problem the time domain is split into a finite number of time increments. Then, attention is focussed on the increment with $t_n < t < t_{n+1} = t_n + \Delta t$. Assuming that the solution u_n at time t_n is known, the unknown u_{n+1} at time t_{n+1} has to be determined. The θ-scheme approximates these differential equations by

$$\frac{\underset{\sim}{u}_{n+1} - \underset{\sim}{u}_n}{\Delta t} + \theta\underline{A}\,\underset{\sim}{u}_{n+1} + (1-\theta)\underline{A}\,\underset{\sim}{u}_n = \theta\underset{\sim}{f}_{n+1} + (1-\theta)\underset{\sim}{f}_n. \tag{15.12}$$

For $\theta = 0$ this scheme reduces to the **Euler explicit** or **forward Euler** scheme, while for $\theta = 1$ the **Euler implicit** or **backward Euler** scheme results. Both of these schemes are first-order accurate, i.e. $\mathcal{O}(\Delta t)$. This means that the accuracy of the solution is linearly related to the size of the time step Δt. The accuracy improves when the time step becomes smaller. For $\theta = 0.5$ the **Crank–Nicholson** scheme results which is second-order accurate, i.e. $\mathcal{O}(\Delta t^2)$.

To illustrate the stability properties of the θ-method, consider a single variable model problem:

$$\frac{du}{dt} + \lambda u = f, \tag{15.13}$$

with $\lambda > 0$. This differential equation has the property that any perturbation to the solution (for example induced by a perturbation of the initial value of u) decays exponentially as a function of time. Assume that $\hat{u}$ satisfies the differential Eq. (15.13) exactly, and let $\tilde{u}$ be a perturbation of $\hat{u}$, hence $u = \hat{u} + \tilde{u}$. Consequently, the perturbation $\tilde{u}$ must obey

$$\frac{d\tilde{u}}{dt} + \lambda\tilde{u} = 0. \tag{15.14}$$

If at $t = 0$, the perturbation equals $\tilde{u} = \tilde{u}_0$, the solution to this equation is

$$\tilde{u} = e^{-\lambda t}\tilde{u}_0. \tag{15.15}$$

Clearly, if $\lambda > 0$, the perturbation decays exponentially as a function of time.

Application of the θ-scheme to the single variable model problem (15.13) yields

$$\frac{u_{n+1} - u_n}{\Delta t} + \theta\lambda u_{n+1} + (1-\theta)\lambda u_n = \theta f_{n+1} + (1-\theta)f_n. \tag{15.16}$$

Now, as before, if $\tilde{u}_n$ is a perturbation of $\hat{u}_n$, this perturbation satisfies

$$\frac{\tilde{u}_{n+1} - \tilde{u}_n}{\Delta t} + \theta \lambda \tilde{u}_{n+1} + (1 - \theta) \lambda \tilde{u}_n = 0. \tag{15.17}$$

Clearly, the perturbation at $t = t_{n+1}$ can be expressed as

$$\tilde{u}_{n+1} = \underbrace{\frac{1-(1 - \theta) \lambda \Delta t}{1 + \theta \lambda \Delta t}}_{A} \tilde{u}_n. \tag{15.18}$$

The factor A is called the amplification factor. To have a stable integration scheme the magnitude of $\tilde{u}_{n+1}$ should be smaller than the magnitude of $\tilde{u}_n$, i.e. the perturbation should not grow as time proceeds. Hence, stability requires $|\tilde{u}_{n+1}| \leq |\tilde{u}_n|$, which holds if the amplification factor $|A| \leq 1$.

Fig. 15.2 shows the amplification factor A as a function of $\lambda \Delta t$ with θ as a parameter. For $0 \leq \theta < 0.5$ the integration scheme is conditionally stable, meaning that the time step Δt has to be chosen sufficiently small related to λ. In the multi-variable case the above corresponds to the requirement that, in case $0 \leq \theta < 0.5$, $\lambda \Delta t$ should be small compared to the eigenvalues of the matrix $\underline{A}$. For $0.5 \leq \theta \leq 1$ the scheme is unconditionally stable, hence for any choice of Δt a stable integration process results.

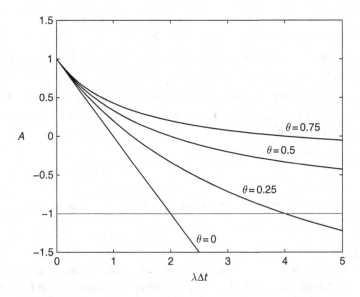

Figure 15.2

Amplification factor A as a function of $\lambda \Delta t$ for various values of θ.

15.4   Spatial discretization

Following a similar derivation as in the previous chapter, the weak form is obtained by multiplication of Eq. (15.1) with a suitable weighting function w, performing an integration over $\Omega = [a, b]$, followed by an integration by parts:

$$\int_\Omega w \frac{\partial u}{\partial t} \, dx + \int_\Omega wv \frac{\partial u}{\partial x} \, dx + \int_\Omega \frac{dw}{dx} c \frac{\partial u}{\partial x} \, dx = B, \tag{15.19}$$

where the right-hand side term B results from the integration by parts:

$$B = w(b) \, c \frac{\partial u}{\partial x}\Big|_{x=b} - w(a) \, c \frac{\partial u}{\partial x}\Big|_{x=a}. \tag{15.20}$$

Notice that no partial integration of the convective term has been performed. The discrete set of equations, according to Eq. (15.19), is derived by subdivision of the domain in elements and by discretization at element level according to

$$u_h(x, t)\,|_{\Omega_e} = \underset{\sim}{N}^{\mathrm{T}}(x)\,\underset{\sim}{u}_e(t), \qquad w_h(x)\,|_{\Omega_e} = \underset{\sim}{N}^{\mathrm{T}}(x)\,\underset{\sim}{w}_e(t). \tag{15.21}$$

Note that the shape functions $\underset{\sim}{N}$ are a function of the spatial coordinate x only and not of the time t. The nodal values of u_h, at element level stored in the column $\underset{\sim}{u}_e$, however, do depend on the time t. Substitution of Eq. (15.21) into Eq. (15.19) yields

$$\sum_{e=1}^{N_{el}} \left(\underset{\sim}{w}_e^{\mathrm{T}} \int_{\Omega_e} \underset{\sim}{N}\,\underset{\sim}{N}^{\mathrm{T}} \, dx \, \frac{d\underset{\sim}{u}_e}{dt} + \underset{\sim}{w}_e^{\mathrm{T}} \int_{\Omega_e} \underset{\sim}{N}\, a \frac{d\underset{\sim}{N}^{\mathrm{T}}}{dx} \, dx \, \underset{\sim}{u}_e \right.$$

$$\left. + \underset{\sim}{w}_e^{\mathrm{T}} \int_{\Omega_e} \frac{d\underset{\sim}{N}}{dx} c \frac{d\underset{\sim}{N}^{\mathrm{T}}}{dx} \, dx \, \underset{\sim}{u}_e \right) = B. \tag{15.22}$$

With

$$\underline{M}_e = \int_{\Omega_e} \underset{\sim}{N}\,\underset{\sim}{N}^{\mathrm{T}} \, dx, \tag{15.23}$$

and

$$\underline{K}_e = \int_{\Omega_e} \underset{\sim}{N}\, a \frac{d\underset{\sim}{N}^{\mathrm{T}}}{dx} \, dx + \int_{\Omega_e} \frac{d\underset{\sim}{N}}{dx} c \frac{d\underset{\sim}{N}^{\mathrm{T}}}{dx} \, dx, \tag{15.24}$$

Eq. (15.22) can be written as

$$\sum_{e=1}^{N_{el}} \underset{\sim}{w}_e^{\mathrm{T}} \left(\underline{M}_e \frac{d\underset{\sim}{u}_e}{dt} + \underline{K}_e \underset{\sim}{u}_e \right) = B. \tag{15.25}$$

After the usual assembly process this is written in global quantities:

$$\underset{\sim}{w}^{\mathrm{T}} \left(\underline{M} \frac{d\underset{\sim}{u}}{dt} + \underline{K}\,\underset{\sim}{u} \right) = \underset{\sim}{w}^{\mathrm{T}} \underset{\sim}{f}, \tag{15.26}$$

where f results from the contribution of B (see Section 14.5). This equation has to be satisfied for all $\underset{\sim}{w}$, hence

$$\underline{M}\frac{d\underset{\sim}{u}}{dt} + \underline{K}\,\underset{\sim}{u} = \underset{\sim}{f}. \tag{15.27}$$

This is a set of first-order differential equations having a similar structure as Eq. (15.11). Therefore, application of the θ-scheme for temporal discretization yields

$$\left(\frac{1}{\Delta t}\underline{M} + \theta\underline{K}\right)\underset{\sim}{u}_{n+1} = \left(\frac{1}{\Delta t}\underline{M} - (1-\theta)\underline{K}\right)\underset{\sim}{u}_n + \underset{\sim}{f}^\theta, \tag{15.28}$$

with

$$\underset{\sim}{f}^\theta = \theta\underset{\sim}{f}_{n+1} + (1-\theta)\underset{\sim}{f}_n. \tag{15.29}$$

Clearly, in the steady case the set of equations, Eqs. (15.27), reduces to

$$\underline{K}\,\underset{\sim}{u} = \underset{\sim}{f}. \tag{15.30}$$

Example 15.2   Consider the steady convection-diffusion problem, according to

$$v\frac{du}{dx} = \frac{d}{dx}\left(c\frac{du}{dx}\right),$$

with the following parameter setting:

$$\Omega = [0\ \ 1]$$
$$\Gamma_u\ :\ x = 0\ \ \text{and}\ \ x = 1$$
$$\Gamma_p = \emptyset$$
$$c = 1$$
$$u(x = 0) = 0$$
$$u(x = 1) = 1$$

The convective velocity v will be varied. For $v = 0$, the diffusion limit, the solution is obvious: u varies linearly in x from $u = 0$ at $x = 0$ to $u = 1$ at $x = 1$. Figs. 15.3(a) to (d) show the solution for $v = 1$, 10, 25 and 100, respectively, using a uniform element distribution with ten linear elements. For $v = 1$ and $v = 10$ the approximate solution u_h (solid line) closely (but not exactly) follows the exact solution (dashed line). However, for $v = 25$ the numerical solution starts to demonstrate an oscillatory behaviour that is more prominent for $v = 100$. Careful analysis of the discrete set of equations shows that the so-called element Peclet number governs this oscillatory behaviour. The element Peclet number is defined as

$$Pe_h = \frac{vh}{2c},$$

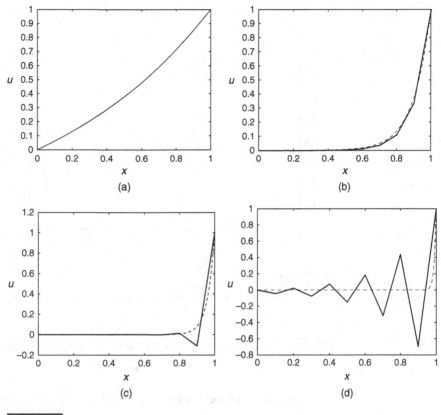

Figure 15.3

Solution of the steady convection-diffusion equation for $v = 1, 10, 25$ and $v = 100$, respectively using ten linear elements; solid line: approximate solution u_h, dashed line: exact solution u.

where h is the element length. Above a certain critical value of Pe_h the solution behaves in an oscillatory fashion. To reduce possible oscillations the element Peclet number should be reduced. For fixed v and c this can only be achieved by reducing the element size h. For example, doubling the number of elements from 10 to 20 eliminates the oscillations at $v = 25$, see Fig. 15.4.

The oscillations that appear in the numerical solution of the steady convection-diffusion equation may be examined as follows. Consider a domain that is subdivided in two linear elements, each having a length equal to h. At one end of the domain the solution is fixed to $u = 0$, while at the other end the solution is set to $u = 1$, or any other arbitrary non-zero value. For constant v and c the governing differential equation may be rewritten as

$$\frac{v}{c}\frac{du}{dx} - \frac{d^2u}{dx^2} = 0.$$

Solution of the steady convection-diffusion problem using 20 elements at $v = 25$.

The set of equations that results after discretization is, as usual:

$$\underline{K}\underset{\sim}{u} = \underset{\sim}{f}.$$

If only two linear elements of equal length h are used the coefficient matrix $\underline{K}$ may be written as

$$\underline{K} = \frac{v}{2c}\begin{bmatrix} -1 & 1 & 0 \\ -1 & 0 & 1 \\ 0 & -1 & 1 \end{bmatrix} + \frac{1}{h}\begin{bmatrix} 1 & -1 & 0 \\ -1 & 2 & -1 \\ 0 & -1 & 1 \end{bmatrix},$$

where the first, asymmetric, part corresponds to the convective term and the second, symmetric, part to the diffusion term. In the absence of a source term the second component of f is zero. Let u_1 and u_3 be located at the ends of the domain such that $u_1 = 0$ and $u_3 = 1$, then u_2 is obtained from

$$2u_2 = 1 - \frac{vh}{2c}.$$

An oscillation becomes manifest if $u_2 < 0$. To avoid this, the **element Peclet number** should be smaller than one:

$$Pe_h = \frac{vh}{2c} < 1.$$

Consequently, at a given convective velocity v and diffusion constant c, the mesh size h can be chosen such that an oscillation-free solution results. In particular for large values of v/c this may result in very fine meshes. To avoid the use of

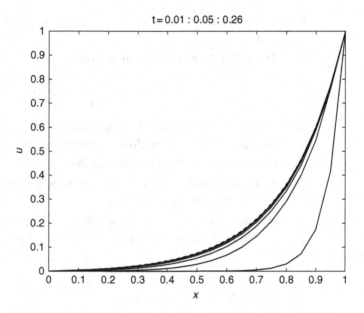

Figure 15.5

Solution of the unsteady convection-diffusion problem using 20 elements at $v = 10$.

very fine meshes, an alternative, stabilized formulation has been developed: the so-called SUPG (Streamline-Upwind/Petrov–Galerkin) formulation. A discussion of this method however, is beyond the scope of the present book.

Example 15.3 Let us consider the instationary convection-diffusion problem. For this problem the same outline as in the previous example for $v = 10$ is chosen and we use a uniform distribution of 20 linear elements. The initial condition is $u(x, t = 0) = 0$ throughout the domain. At the first time step the boundary condition $u(x = 1, t) = 1$ is imposed. The unsteady solution is obtained using a time step of $\Delta t = 0.01$, while $\theta = 0.5$ is selected for the θ-scheme. Fig. 15.5 shows the time-evolution of the solution towards the steady state value (denoted by the dashed line) for $v = 10$.

Exercises

15.1 Consider the domain $\Omega = [\,0\ 1\,]$. On the domain the one-dimensional steady convection-diffusion equation:

$$v\frac{du}{dx} = c\frac{d^2u}{dx^2}$$

holds. As boundary conditions, at $x = 0$, $u = 0$ and at $x = 1$, $u = 1$ are specified.

(a) Prove that the exact solution is given by

$$u = \frac{1}{1 - e^{\frac{v}{c}}}(1 - e^{\frac{v}{c}x}).$$

(b) Verify this by means of the script demo_fem1dcd, to solve the one-dimensional convection-diffusion problem, which can be found in the directory oned of the programme library mlfem_nac. Use five elements and select $c = 1$, while v is varied. Choose $v = 0$, $v = 1$, $v = 10$ and $v = 20$. Explain the results.

(c) According to Section 15.4, the solution is expected to be oscillation free if the element Peclet number is smaller than 1:

$$Pe_h = \frac{ah}{2c} < 1.$$

Verify that this is indeed the case.

15.2 Investigate the unsteady convection-diffusion problem:

$$\frac{\partial u}{\partial t} + v\frac{\partial u}{\partial x} = \frac{\partial}{\partial x}\left(c\frac{\partial u}{\partial x}\right),$$

on the domain $\Omega = [0\ 1]$ subject to the initial condition:

$$u_{ini}(x, t = 0) = 0,$$

inside the domain Ω and the boundary conditions:

$$u = 0 \text{ at } x = 0, \qquad u = 1 \text{ at } x = 1.$$

The θ-scheme for time integration is applied. Modify the m-file demo_fem1dcd accordingly. Use ten linear elements.

(a) Choose $v = c = 1$ and solve the problem with different values of θ. Use $\theta = 0.5$, $\theta = 0.4$ and $\theta = 0.25$. For each problem start with a time step of 0.01 and increase the time step with 0.01 until a maximum of 0.05. Describe what happens with the solution.

(b) In the steady state case, what is the maximum value of the convective velocity v such that the solution is oscillation free for $c = 1$?

(c) Does the numerical solution remain oscillation free in the unsteady case for $\theta = 0.5$ and $\Delta t = 0.001, 0.01, 0.1$? What happens?

(d) What happens at $\Delta t = 0.001$, if the convective velocity is reduced?

15.3 Investigate the unsteady convection problem:

$$\frac{\partial u}{\partial t} + v\frac{\partial u}{\partial x} = \frac{\partial}{\partial x}\left(c\frac{\partial u}{\partial x}\right)$$

on the spatial domain $\Omega = [0\ 1]$ and a temporal domain that spans $t = [0\ 0.5]$. At $x = 0$ the boundary condition $u = 0$ is prescribed, while at $x = 1$, $c\ du/dx = 0$ is selected. The convection dominated case is investigated, with $v = 1$ and $c = 0.01$. The problem is solved using the θ-scheme with $\theta = 0.5$, a time step $\Delta t = 0.05$ and 40 linear elements.

(a) First, the initial condition:

$$u(x, t = 0) = \sin(2\pi x)$$

is considered. Adapt the `demo_fem1dcd` and the FEM program accordingly. In particular, make sure that the initial condition is handled properly in `fem1dcd`. (Hint: the initial condition is specified in `sol(:,1)`.) Solve this problem. Plot the solution for all time steps.

(b) Second, let the initial condition be given by

$$u(x, t = 0) = 0 \text{ for } x < 0.25 \text{ and } x > 0.5,$$

while

$$u(x, t = 0) = 1 \text{ for } 0.25 \leq x \leq 0.5.$$

Solve this problem in the same way.

15.4 Many biological materials can be described as a mixture of a porous solid and a fluid, like articular cartilage [13], skin [14], intervertebral disk [6], heart tissue [11], etc. A confined compression test can be used to determine the material parameters. A circular specimen is placed in a confining ring. A porous filter supports the solid phase, while fluid from the specimen can be expelled. On top a tight-fitting indenter is placed to mechanically load the specimen with a constant force. Because of the confining ring the specimen can only deform in one direction. When a step load is applied to the

tissue, at first a pressure will be built up in the fluid. Immediately after loading the fluid will start moving through the filter and the solid will gradually take over the load. After some time the pressure in the fluid is zero and all load is taken by the solid. It can be derived from the theory of mixtures that the pressure in the fluid is described by the instationary diffusion equation:

$$\frac{\partial p}{\partial t} = HK\frac{\partial^2 p}{\partial^2 x},$$

with K the confined compression modulus (also called aggregate modulus) and H the permeability of the porous solid. The boundary conditions are:

- $p = 0$ at $x = 0$; free drainage at the porous filter
- $\partial p/\partial x = 0$ at $x = 10^{-2}$ [m]; no flow through the surface where the indenter makes contact with the specimen

The initial condition is $p = 1000$ [Nm^{-2}] at every point in the specimen at $t = 0$ (except at $x = 0$). This means that at $t = 0$ all load is carried by the fluid and not by the solid. The aggregate modulus satisfies $K = 10^5$ [Nm^{-2}]; the permeability is given by $H = 10^{-14}$ [m^{4}N^{-1}s^{-1}].

(a) Adjust `demo_fem1dcd` to calculate the pressure as a function of time for the above given problem. Choose $v = 0$ and take as a time step $\Delta t = 2500$ [s]. Use $\theta = 1$ for the time integration scheme.

(b) Determine the fluid pressure near the contact surface with the indenter and plot this pressure as a function of time.

16 Solution of the three-dimensional convection-diffusion equation by means of the Finite Element Method

16.1 Introduction

The two- and three-dimensional convection-diffusion equation plays an important role in many applications in biomedical engineering. One typical example from recent research is the analysis of the effectiveness of different types of bioreactors for tissue engineering. Tissue engineering is a rapidly evolving interdisciplinary research area aiming at the replacement or restoration of diseased or damaged tissue. In many cases devices made of artificial materials are only capable of partially restoring the original function of native tissues, and may not last for the full lifetime of a patient. In addition, there is no artificial replacement for a large number of tissues and organs. In tissue engineering new, autologous tissues are grown outside the human body by seeding cultured cells on scaffolds and further developed in a bioreactor for later implantation. The tissue proliferation and differentiation process is strongly affected by mechanical stimuli and transport of oxygen, minerals, nutrients and growth factors. To optimize bioreactor systems it is necessary to analyse how these systems behave. The convection-diffusion equation plays an important role in this kind of simulating analysis.

Fig. 16.1 shows two different bioreactor configurations, which both have been used in the past to tissue engineer articular cartilage. The work was especially focussed on glucose, oxygen and lactate, because these metabolites play a major role in the chondrocyte biosynthesis and survival. Questions ranged from: 'Does significant nutrient depletion occur at the high cell concentrations required for chondrogenesis?' to 'Do increasing transport limitations due to matrix accumulation significantly affect metabolite distributions?' Fig. 16.2 shows a typical result of the calculated oxygen distributions in the two bioreactor configurations.

This chapter explains the discretization of the convection-diffusion equation in two or three dimensions. First the diffusion equation is discussed, thereafter the convection-diffusion equation is elaborated. The spatial discretization of the weighting function is based on the Galerkin method.

(a)

(b)

Figure 16.1

Culture configurations. (a) Petri dish (b) Compression set-up. Adapted from [16].

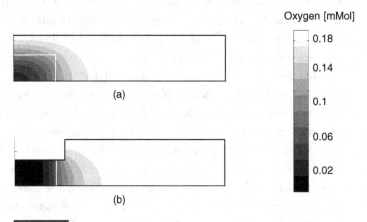

Oxygen [mMol]

0.18

0.14

0.1

0.06

0.02

(a)

(b)

Figure 16.2

Oxygen distribution in the static case at 48 h. Because of axisymmetry only the right half of the domain cross section is shown. The construct position is indicated with a white line. (a) Petri dish (b) Compression set-up. Adapted from [16].

16.2 Diffusion equation

Consider a two- or three-dimensional domain Ω with boundary Γ. As in the one-dimensional model problem, the boundary Γ is split into a part Γ_u along which the essential boundary conditions are specified, and a part Γ_p along which the natural boundary conditions may be specified. The generic form of the diffusion equation is given by

$$\vec{\nabla} \cdot (c\vec{\nabla}u) + f = 0, \tag{16.1}$$

where c denotes the diffusion coefficient and f a source term. A more general form of Eq. (16.1) is obtained by replacing the scalar c with a second-order tensor:

$$\vec{\nabla} \cdot (\mathbf{C} \cdot \vec{\nabla}u) + f = 0. \tag{16.2}$$

However, currently attention is restricted to Eq. (16.1).

The essential boundary conditions along Γ_u read:

$$u = U \text{ at } \Gamma_u, \tag{16.3}$$

while the natural boundary conditions along Γ_p are given by

$$\vec{n} \cdot c\vec{\nabla}u = P \text{ at } \Gamma_p, \tag{16.4}$$

with $\vec{n}$ the unit outward normal vector to the boundary Γ.

16.3 Divergence theorem and integration by parts

Let $\vec{n}$ be the unit outward normal to the boundary Γ of the domain Ω, and ϕ a sufficiently smooth function on Ω, then

$$\int_{\Omega} \vec{\nabla}\phi \, d\Omega = \int_{\Gamma} \vec{n}\phi \, d\Gamma. \tag{16.5}$$

If the function ϕ is replaced by a vector it can easily be derived:

$$\int_{\Omega} \vec{\nabla} \cdot \vec{\phi} \, d\Omega = \int_{\Gamma} \vec{n} \cdot \vec{\phi} \, d\Gamma. \tag{16.6}$$

Eq. (16.6) is known as the **divergence theorem**. For a proof of these equations, see for example Adams [1].

Let both ϕ and ψ be sufficiently smooth functions on Ω, then

$$\int_{\Omega} (\vec{\nabla}\phi)\psi \, d\Omega = \int_{\Gamma} \vec{n}\phi\psi \, d\Gamma - \int_{\Omega} \phi\vec{\nabla}\psi \, d\Omega. \tag{16.7}$$

This is called integration by parts. To prove this we must integrate the product rule of differentiation:

$$\vec{\nabla}(\phi\psi) = (\vec{\nabla}\phi)\psi + \phi\vec{\nabla}\psi, \tag{16.8}$$

to obtain

$$\int_{\Omega} \vec{\nabla}(\phi\psi) \, d\Omega = \int_{\Omega} (\vec{\nabla}\phi)\psi \, d\Omega + \int_{\Omega} \phi\vec{\nabla}\psi \, d\Omega. \tag{16.9}$$

Subsequently, use the divergence theorem to convert the left-hand side into the boundary integral:

$$\int_\Omega \vec{\nabla}(\phi\psi)\, d\Omega = \int_\Gamma \vec{n}\phi\psi\, d\Gamma. \tag{16.10}$$

This yields the desired result.

16.4 Weak form

Following the same steps as in Chapter 14, the differential equation Eq. (16.1) is multiplied with a weighting function w and integrated over the domain Ω:

$$\int_\Omega w\left(\vec{\nabla}\cdot(c\vec{\nabla}u)+f\right) d\Omega = 0, \quad \text{for all} \quad w. \tag{16.11}$$

Next the integration by parts rule according to Eq. (16.7) is used:

$$\int_\Gamma w\,\vec{n}\cdot c\vec{\nabla}u\, d\Gamma - \int_\Omega \vec{\nabla}w\cdot(c\vec{\nabla}u)\, d\Omega + \int_\Omega wf\, d\Omega = 0. \tag{16.12}$$

The boundary integral can be split into two parts, depending on the essential and natural boundary conditions:

$$\int_\Gamma w\,\vec{n}\cdot c\vec{\nabla}u\, d\Gamma = \int_{\Gamma_u} w\,\vec{n}\cdot c\vec{\nabla}u\, d\Gamma + \int_{\Gamma_p} wP\, d\Gamma, \tag{16.13}$$

where $c\vec{\nabla}u\cdot\vec{n} = P$ at Γ_p is used. It will be clear that, similar to the derivations in Chapter 14, the first integral on the right-hand side of Eq. (16.13) is unknown, while the second integral offers the possibility to incorporate the natural boundary conditions. For the time being we keep both integrals together (to limit the complexity of the equations and rewrite Eq. (16.12) according to:

$$\int_\Omega \vec{\nabla}w\cdot(c\vec{\nabla}u)\, d\Omega = \int_\Omega wf\, d\Omega + \int_\Gamma w\,\vec{n}\cdot c\vec{\nabla}u\, d\Gamma. \tag{16.14}$$

16.5 Galerkin discretization

Step 1 Introduce a mesh by splitting the domain Ω into a number of non-overlapping elements Ω_e. In a two-dimensional configuration the elements typically have either a triangular (in this case the mesh is sometimes referred to as a triangulation) or a quadrilateral shape. A typical example of triangulation is given in Fig. 16.3. Each triangle corresponds to an element.

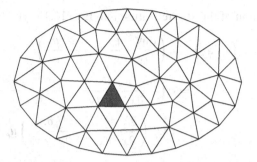

Figure 16.3

Example of a two-dimensional finite element mesh using triangular elements. One element has been highlighted.

The integration over the domain Ω can be performed by a summation of the integrals over each element:

$$\sum_{e=1}^{N_{el}} \int_{\Omega_e} \vec{\nabla}w \cdot (c\vec{\nabla}u) \; d\Omega = \sum_{e=1}^{N_{el}} \left(\int_{\Omega_e} wf \; d\Omega + \int_{\Gamma_e} w\vec{n} \cdot c\vec{\nabla}u \; d\Gamma \right). \quad (16.15)$$

The boundary part Γ_e denotes the intersection of element Ω_e with the boundary Γ, hence $\Gamma_e = \Omega_e \cap \Gamma$. Clearly, not each element will have an intersection with Γ.

Step 2 If a Cartesian coordinate system is used for two-dimensional problems (extension to the three-dimensional case is straightforward), the inner product $\vec{\nabla}w \cdot (c\vec{\nabla}u)$ yields

$$\vec{\nabla}w \cdot (c\vec{\nabla}u) = \left(\frac{\partial w}{\partial x}\vec{e}_x + \frac{\partial w}{\partial y}\vec{e}_y \right) \cdot c \left(\frac{\partial u}{\partial x}\vec{e}_x + \frac{\partial u}{\partial y}\vec{e}_y \right)$$

$$= c \left(\frac{\partial w}{\partial x}\frac{\partial u}{\partial x} + \frac{\partial w}{\partial y}\frac{\partial u}{\partial y} \right). \quad (16.16)$$

Step 3 Introduce a discretization for both the weighting function w and the unknown u, so within each element

$$w_h|_{\Omega_e} = \sum_{i=1}^{n} N_i(x,y)\, w_{e,i} = \underset{\sim}{N}^{\mathrm{T}}(x,y)\, \underset{\sim}{w}_e \quad (16.17)$$

$$u_h|_{\Omega_e} = \sum_{i=1}^{n} N_i(x,y)\, u_{e,i} = \underset{\sim}{N}^{\mathrm{T}}(x,y)\, \underset{\sim}{u}_e. \quad (16.18)$$

Step 4 Substitution of this discretization into Eq. (16.16) yields

$$
\begin{aligned}
\vec{\nabla} w_h \cdot \left(c \vec{\nabla} u_h \right) &= c \left(\frac{\partial w_h}{\partial x} \frac{\partial u_h}{\partial x} + \frac{\partial w_h}{\partial y} \frac{\partial u_h}{\partial y} \right) \\
&= c \left(\frac{\partial \underset{\sim}{N}^T \underset{\sim}{w}_e}{\partial x} \frac{\partial \underset{\sim}{N}^T \underset{\sim}{u}_e}{\partial x} + \frac{\partial \underset{\sim}{N}^T \underset{\sim}{w}_e}{\partial y} \frac{\partial \underset{\sim}{N}^T \underset{\sim}{u}_e}{\partial y} \right) \\
&= c \underset{\sim}{w}_e^T \left(\frac{\partial \underset{\sim}{N}}{\partial x} \frac{\partial \underset{\sim}{N}^T}{\partial x} + \frac{\partial \underset{\sim}{N}}{\partial y} \frac{\partial \underset{\sim}{N}^T}{\partial y} \right) \underset{\sim}{u}_e.
\end{aligned}
\tag{16.19}
$$

Step 5 Using the discretization in Eq. (16.15) yields

$$
\sum_{e=1}^{N_{el}} \underset{\sim}{w}_e^T \int_{\Omega_e} c \left(\frac{\partial \underset{\sim}{N}}{\partial x} \frac{\partial \underset{\sim}{N}^T}{\partial x} + \frac{\partial \underset{\sim}{N}}{\partial y} \frac{\partial \underset{\sim}{N}^T}{\partial y} \right) d\Omega \, \underset{\sim}{u}_e
$$

$$
= \sum_{e=1}^{N_{el}} \left(\underset{\sim}{w}_e^T \int_{\Omega_e} \underset{\sim}{N} f \, d\Omega + \underset{\sim}{w}_e^T \int_{\Gamma_e} \underset{\sim}{N} \vec{n} \cdot c \vec{\nabla} u \, d\Gamma \right).
\tag{16.20}
$$

The element matrix is given by

$$
\underline{K}_e = \int_{\Omega_e} c \left(\frac{\partial \underset{\sim}{N}}{\partial x} \frac{\partial \underset{\sim}{N}^T}{\partial x} + \frac{\partial \underset{\sim}{N}}{\partial y} \frac{\partial \underset{\sim}{N}^T}{\partial y} \right) d\Omega,
\tag{16.21}
$$

and the element column by

$$
\underset{\sim}{f}_e = \int_{\Omega_e} \underset{\sim}{N} f \, d\Omega + \int_{\Gamma_e} \underset{\sim}{N} \vec{n} \cdot c \vec{\nabla} u \, d\Gamma.
\tag{16.22}
$$

Using this notation, Eq. (16.20) may be written as

$$
\sum_{e=1}^{N_{el}} \underset{\sim}{w}_e^T \underline{K}_e \, \underset{\sim}{u}_e = \sum_{e=1}^{N_{el}} \underset{\sim}{w}_e^T \underset{\sim}{f}_e.
\tag{16.23}
$$

Following a similar procedure as outlined in Chapter 14, this may be rearranged into

$$
\underset{\sim}{w}^T \underline{K} \, \underset{\sim}{u} = \underset{\sim}{w}^T \underset{\sim}{f},
\tag{16.24}
$$

which should hold for all $\underset{\sim}{w}$. This finally leads to:

$$
\underline{K} \underset{\sim}{u} = \underset{\sim}{f}.
\tag{16.25}
$$

Similar to the situation described in Section 14.6 the column $\underset{\sim}{u}$ contains an unknown and a known part depending on the essential boundary conditions. In the nodes, where essential boundary conditions are prescribed, the associated external loads in the right-hand side integrals are unknown. However, the set equations can be partitioned as has been done in Chapter 14 to arrive at a set that can be solved.

16.6 Convection-diffusion equation

Assuming isotropic diffusion, the convection-diffusion equation is given by

$$\frac{\partial u}{\partial t} + \vec{v} \cdot \vec{\nabla} u = \vec{\nabla} \cdot (c\vec{\nabla} u) + f, \tag{16.26}$$

with $\vec{v}$ the convective velocity. This equation should hold on the spatial domain Ω during a certain period of time, say $S = [0, T]$. Initial boundary conditions must be specified:

$$u(\vec{x}, t = 0) = u_{\text{ini}}(\vec{x}) \text{ in } \Omega, \tag{16.27}$$

as well as essential and natural boundary conditions:

$$u = U \text{ at } \Gamma_u \tag{16.28}$$

$$\vec{n} \cdot c\vec{\nabla} u = P \text{ at } \Gamma_p. \tag{16.29}$$

The weak form is obtained analogously to the procedure of Section 16.4, giving

$$\int_{\Omega} \left(w\frac{\partial u}{\partial t} + w\vec{v} \cdot \vec{\nabla} u + \vec{\nabla} w \cdot (c\vec{\nabla} u) \right) d\Omega$$
$$= \int_{\Omega} wf \, d\Omega + \int_{\Gamma} w\vec{n} \cdot c\vec{\nabla} u \, d\Gamma. \tag{16.30}$$

Spatial discretization is performed in a two-dimensional configuration by introducing

$$w_h|_{\Omega_e} = \underline{N}^{\mathrm{T}}(x, y) \, \underline{w}_e, \tag{16.31}$$

and

$$u_h|_{\Omega_e} = \underline{N}^{\mathrm{T}}(x, y) \, \underline{u}_e. \tag{16.32}$$

For a particular element Ω_e, the individual integrals of Eq. (16.30) can be converted to:

$$\int_{\Omega_e} w\frac{\partial u}{\partial t} \, d\Omega = \underline{w}_e^{\mathrm{T}} \underline{M}_e \frac{d\underline{u}_e}{dt}, \tag{16.33}$$

with:

$$\underline{M}_e = \int_{\Omega_e} \underline{N}\underline{N}^{\mathrm{T}} \, d\Omega. \tag{16.34}$$

Further, by using

$$\vec{v} = v_x\vec{e}_x + v_y\vec{e}_y, \tag{16.35}$$

it can be written:

$$\int_{\Omega_e} w\vec{v} \cdot \vec{\nabla} u \, d\Omega = \underline{w}_e^{\mathrm{T}} \underline{C}_e \underline{u}_e, \tag{16.36}$$

with

$$
\underline{C}_e = \int_{\Omega_e} \underset{\sim}{N} \left(v_x \frac{\partial \underset{\sim}{N^{\mathrm{T}}}}{\partial x} + v_y \frac{\partial \underset{\sim}{N^{\mathrm{T}}}}{\partial y} \right) d\Omega. \tag{16.37}
$$

The remaining integrals from Eq. (16.30) follow from the previous section. Therefore, Eq. (16.30) may be formulated as

$$
\sum_{e=1}^{N_{el}} \underset{\sim}{w_e^{\mathrm{T}}} \left(\underline{M}_e \frac{d\underset{\sim}{u}_e}{dt} + (\underline{C}_e + \underline{K}_e)\, \underset{\sim}{u}_e \right) = \sum_{e=1}^{N_{el}} \underset{\sim}{w_e^{\mathrm{T}}} \underset{\sim}{f}_e, \tag{16.38}
$$

which must be satisfied for all admissible weighting values, hence after the assembly process the following set of equations results:

$$
\underline{M} \frac{d\underset{\sim}{u}}{dt} + (\underline{C} + \underline{K})\, \underset{\sim}{u} = \underset{\sim}{f}. \tag{16.39}
$$

Application of the θ-scheme for temporal discretization in the time increment $[t_n, t_{n+1}]$ results in:

$$
\left(\underline{M} \frac{1}{\Delta t} + \theta \underline{C} + \theta \underline{K} \right) \underset{\sim}{u}_{n+1}
$$
$$
= \left(\underline{M} \frac{1}{\Delta t} - (1-\theta)\underline{C} - (1-\theta)\underline{K} \right) \underset{\sim}{u}_n + \underset{\sim}{f}^{\theta}, \tag{16.40}
$$

with $\Delta t = t_{n+1} - t_n$ and

$$
\underset{\sim}{f}^{\theta} = \theta \underset{\sim}{f}_{n+1} + (1-\theta) \underset{\sim}{f}_n. \tag{16.41}
$$

16.7 Isoparametric elements and numerical integration

To compute the element matrices $\underline{M}_e$, $\underline{C}_e$ and $\underline{K}_e$ and the element array $\underset{\sim}{f}_e$ the shape functions $\underset{\sim}{N}$ and the shape function derivatives with respect to the coordinates x and y, need to be available. To define the shape functions it is convenient to introduce a local coordinate system.

In particular for an arbitrarily shaped quadrilateral element, for instance as depicted in Fig. 16.4(a) it is difficult or impossible to define explicitly the shape functions with respect to the global coordinates x and y. However, defining the shape functions in a unit square domain, as depicted in Fig. 16.4(b) is straightforward. The local coordinate system defined in Fig. 16.4(b) is chosen such that $-1 \leq \xi \leq 1$ and $-1 \leq \eta \leq 1$. Node one, for instance, has local coordinates $\xi = -1$ and $\eta = -1$. Remember that the shape function of a certain node has to be one at the spatial location of this node and zero at the spatial location of all other nodes. Using the local coordinate system allows an elegant definition of the shape functions of a four-node quadrilateral element:

(a) Quadrilateral element with respect to a **global** coordinate system

(b) Quadrilateral element with respect to a **local** coordinate system

Figure 16.4

Quadrilateral element with respect to global and local coordinate systems.

$$N_1 = \frac{1}{4}(1 - \xi)(1 - \eta)$$

$$N_2 = \frac{1}{4}(1 + \xi)(1 - \eta)$$

$$N_3 = \frac{1}{4}(1 + \xi)(1 + \eta)$$

$$N_4 = \frac{1}{4}(1 - \xi)(1 + \eta).$$

(16.42)

An element having these shape functions is called a **bi-linear** element. Along the edges of the element the shape functions are linear with respect to either ξ or η. Within the element, however, the shape functions are bi-linear with respect to ξ and η. For instance:

$$N_1 = \frac{1}{4}(1 - \xi - \eta + \xi\eta).$$

(16.43)

Fig. 16.5 shows N_1 visualized as a contour plot.

The shape function derivatives with respect to the local coordinates ξ and η are easily computed. However, the shape function derivatives with respect to the global coordinates x and y are needed. For this purpose the concept of **isoparametric elements** is used.

For isoparametric elements the global coordinates within an element are interpolated based on the nodal coordinates using the shape functions

$$x|_{\Omega_e} = \underset{\sim}{N}^{\mathrm{T}}(\xi, \eta)\,\underset{\sim}{x}_e, \qquad y|_{\Omega_e} = \underset{\sim}{N}^{\mathrm{T}}(\xi, \eta)\,\underset{\sim}{y}_e,$$

(16.44)

where $\underset{\sim}{x}_e$ and $\underset{\sim}{y}_e$ contain the nodal x- and y-coordinates, respectively. These equations reflect the transformation from the local coordinates (ξ, η) to the global coordinates (x, y). The derivatives of N_i with respect to the Cartesian coordinates x and y can be evaluated with the aid of the chain rule:

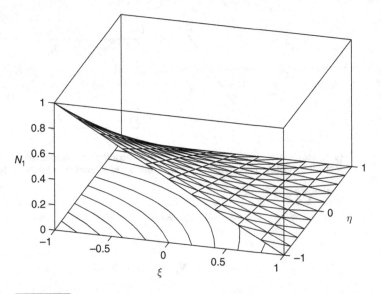

Shape function associated with node 1.

$$\frac{\partial N_i}{dx} = \frac{\partial N_i}{\partial \xi}\frac{\partial \xi}{\partial x} + \frac{\partial N_i}{\partial \eta}\frac{\partial \eta}{\partial x}$$

$$\frac{\partial N_i}{dy} = \frac{\partial N_i}{\partial \xi}\frac{\partial \xi}{\partial y} + \frac{\partial N_i}{\partial \eta}\frac{\partial \eta}{\partial y}. \tag{16.45}$$

These relations can be rewritten in the following matrix form:

$$\begin{bmatrix} \dfrac{\partial N_i}{\partial x} \\ \dfrac{\partial N_i}{\partial y} \end{bmatrix} = \begin{bmatrix} \dfrac{\partial \xi}{\partial x} & \dfrac{\partial \eta}{\partial x} \\ \dfrac{\partial \xi}{\partial y} & \dfrac{\partial \eta}{\partial y} \end{bmatrix} \begin{bmatrix} \dfrac{\partial N_i}{\partial \xi} \\ \dfrac{\partial N_i}{\partial \eta} \end{bmatrix}. \tag{16.46}$$

The derivatives $\partial N_i/\partial \xi$ and $\partial N_i/\partial \eta$ are readily available, but the terms in the matrix cannot be directly computed since the explicit expressions $\xi = \xi(x, y)$ and $\eta = \eta(x, y)$ are not known. However, due to the isoparametric formulation the inverse relations are known, so the following matrix can be calculated easily:

$$\underline{x}_{,\xi} = \begin{bmatrix} \dfrac{\partial x}{\partial \xi} & \dfrac{\partial y}{\partial \xi} \\ \dfrac{\partial x}{\partial \eta} & \dfrac{\partial y}{\partial \eta} \end{bmatrix}, \tag{16.47}$$

with components:

$$\frac{\partial x}{\partial \xi} = \frac{\partial \underset{\sim}{N}^{\mathrm{T}}}{\partial \xi} x_e$$

$$\frac{\partial x}{\partial \eta} = \frac{\partial \underset{\sim}{N}^{\mathrm{T}}}{\partial \eta} x_e$$

$$\frac{\partial y}{\partial \xi} = \frac{\partial \underset{\sim}{N}^{\mathrm{T}}}{\partial \xi} y_e \qquad (16.48)$$

$$\frac{\partial y}{\partial \eta} = \frac{\partial \underset{\sim}{N}^{\mathrm{T}}}{\partial \eta} y_e.$$

Matrix (16.47) is the inverse of the matrix in Eq. (16.46) (this can be checked by multiplying the two matrices, which gives the unit matrix). Accordingly

$$\begin{bmatrix} \frac{\partial \xi}{\partial x} & \frac{\partial \eta}{\partial x} \\ \frac{\partial \xi}{\partial y} & \frac{\partial \eta}{\partial y} \end{bmatrix} = (\underline{x}_{,\xi})^{-1} = \frac{1}{j} \begin{bmatrix} \frac{\partial y}{\partial \eta} & -\frac{\partial y}{\partial \xi} \\ -\frac{\partial x}{\partial \eta} & \frac{\partial x}{\partial \xi} \end{bmatrix} \qquad (16.49)$$

where

$$j = \det(\underline{x}_{,\xi}) = \frac{\partial x}{\partial \xi}\frac{\partial y}{\partial \eta} - \frac{\partial x}{\partial \eta}\frac{\partial y}{\partial \xi}. \qquad (16.50)$$

It is an elaborate process and usually not possible to analytically compute the integrals in the expressions for the element matrices or arrays, so generally they are approximated by numerical integration. Each of the components of the matrices, such as $\underline{K}_e$ etc., that need to be computed consists of integrals of a given function, say $g(x,y)$, over the domain of the element Ω_e. These may be transformed to an integral over the unit square $-1 \leq \xi \leq 1, -1 \leq \eta \leq 1$, according to

$$\int_{\Omega_e} f(x,y)\,d\Omega = \int_{-1}^{1} \int_{-1}^{1} f(x(\xi,\eta),y(\xi,\eta))j(\xi,\eta)\,d\xi\,d\eta, \qquad (16.51)$$

with $j(\xi,\eta)$ defined by Eq. (16.50).

The integral over the unit square may be approximated by a numerical integration (quadrature) rule, giving

$$\int_{-1}^{1} \int_{-1}^{1} f(\xi,\eta)\,j(\xi,\eta)\,d\xi\,d\eta \approx \sum_{i=1}^{n_{\mathrm{int}}} f(\xi_i,\eta_i)j(\xi_i,\eta_i)W(\xi_i,\eta_i). \qquad (16.52)$$

For example, in case of a two-by-two Gaussian integration rule the location of the integration points have ξ,η-coordinates and associated weights:

$$\xi_1 = \frac{-1}{\sqrt{3}}, \qquad \eta_1 = \frac{-1}{\sqrt{3}}, \qquad W_1 = 1$$

$$\xi_2 = \frac{1}{\sqrt{3}}, \qquad \eta_2 = \frac{-1}{\sqrt{3}}, \qquad W_2 = 1$$

$$(16.53)$$

$$\xi_3 = \frac{1}{\sqrt{3}}, \quad \eta_3 = \frac{1}{\sqrt{3}}, \quad W_3 = 1$$

$$\xi_4 = \frac{-1}{\sqrt{3}}, \quad \eta_4 = \frac{1}{\sqrt{3}}, \quad W_4 = 1.$$

16.8 Example

One of the treatments for coronary occlusions that may lead to an infarct is to put a stent at the location of the occlusion. One of the problems with this intervention is that the blood vessels often occlude again quite soon after the stent is placed. One of the solutions may be to design a stent that gradually releases drugs to prevent such an occlusion from occurring again, see Fig. 16.6. How these drugs propagate through the vascular tree is a convection-diffusion problem.

Consider the domain as sketched in Fig. 16.7. It represents a section of a long channel. For reasons of simplicity the three-dimensional problem is modelled as a two-dimensional problem: the relevant fields in the configuration are assumed to be independent of the coordinate perpendicular to the xy-plane. A so-called Newtonian fluid (modelling blood in a first approximation) flows through the channel as indicated in the figure.

Let u denote the concentration of a certain drug. Along the entrance of the domain the drug concentration is zero. Along the small part of the wall indicated

Figure 16.6

Schematic of a stent in a blood vessel.

Figure 16.7

Specification of the convection-diffusion problem.

Figure 16.8

Computational domain of the convection-diffusion problem.

with a thick line, the drug concentration is prescribed, say $u = 1$. The drug diffuses into the liquid with a diffusion constant c, but is also convected by the fluid. The aim is to compute the concentration profile in the two-dimensional channel for a number of fluid velocities. The computational domain is indicated by the dashed line, and is further outlined in Fig. 16.8. Because of symmetry only the top half of the vessel is modelled.

For stationary flow conditions, the velocity field $\vec{v}$ is described by means of a parabolic profile (Poisseuille flow) according to

$$\vec{v} = a(1 - y^2)\,\vec{e}_x.$$

As mentioned before, along boundary C1 the fluid flows into the domain with a concentration $u = 0$, while along boundary C2 the concentration $u = 1$ is prescribed. Along the remaining parts of the boundary the natural boundary condition:

$$\vec{n} \cdot c\vec{\nabla}u = 0,$$

is imposed. This means that the top wall is impenetrable for the drug, while this condition must also be enforced along the symmetry line $y = 0$. Specification of this condition on the outflow boundary is somewhat disputable, but difficult to avoid, because only a small part of the circulation system is modelled. By choosing the outflow boundary far away from the source of the drug the influence of this boundary condition is small.

The corresponding mesh is shown in Fig. 16.9. The problem is discretized using bi-quadratic elements.

The steady convection diffusion problem is solved, for $c = 1$ and a sequence 0,1,10,25,100 of parameter a. Clearly, with increasing a the velocity in the x-direction increases proportionally, hence convection becomes increasingly important. For increasing a contours of constant u are depicted in Fig. 16.10. In all

Figure 16.9

Mesh of the convection-diffusion problem.

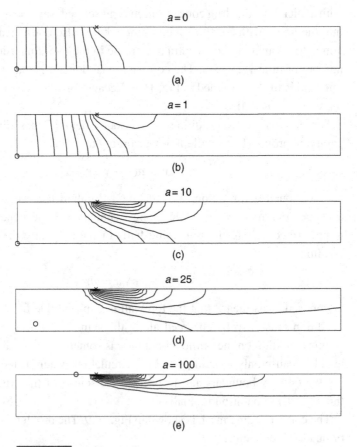

Figure 16.10

Contour lines of constant u values for a range of values of a. With increasing a the effect of convection increases.

cases 10 contour lines are shown ranging from 0.1 to 1 with increments of 0.1. The effect of an increasing velocity is clearly demonstrated.

Exercises

16.1   The weak form of the two-dimensional convection-diffusion equation is given by

$$\int_\Omega \left(w\frac{\partial u}{\partial t} + w\vec{v} \cdot \vec{\nabla}u + \vec{\nabla}w \cdot (c\vec{\nabla}u) \right) d\Omega$$

$$= \int_\Omega wf \, d\Omega + \int_\Gamma w\vec{n} \cdot c\vec{\nabla}u = \, d\Gamma.$$

After discretization, the element inertia matrix is defined as

$$\underline{M}_e = \int_{\Omega_e} \underset{\sim}{N} \, \underset{\sim}{N}^{\mathrm{T}} \, d\Omega,$$

while the element matrix related to the convective part is given by

$$\underline{C}_e = \int_{\Omega_e} \underset{\sim}{N} \left(v_x \frac{d\underset{\sim}{N}^{\mathrm{T}}}{dx} + v_y \frac{d\underset{\sim}{N}^{\mathrm{T}}}{dy} \right) d\Omega.$$

Consider the element as depicted in the figure below.

This is a bi-linear element. The element spans the spatial domain $-2 \le x \le 2$ and $2 \le y \le 2$.

(a)   Compute the element inertia matrix $\underline{M}_e$ using a 2×2 Gauss integration rule. It is recommended to use MATLAB for this computation.

(b)   Suppose that the location of the integration points coincides with the nodes of the element. What is the element inertia matrix $\underline{M}_e$ in this case?

(c)   Compute the matrix $\underline{C}_e$ if $v_x = 1$ and $v_y = 0$, using a 2 × 2 Gauss integration rule.

(d)   Suppose that along the edge of the element located at $x = 2$ a constant flux $q = \vec{n} \cdot c\vec{\nabla}u$ is prescribed. Then compute the column

$$\int_{\Gamma_e} N q \, d\Gamma,$$

for this edge, contributing to the element column $\underset{\sim}{f}_e$.

16.2   Consider the mesh depicted in the figure below. The solution vector $\underset{\sim}{u}$ contains the nodal solutions in the sequence defined by the node numbers, hence:

$$\underset{\sim}{u}^T = [\, u_1 \; u_2 \; \cdots \; u_{16} \,] \; .$$

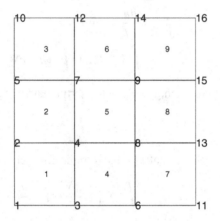

Let the element topology array of the third element be given by
`top(3,:)=[5 7 12 10 1 1]`.

(a)   Is this topology array unique?

(b)   What is the `pos` array for this element?

(c)   Define the `dest` array.

(d)   The solution $\underset{\sim}{u}$ is stored in the array `sol`. How are the nodal solutions of element 8 extracted from `sol` using the array `pos`?

(e)   Suppose that the array `nodes` contains the node numbers of the left edge, say `nodes=[1 2 5 10]`. How are the nodal solutions of these nodes extracted from the solution array `sol` via the array `dest`?

(f)   Consider the third element whose element topology array is given by `top(3,:)=[5 7 12 10 1 1]`. If the nodal solution array $\underset{\sim}{u}_e$ for this element is given by:

$$\underset{\sim}{u}_e^T = [\, 1 \; 4 \; 3 \; 7 \,],$$

compute the solution u for $\xi = \frac{1}{4}$ and $\eta = \frac{3}{4}$.

(i)   Suppose the element topology array is given by `top(3,:)=[12 10 5 7 1 1]`. What is the corresponding element nodal solution array $\underset{\sim}{u}_e$?

16.3   Consider the square domain $0 \leq x \leq 1, 0 \leq y \leq 1$. Along the line $x = 0$ the boundary condition $u = 0$ is imposed, and along the line $x = 1$ the boundary condition $u = 1$ is prescribed. Along the other boundaries the natural boundary condition $\vec{n} \cdot c\vec{\nabla}u = 0$ is specified. Consider the steady convection diffusion problem on this domain.

(a)   Suppose that the convection-diffusion problem represents the temperature equation. What physical meaning does the above natural boundary condition have?

(b)   Adjust the m-file demo_cd such that the above problem is solved using 5×5 linear elements. Select the diffusion constant $c = 1$ and the convective velocity $\vec{v} = \vec{e}_x$ (Remark: see the element m-file elcd and the associated m-file elcd_a). Specify the structured array mat according to this.

(c)   Extract the solution along the line $y = 0$. Hint, first select the nodes using usercurves containing the nodes along all the curves that have been defined for the mesh generator crmesh. Then use dest to extract the relevant solution components.

(d)   Compare the solution along $y = 0$ with the one-dimensional solution, using the program fem1dcd.

16.4   Consider the domain as sketched in the figure. Along the boundary Γ_1 the specification $u = 1$ is chosen, while along $\Gamma_7\, u = 0$ is chosen as a boundary condition. Along all other boundaries $\vec{n} \cdot c\vec{\nabla}u = 0$ is imposed. Choose the convective velocity $\vec{v}$ such that:

$$y < 0: \quad \vec{v} = \vec{0}$$
$$y \geq 0: \quad \vec{v} = y\vec{e}_x$$

Furthermore, choose $c = 1$ throughout the domain. Model this problem using femlin_cd. Approximately beyond which value of x along the boundary Γ_6 holds $u > 0.8$?

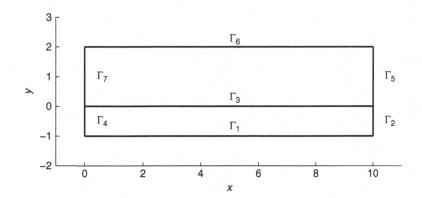

16.5   If the flux $\vec{n} \cdot c\vec{\nabla}u = P$ along Γ_p is prescribed, the integral (element level):

$$\int_{\Gamma_e} \underset{\sim}{N} P \, d\Gamma,$$

needs to be computed. It may be transformed to the local coordinate system $\xi \in [-1 \; 1]$, such that:

$$\int_{\Gamma_e} \underset{\sim}{N} P \, d\Gamma = \int_{-1}^{1} \underset{\sim}{N} P \frac{\partial x}{\partial \xi} \, d\xi,$$

where it is assumed that Γ_e is oriented in x-direction.

If linear elements are chosen, the shape functions along the boundary are given by:

$$\underset{\sim}{N} = \begin{bmatrix} \frac{1}{2}(1 - \xi) \\ \frac{1}{2}(1 + \xi) \end{bmatrix}.$$

Use an isoparametric element and let L denote the length of Γ_e, then demonstrate that for *constant P*:

$$\int_{\Gamma_e} \underset{\sim}{N} P \, d\Gamma = \frac{LP}{2} \begin{bmatrix} 1 \\ 1 \end{bmatrix}.$$

17 Shape functions and numerical integration

17.1 Introduction

In the previous chapter the shape functions N_i have hardly been discussed in any detail. The key purpose of this chapter is first to introduce isoparametric shape functions, and second to outline numerical integration of the integrals appearing in the element coefficient matrices and element column. Before this can be done it is useful to understand the minimum requirements to be imposed on the shape functions. The key question involved is, what conditions should at least be satisfied such that the approximate solution of the boundary value problems, dealt with in the previous chapter, generated by a finite element analysis, converges to the exact solution at mesh refinement. The answer is:

(i)   The shape functions should be smooth within each element Ω_e, i.e. shape functions are not allowed to be discontinuous within an element.

(ii)   The shape functions should be continuous across each element boundary. *This condition does not always have to be satisfied, but this is beyond the scope of the present book.*

(iii)   The shape functions should be complete, i.e. at element level the shape functions should enable the representation of uniform gradients of the field variable(s) to be approximated.

Conditions (i) and (ii) allow that the gradients of the shape functions show finite jumps across the element interface. However, smoothness in the element interior assures that all integrals in which gradients of the unknown function, say u, occur can be evaluated. In Fig. 17.1(a) an example is given of an admissible shape function. In this case the derivative of the shape function is discontinuous over the element boundary, however the jump is finite. In Fig. 17.1(b) the discontinuous shape function at the element boundary leads to an infinite derivative and the integrals in the weighted residual equations can no longer be evaluated.

Completeness When the finite element mesh is refined further and further, at the element level the exact solution becomes more and more linear in the coordinates and its derivatives approach constant values in each element. To ensure that

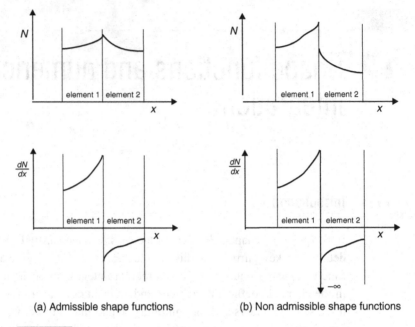

(a) Admissible shape functions   (b) Non admissible shape functions

Figure 17.1

Continuity of shape functions.

an adequate approximation can be achieved, the shape functions have to contain
all constant and linear functions. Assuming that the exact solution is described by
an arbitrary linear polynomial, the element interpolation has to be able to exactly
describe this field. In mathematical terms this means the following. Assume u to
be approximated by

$$u_h(\vec{x}) = \sum_{i=1}^{n} N_i(\vec{x})\, u_i, \tag{17.1}$$

with $N_i(\vec{x})$ the interpolation functions and u_i the nodal values of $u_h(\vec{x})$. Consider the case, where the nodal values u_i are selected to be related to the nodal
coordinates x_i, y_i, z_i by

$$u_i = c_0 + c_1 x_i + c_2 y_i + c_3 z_i, \tag{17.2}$$

according to a linear field. Substitution of Eq. (17.2) into (17.1) reveals:

$$u_h = c_0 \sum_{i=1}^{n} N_i(\vec{x}) + c_1 \sum_{i=1}^{n} N_i(\vec{x})\, x_i + c_2 \sum_{i=1}^{n} N_i(\vec{x})\, y_i + c_3 \sum_{i=1}^{n} N_i(\vec{x})\, z_i. \tag{17.3}$$

For the element types described in the current chapter the coordinates within the elements are interpolated in the following way:

$$x = \sum_{i=1}^{n} N_i(\vec{x}) x_i$$

$$y = \sum_{i=1}^{n} N_i(\vec{x}) y_i$$

$$z = \sum_{i=1}^{n} N_i(\vec{x}) z_i. \tag{17.4}$$

Completeness implies that Eq. (17.3) has to lead to

$$u_h(x, y, z) = c_0 + c_1 x + c_2 y + c_3 z, \tag{17.5}$$

for arbitrary $c_0, \ldots, c_3$. Hence

$$\sum_{i=1}^{n} N_i(\vec{x}) = 1. \tag{17.6}$$

The last equation can be used to check whether the shape functions may have been specified in an adequate way.

17.2 Isoparametric, bilinear quadrilateral element

Consider a four-noded, straight-edged, element $\Omega_e \in \mathbf{R}^2$ as depicted in Fig. 17.2(a). The nodal points are assumed to be numbered counterclockwise from 1 to 4. This element can be mapped onto a square, with local coordinates ξ and

(a) Quadrilateral element with respect to global coordinates

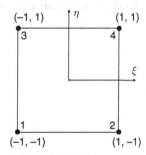

(b) Quadrilateral element with respect to local coordinates

Figure 17.2

Quadrilateral element with respect to a global and a local coordinate system.

η, with $\xi, \eta \in [-1, 1]$. A point with coordinates $x, y \in \Omega_e$ is related to a point $\xi, \eta \in [-1, 1]$ by the mapping

$$x(\xi, \eta) = \sum_{i=1}^{4} N_i(\xi, \eta) x_i \qquad (17.7)$$

$$y(\xi, \eta) = \sum_{i=1}^{4} N_i(\xi, \eta) y_i. \qquad (17.8)$$

The shape functions N_i may be determined by assuming the 'bi-linear' expansion:

$$x(\xi, \eta) = \alpha_0 + \alpha_1 \xi + \alpha_2 \eta + \alpha_3 \xi \eta$$
$$y(\xi, \eta) = \beta_0 + \beta_1 \xi + \beta_2 \eta + \beta_3 \xi \eta. \qquad (17.9)$$

The parameters α_j and β_j ($j = 0, 1, 2, 3$) must be calculated such that the relations

$$x(\xi_i, \eta_i) = x_i$$
$$y(\xi_i, \eta_i) = y_i, \qquad (17.10)$$

are satisfied, where ξ_i and η_i refer to the local coordinates of the nodes. Applying the restriction (17.10) to Eq. (17.9) leads to

$$\begin{pmatrix} x_1 \\ x_2 \\ x_3 \\ x_4 \end{pmatrix} = \begin{pmatrix} 1 & -1 & -1 & 1 \\ 1 & 1 & -1 & -1 \\ 1 & 1 & 1 & 1 \\ 1 & -1 & 1 & -1 \end{pmatrix} \begin{pmatrix} \alpha_0 \\ \alpha_1 \\ \alpha_2 \\ \alpha_3 \end{pmatrix}. \qquad (17.11)$$

From this set of equations the coefficients α_j can be linearly expressed in the nodal coordinates x_1, x_2, x_3 and x_4. By reorganizing, the eventual expressions for the shape functions become:

$$N_1(\xi, \eta) = \frac{1}{4}(1 - \xi)(1 - \eta)$$

$$N_2(\xi, \eta) = \frac{1}{4}(1 + \xi)(1 - \eta)$$

$$N_3(\xi, \eta) = \frac{1}{4}(1 + \xi)(1 + \eta)$$

$$N_4(\xi, \eta) = \frac{1}{4}(1 - \xi)(1 + \eta). \qquad (17.12)$$

The element is called *isoparametric* as both the spatial coordinates as well as the element interpolation function u_h are interpolated with the same shape functions, that is

$$u_h(\xi, \eta) = \sum_{i=1}^{n} N_i(\xi, \eta) \, u_i. \tag{17.13}$$

17.3 Linear triangular element

There are two ways to arrive at a triangular element. First, it is possible to coalesce two nodes of the quadrilateral element, for instance node 3 and 4. This is done by setting $\underset{\sim}{x}_4 = \underset{\sim}{x}_3$ (with $\underset{\sim}{x}_3^{\mathrm{T}} = [x_3, y_3]$ and $\underset{\sim}{x}_4^{\mathrm{T}} = [x_4, y_4]$) and by defining new shape functions $\hat{N}_i$ according to

$$
\begin{aligned}
\underset{\sim}{x} &= \sum_{i=1}^{4} N_i \underset{\sim}{x}_i \\
&= \underbrace{N_1}_{\hat{N}_1} \underset{\sim}{x}_1 + \underbrace{N_2}_{\hat{N}_2} \underset{\sim}{x}_2 + \underbrace{(N_3 + N_4)}_{\hat{N}_3} \underset{\sim}{x}_3 \\
&= \sum_{i=1}^{3} \hat{N}_i \underset{\sim}{x}_i.
\end{aligned}
\tag{17.14}
$$

Fig. 17.3 illustrates this operation.

The second method is based on using so-called triangle coordinates. A convenient set of coordinates $\lambda_1, \lambda_2, \lambda_3$ for a triangle can be defined by means of the following equations:

$$x = \lambda_1 x_1 + \lambda_2 x_2 + \lambda_3 x_3 \tag{17.15}$$

$$y = \lambda_1 y_1 + \lambda_2 y_2 + \lambda_3 y_3 \tag{17.16}$$

$$1 = \lambda_1 + \lambda_2 + \lambda_3. \tag{17.17}$$

To every set $\lambda_1, \lambda_2, \lambda_3 \in [0, 1]$ corresponds a unique set of Cartesian coordinates x, y. The triangle coordinates are not independent, but related by Eq. (17.17).

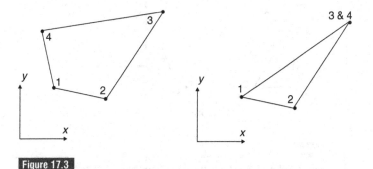

Figure 17.3

Degeneration from quadrilateral element to triangular element.

Solving Eqs. (17.15) to (17.17) for λ_i leads to

$$\lambda_1 = \frac{1}{\Delta}\left((x_2 y_3 - x_3 y_2) + (y_2 - y_3)x + (x_3 - x_2)y\right)$$

$$\lambda_2 = \frac{1}{\Delta}\left((x_3 y_1 - x_1 y_3) + (y_3 - y_1)x + (x_1 - x_3)y\right)$$

$$\lambda_3 = \frac{1}{\Delta}\left((x_1 y_2 - x_2 y_1) + (y_1 - y_2)x + (x_2 - x_1)y\right), \qquad (17.18)$$

where

$$\Delta = (x_3 - x_2)(y_1 - y_2) - (y_2 - y_3)(x_2 - x_1). \qquad (17.19)$$

Note that $|\Delta|$ is 2 times the square of the triangle surface. Eqs. (17.15) to (17.17) shows that λ_i can be (linearly) expressed in x and y: $\lambda_i = \lambda_i(x, y)$. The relation satisfies

$$\lambda_i(x_j, y_j) = \delta_{ij}. \qquad (17.20)$$

Now, the shape functions for the 3-node, 6-node and 7-node triangular elements are:

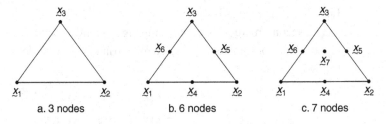

a. 3 nodes   b. 6 nodes   c. 7 nodes

Figure 17.4

Triangular elements with 3, 6 or 7 nodes.

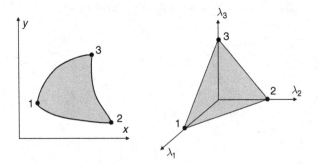

Figure 17.5

The mapping of a triangle from the global coordinate system to the local coordinate system with triangle coordinates.

(a) 3-node triangular element, linear interpolation:

$$N_1 = \lambda_1$$
$$N_2 = \lambda_2$$
$$N_3 = \lambda_3 \qquad (17.21)$$

(b) 6-node triangular element, quadratic interpolation:

$$N_1 = \lambda_1(2\lambda_1 - 1)$$
$$N_2 = \lambda_2(2\lambda_2 - 1)$$
$$N_3 = \lambda_3(2\lambda_3 - 1)$$
$$N_4 = 4\lambda_1\lambda_2$$
$$N_5 = 4\lambda_2\lambda_3$$
$$N_6 = 4\lambda_1\lambda_3 \qquad (17.22)$$

(c) 7-node triangular element, bi-quadratic interpolation:

$$N_1 = \lambda_1(2\lambda_1 - 1) + 3\lambda_1\lambda_2\lambda_3$$
$$N_2 = \lambda_2(2\lambda_2 - 1) + 3\lambda_1\lambda_2\lambda_3$$
$$N_3 = \lambda_3(2\lambda_3 - 1) + 3\lambda_1\lambda_2\lambda_3$$
$$N_4 = 4\lambda_1\lambda_2 - 12\lambda_1\lambda_2\lambda_3$$
$$N_5 = 4\lambda_2\lambda_3 - 12\lambda_1\lambda_2\lambda_3$$
$$N_6 = 4\lambda_1\lambda_3 - 12\lambda_1\lambda_2\lambda_3$$
$$N_7 = 27\lambda_1\lambda_2\lambda_3. \qquad (17.23)$$

The factor $\lambda_1\lambda_2\lambda_3$ is called a 'bubble' function giving zero contributions along the boundaries of the element.

There are two major differences between the method used in Section 17.2 and the method with triangle coordinates:

- Determining the derivatives of these shape functions with respect to the global coordinates is not trivial. Consider the derivative of a shape function N_i with respect to x. Applying the chain rule for differentiation would lead to

$$\frac{\partial N_i}{\partial x} = \frac{\partial N_i}{\partial \lambda_1}\frac{\partial \lambda_1}{\partial x} + \frac{\partial N_i}{\partial \lambda_2}\frac{\partial \lambda_2}{\partial x} + \frac{\partial N_i}{\partial \lambda_3}\frac{\partial \lambda_3}{\partial x}. \qquad (17.24)$$

But, by definition a partial derivative as to one variable implies that the other variables have to be considered as constant. In this case a partial derivative with respect to λ_1 means that, when this derivative is determined, λ_2 and λ_3 have to be considered constant. However, the λ_i's are related by Eq. (17.17). Only two variables can be considered as independent. A solution for this dilemma is to eliminate λ_3 from the shape functions by using:

$$\lambda_3 = 1 - \lambda_1 - \lambda_2. \tag{17.25}$$

A more obvious way of solving the problem is to substitute Eq. (17.18) into the equations for the shape functions, thus eliminating all triangular coordinates and directly determine $\partial N_i / \partial x$ and $\partial N_i / \partial y$.

- The second issue is the difference in integration limits which have to correspond with a triangle. For the square element in Section 17.2 the domain for integration is simple, meaning that the surface integral can be split into two successive single integrals with ξ and η as variables. The integration limits are -1 to $+1$. For the triangle this is more complicated and the limits of integration now involve the coordinate itself. This item will be discussed shortly in Section 17.5.

17.4 Lagrangian and Serendipity elements

In principle higher-order elements are more accurate than the linear ones discussed so far. However, the computation of element coefficient matrices and element arrays is more expensive for higher-order elements, and the cost-effectiveness depends on the particular problem investigated. Cost-effectiveness in the sense that there is a trade-off between the accuracy, using a smaller number of higher-order elements, versus using more linear elements.

Higher-order elements can systematically be derived from Lagrange polynomials. A (one-dimensional) set of Lagrange polynomials on an element with domain $\xi_1 \leq \xi \leq \xi_n$ is defined by

$$l_a^{n-1}(\xi) = \frac{\prod_{b=1, b \neq a}^{n} (\xi - \xi_b)}{\prod_{b=1, b \neq a}^{n} (\xi_a - \xi_b)} \tag{17.26}$$

$$= \frac{(\xi - \xi_1) \ldots (\xi - \xi_{a-1})(\xi - \xi_{a+1}) \ldots (\xi - \xi_n)}{(\xi_a - \xi_1) \ldots (\xi_a - \xi_{a-1})(\xi_a - \xi_{a+1}) \ldots (\xi_a - \xi_n)},$$

with n the number of nodes of the element and with $a = 1, 2, \ldots n$ referring to a node number. Notice that the above polynomial is of the order $(n - 1)$.

For instance, first-order (linear) polynomials are found for $n = 2$, hence

$$l_1^1 = \frac{\xi - \xi_2}{\xi_1 - \xi_2}, \tag{17.27}$$

and

$$l_2^1 = \frac{\xi - \xi_1}{\xi_2 - \xi_1}. \tag{17.28}$$

17.4.1 Lagrangian elements

The shape functions of an element of order $(n-1)$ in one dimension are chosen as

$$N_a = l_a^{n-1}. \tag{17.29}$$

The quadratic shape function associated with node 1 of the element depicted in Fig. 17.6 (with $\xi_1 = -1$, $\xi_2 = 0$, $\xi_3 = 1$) satisfies

$$N_1(\xi) = l_1^2(\xi) = \frac{(\xi - \xi_2)(\xi - \xi_3)}{(\xi_1 - \xi_2)(\xi_1 - \xi_3)} = \frac{1}{2}\xi(\xi - 1). \tag{17.30}$$

Likewise it follows that

$$N_2(\xi) = -(\xi + 1)(\xi - 1) = 1 - \xi^2, \tag{17.31}$$

and

$$N_3(\xi) = \frac{1}{2}\xi(\xi + 1). \tag{17.32}$$

In two dimensions, the shape functions for the 9-node element as visualized in Fig. 17.7 are formed by multiplication of two Lagrangian polynomials. Leading to:

$$N_1(\xi, \eta) = \frac{1}{4}\xi(\xi - 1)\eta(\eta - 1)$$

$$N_2(\xi, \eta) = -\frac{1}{2}(\xi + 1)(\xi - 1)\eta(\eta - 1)$$

A one-dimensional quadratic element, $-1 \leq \xi \leq 1$.

A two-dimensional quadratic element, $-1 \leq \xi \leq 1$, $-1 \leq \eta \leq 1$.

$$N_3(\xi, \eta) = -\frac{1}{4}\xi(\xi + 1)\eta(\eta - 1)$$

$$N_4(\xi, \eta) = -\frac{1}{2}\xi(\xi + 1)(\eta + 1)(\eta - 1)$$

$$N_5(\xi, \eta) = \frac{1}{4}\xi(\xi + 1)(\eta + 1)\eta$$

$$N_6(\xi, \eta) = -\frac{1}{2}(\xi + 1)(\xi - 1)(\eta + 1)\eta$$

$$N_7(\xi, \eta) = -\frac{1}{4}\xi(\xi - 1)(\eta + 1)\eta$$

$$N_8(\xi, \eta) = -\frac{1}{2}\xi(\xi - 1)(\eta + 1)(\eta - 1)$$

$$N_9(\xi, \eta) = (\xi + 1)(\xi - 1)(\eta + 1)(\eta - 1). \tag{17.33}$$

17.4.2 Serendipity elements

For serendipity elements no internal nodes are used. Consider a 'quadratic element', as depicted in Fig. 17.8 (right). The shape functions of the corner nodes are defined by

$$N_1 = \frac{1}{4}(1 - \xi)(1 - \eta)(-\xi - \eta - 1)$$

$$N_2 = \frac{1}{4}(1 + \xi)(1 - \eta)(+\xi - \eta - 1)$$

$$N_3 = \frac{1}{4}(1 + \xi)(1 + \eta)(+\xi + \eta - 1)$$

$$N_4 = \frac{1}{4}(1 - \xi)(1 + \eta)(-\xi + \eta - 1), \tag{17.34}$$

while the shape functions for the mid-side nodes read

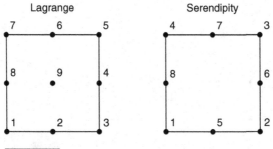

Figure 17.8

Example of a Serendipity element compared to the 'equal order' Lagrangian element.

$$N_5 = \frac{1}{2}(1 - \xi^2)(1 - \eta)$$

$$N_6 = \frac{1}{2}(1 + \xi)(1 - \eta^2)$$

$$N_7 = \frac{1}{2}(1 - \xi^2)(1 + \eta)$$

$$N_8 = \frac{1}{2}(1 - \xi)(1 - \eta^2). \qquad (17.35)$$

Other examples may be found in Zienkiewicz [18] and Hughes [10].

17.5 Numerical integration

Let $f : \Omega_e \mapsto \mathbf{R}$ be some function, and assume that the integral:

$$\int_{\Omega_e} f(x) \, dx, \qquad (17.36)$$

over the domain Ω_e of an element is to be computed. In finite element computations there is a mapping from the x-space to the ξ-space, such that (see Section 16.7, on isoparametric elements)

$$\int_{\Omega_e} f(x) \, dx = \int_{-1}^{1} \underbrace{f(\xi) \frac{dx}{d\xi}(\xi)}_{\phi(\xi)} \, d\xi. \qquad (17.37)$$

This integral can be approximated with a numerical integration rule:

$$\int_{-1}^{1} g(\xi) \, d\xi \approx \sum_{i=1}^{n_{\text{int}}} g(\xi_i) \, W_i, \qquad (17.38)$$

where ξ_i denotes the location of an integration point and W_i the associated weight factor.

In Fig. 17.9 an interpretation is given of the above numerical integration rule. At a discrete number of points ξ_i within the interval $\xi \in [-1, +1]$ the function value $g(\xi_i)$ is evaluated. Related to each point ξ_i a rectangle is defined with height $g(\xi_i)$ and width W_i. Note that it is not necessary that the point ξ_i is located on the symmetry line of the rectangle. By adding up the surfaces $g(\xi_i) W_i$ of all rectangles an approximation is obtained of the total surface underneath the function, which is the integral. It is clear that the weight factor W_i in Eq. (17.38) can be interpreted as the width of the interval around ξ_i.

The integration rule that is mostly used is the **Gaussian quadrature**. In that case the locations of the integration points and weight factors are chosen so as

Table 17.1   Gaussian quadrature up to n_{int}=3.

n_{int}	ξ_i	W_i
1	$\xi_1 = 0$	$W_1 = 2$
2	$\xi_1 = \frac{-1}{\sqrt{3}}, \xi_2 = \frac{1}{\sqrt{3}}$	$W_1 = W_2 = 1$
3	$\xi_1 = -\sqrt{\frac{3}{5}}, \xi_2 = 0, \xi_3 = \sqrt{\frac{3}{5}}$	$W_1 = W_3 = \frac{5}{9}, W_2 = \frac{8}{9}$

Figure 17.9

Numerical integration of a function $\phi(\xi)$.

to obtain optimal accuracy for polynomial expressions of $g(\xi)$. In Table 17.1 the location of the integration points and the associated weight factors are given up to $n_{int} = 3$. For two-dimensional problems the above generalizes to

$$\int_{\Omega_e} f(x,y) \, d\Omega = \int_{-1}^{1} \int_{-1}^{1} f(x(\xi,\eta), y(\xi,\eta)) j(\xi,\eta) \, d\xi d\eta$$

$$= \int_{-1}^{1} \int_{-1}^{1} g(\xi,\eta) \, d\xi d\eta, \tag{17.39}$$

with $j(\xi,\eta)$ according to Eq. (16.50) and

$$\int_{-1}^{1} \int_{-1}^{1} g(\xi,\eta) \, d\xi \, d\eta \approx \sum_{i=1}^{n_{int}} \sum_{j=1}^{n_{int}} g(\xi_i, \eta_j) W_i W_j. \tag{17.40}$$

The above integration scheme can be elaborated for the 9-node rectangular Lagrangian element in Fig. 17.10 using Table 17.2.

As was already remarked in Section 17.3 integration over a triangular domain is not trivial. For a triangular element that is formed by degeneration from a quadrilateral element the integration can be performed in the same way, with the same integration points, as given above.

In the case that triangular coordinates are used, the evaluation of the integrals is far from trivial. If the triangle coordinates λ_1 and λ_2 are maintained by eliminating

Table 17.2 Integration points in the square and associated weight factors.

n_{int}	point	location of the integration points		W_i
		ξ	η	
9	1	−0.7745966692	−0.7745966692	0.3086420047
	2	0.7745966692	−0.7745966692	0.3086420047
	3	0.7745966692	0.7745966692	0.3086420047
	4	−0.7745966692	0.7745966692	0.3086420047
	5	0	−0.7745966692	0.4938271818
	6	0.7745966692	0	0.4938271818
	7	0	0.7745966692	0.4938271818
	8	−0.7745966692	0	0.4938271818
	9	0	0	0.7901234686

Figure 17.10

Position of the integration points in the square.

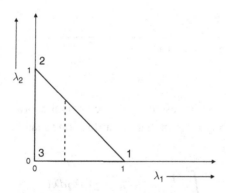

Figure 17.11

Mapping of triangle in a λ_1, λ_2-coordinate system.

Table 17.3 Integration points in the triangles.

n_{int}	point	location of the integration points			weight factors
		λ_1	λ_2	λ_3	W_i
3	1	0.5	0.5	0	0.16667
	2	0	0.5	0.5	0.16667
	3	0.5	0	0.5	0.16667
7	1	0.3333333333	0.3333333333	0.3333333333	0.11250
	2	0.0597158717	0.4701420641	0.4701420641	0.00662
	3	0.4701420641	0.0597158717	0.4701420641	0.00662
	4	0.4701420641	0.4701420641	0.0597158717	0.00662
	5	0.7974269853	0.1012865073	0.1012865073	0.00630
	6	0.1012865073	0.7974269853	0.1012865073	0.00630
	7	0.1012865073	0.1012865073	0.7974269853	0.00630

Figure 17.12

Position of the integration points in the triangles.

the coordinate λ_3 these can represented in a rectangular coordinate system as given in Fig. 17.11. It can easily be seen that the surface integral of a function $\phi(\lambda_1, \lambda_2)$ can be written as

$$\int_0^1 \int_0^{1-\lambda_1} \phi(\lambda_1, \lambda_2)\, d\lambda_2 d\lambda_1. \tag{17.41}$$

Despite a problem with a variable integral limit, it appeared possible to derive numerical integration rules for triangles. In Table 17.3 the position of integration points in triangular coordinates as well as weight factors are given for two higher-order elements.

Exercises

17.1 A MATLAB script to calculate the shape functions and the derivatives with respect to the isoparametric coordinates of 4-noded isoparametric bi-linear elements in some point of the elements may consist of the following code:

```
% Shape = program to calculate shape functions
xi= ;
eta= ;
x=[ , , , ]';
y=[ , , , ]';
N=[0.25*(1-xi)*(1-eta);
   0.25*(1+xi)*(1-eta);
   0.25*(1+xi)*(1+eta);
   0.25*(1-xi)*(1+eta)]
dNdxi=[-0.25*(1-eta);
        0.25*(1-eta);
        0.25*(1+eta);
       -0.25*(1+eta)]
dNdeta=[-0.25*(1-xi);
        -0.25*(1+xi);
         0.25*(1+xi);
         0.25*(1-xi)]
```

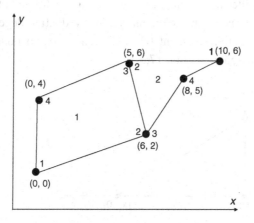

To study some of the properties of the shape functions the script will be completed by considering the elements in the figure above and extended in the following. In this figure the local node numbering has also been indicated.

(a)   Use the program to test whether

$$\sum_{i=1}^{n} N_i = 1,$$

and

$$\sum_{i=1}^{n} \frac{\partial N_i}{\partial \xi} = \sum_{i=1}^{n} \frac{\partial N_i}{\partial \eta} = 0,$$

for a number of combinations of (ξ, η).

(b)   Extend the program to calculate the Jacobian matrix:

$$\underline{x}_{,\xi} = \begin{pmatrix} \frac{\partial x}{\partial \xi} & \frac{\partial x}{\partial \eta} \\ \frac{\partial y}{\partial \xi} & \frac{\partial y}{\partial \eta} \end{pmatrix},$$

and the Jacobian determinant $j = \det(\underline{x}_{,\xi})$.

(c)   Determine the Jacobian determinant in the following points:

$$(\xi_1, \eta_1) = (0, 0)$$
$$(\xi_2, \eta_2) = (0.5, 0.5)$$
$$(\xi_3, \eta_3) = (1, 0)$$
$$(\xi_4, \eta_4) = (1, -1)$$

for both elements which are shown in the figure. What can you conclude from this?

17.2   Consider the element that is given in the figure below. The element shape functions of this element are derived after degeneration of a 4-noded quadrilateral element by coalescence of two nodes in the same way as discussed in Section 17.3.

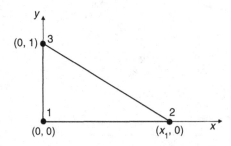

(a)   Give analytical expressions for the shape functions $N_i(\xi, \eta)$,
(b)   Compute the Jacobian determinant for $\xi = \eta = 0$.
(c)   Plot the result as a function of x_1 on the interval $[-2, +2]$ and comment on the result.

17.3 Consider the one-dimensional 4-noded element as given in the figure.

$$\xi=-1 \qquad \xi=-1/3 \qquad \xi=+1/3 \qquad \xi=+1$$

Use Lagrange polynomials to derive the element shape functions.

17.4 After solving the stationary convection-diffusion equation for the 4-noded quadrilateral element e the following element solution vector has been found:

$$\underset{\sim}{u}_e = \begin{bmatrix} 0.4 \\ 0.1 \\ 0.3 \\ 1 \end{bmatrix}.$$

(a) Use the script from Exercise 17.1 as the basis to calculate the u_h at the point $(\xi, \eta) = (0.5, 0.5)$.

(b) The nodal coordinates of the element are given:

$$\text{coord} = \begin{bmatrix} 5 & 2 \\ 15 & 1 \\ 10 & 4 \\ 4 & 4 \end{bmatrix}.$$

Calculate $\partial u_h / \partial x$ and $\partial u_h / \partial y$ for $(\xi, \eta) = (0.5, 0.5)$.

17.5 Consider the quadratic 6-node triangular element in the figure.

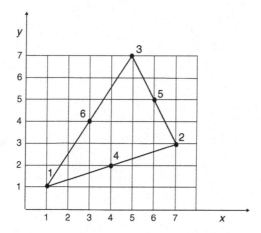

The solution vector for this element after a computation is given as

$$\underset{\sim}{u}_e = \begin{bmatrix} 2 \\ 5 \\ 3 \\ 2 \\ 0 \\ 2 \end{bmatrix}.$$

(a) Write a MATLAB program to calculate the solution u_h at point $(x, y) = (4, 4)$.

(b) Give expressions for $\partial u_h / \partial x$ and $\partial u_h / \partial y$ as a function of x and y.

18 Infinitesimal strain elasticity problems

18.1 Introduction

One of the first applications of the Finite Element Method in biomechanics has been the analysis of the mechanical behaviour of bone [2]. In particular the impact of prosthesis implants has been investigated extensively. An example of a finite element mesh used to analyse the mechanical stresses and strain in a human femur is given in Fig. 18.1(a). In Fig. 18.1(b) the femur head is replaced by a prosthesis. In this case the prosthesis has different mechanical properties compared to the bone, leading to high stress concentrations at some points in the bone and stress shielding (lower stresses than normal) in other parts. This normally leads to a remodelling process in the bone that has to be accounted for when new prostheses are designed. The purpose of this chapter is to introduce the finite element theory that forms the basis of these analyses.

18.2 Linear elasticity

Neglecting inertia, the momentum equation, Eq. (11.9), may be written as

$$\vec{\nabla} \cdot \boldsymbol{\sigma} + \vec{f} = \vec{0}, \tag{18.1}$$

where $\vec{\nabla}$ denotes the gradient operator, $\boldsymbol{\sigma}$ the Cauchy stress tensor and $\vec{f} = \rho\vec{q}$ a given distributed volume load. This equation should hold at each position within the domain of interest Ω, having boundary Γ, and must be supplemented with suitable boundary conditions. Either the displacement field $\vec{u}$ is specified along Γ_u:

$$\vec{u} = \vec{u}_0 \text{ at } \Gamma_u, \tag{18.2}$$

or the external load along Γ_p is specified:

$$\boldsymbol{\sigma} \cdot \vec{n} = \vec{p} \text{ at } \Gamma_p. \tag{18.3}$$

The vector $\vec{n}$ denotes the unit outward normal at the boundary Γ.

Hip implant

(a) (b)

(a) Mesh to analyse stress and strain distributions in a human femur (b) a finite element mesh of a femur, where the femur head has been replaced by a prosthesis. (Image courtesy of Mr Bert van Rietbergen.)

If isotropic, linearly elastic material behaviour according to Hooke's law is assumed, see Section 12.2, and the Cauchy stress tensor is related to the infinitesimal strain tensor $\boldsymbol{\varepsilon}$ by

$$\boldsymbol{\sigma} = K\text{tr}(\boldsymbol{\varepsilon})\boldsymbol{I} + 2G\boldsymbol{\varepsilon}^d, \tag{18.4}$$

where K is the compression modulus and G the shear modulus. The infinitesimal strain tensor is given by:

$$\boldsymbol{\varepsilon} = \frac{1}{2}(\vec{\nabla}\vec{u} + (\vec{\nabla}\vec{u})^{\text{T}}), \tag{18.5}$$

where $\vec{u}$ denotes the displacement field. Furthermore, $\text{tr}(\boldsymbol{\varepsilon})$ is the trace and $\boldsymbol{\varepsilon}^d$ the deviatoric part of the infinitesimal strain tensor. The shear modulus G and the compression modulus K may be expressed in terms of Young's modulus E and Poisson's ratio ν via

$$G = \frac{E}{2(1 + \nu)} \tag{18.6}$$

$$K = \frac{E}{3(1 - 2\nu)}. \tag{18.7}$$

Introduce the Cartesian basis $\{\vec{e}_x, \vec{e}_y, \vec{e}_z\}$. With respect to this basis the Cauchy stress tensor may be written as

$$\boldsymbol{\sigma} = \sigma_{xx}\vec{e}_x\vec{e}_x + \sigma_{xy}\vec{e}_x\vec{e}_y + \cdots + \sigma_{zz}\vec{e}_z\vec{e}_z, \tag{18.8}$$

or, using the Einstein summation convention (the presence of double indices implies summation over these indices):

$$\sigma = \sigma_{ij}\vec{e}_i\vec{e}_j, \tag{18.9}$$

with $i,j = x,y,z$. The Cauchy stress matrix is given by

$$\underline{\sigma} = \begin{bmatrix} \sigma_{xx} & \sigma_{xy} & \sigma_{xz} \\ \sigma_{xy} & \sigma_{yy} & \sigma_{yz} \\ \sigma_{xz} & \sigma_{yz} & \sigma_{zz} \end{bmatrix}, \tag{18.10}$$

where use has been made of the symmetry of the Cauchy stress tensor $\sigma^{\mathrm{T}} = \sigma$. The strain matrix is given by

$$\underline{\varepsilon} = \begin{bmatrix} \varepsilon_{xx} & \varepsilon_{xy} & \varepsilon_{xz} \\ \varepsilon_{xy} & \varepsilon_{yy} & \varepsilon_{yz} \\ \varepsilon_{xz} & \varepsilon_{yz} & \varepsilon_{zz} \end{bmatrix}. \tag{18.11}$$

The trace of the strain tensor is given by

$$\mathrm{tr}(\varepsilon) = \varepsilon_{xx} + \varepsilon_{yy} + \varepsilon_{zz}, \tag{18.12}$$

and the deviatoric part of the strain matrix is given by

$$\underline{\varepsilon}^d = \begin{bmatrix} \frac{2}{3}\varepsilon_{xx} - \frac{1}{3}(\varepsilon_{yy} + \varepsilon_{zz}) & \varepsilon_{xy} & \varepsilon_{xz} \\ \varepsilon_{xy} & \frac{2}{3}\varepsilon_{yy} - \frac{1}{3}(\varepsilon_{xx} + \varepsilon_{zz}) & \varepsilon_{yz} \\ \varepsilon_{xz} & \varepsilon_{yz} & \frac{2}{3}\varepsilon_{zz} - \frac{1}{3}(\varepsilon_{xx} + \varepsilon_{yy}) \end{bmatrix}. \tag{18.13}$$

18.3 Weak formulation

As before, the weak formulation based on the method of weighted residuals is obtained by pre-multiplying Eq. (18.1) with a weight function. Since Eq. (18.1) is a vector equation, the weighting function is chosen to be a vector field as well: $\vec{w}$. Taking the inner product of the weight function $\vec{w}$ with the momentum Eq. (18.1) yields a scalar expression, which upon integration over the domain Ω yields

$$\int_{\Omega} \vec{w} \cdot (\vec{\nabla} \cdot \sigma + \vec{f}) \, d\Omega = 0. \tag{18.14}$$

This integral equation must hold for all weighting functions $\vec{w}$. By application of the product rule, it can be shown that

$$\vec{\nabla} \cdot (\sigma \cdot \vec{w}) = (\vec{\nabla} \cdot \sigma) \cdot \vec{w} + (\vec{\nabla}\vec{w})^{\mathrm{T}} : \sigma. \tag{18.15}$$

In Eq. (18.15) the double dot product $(\vec{\nabla}\vec{w})^{\mathrm{T}} : \boldsymbol{\sigma}$ is used. The definition of the double dot product of two tensors $\boldsymbol{A}$ and $\boldsymbol{B}$ is

$$\boldsymbol{A} : \boldsymbol{B} = \mathrm{tr}(\boldsymbol{A} \cdot \boldsymbol{B}) = A_{ij}B_{ji}, \tag{18.16}$$

where the Einstein summation convention has been used. With respect to a Cartesian basis and using index notation it is straightforward to prove Eq. (18.15). First of all notice that

$$\boldsymbol{\sigma} \cdot \vec{w} = \sigma_{ij}w_j\vec{e}_i. \tag{18.17}$$

Consequently

$$\vec{\nabla}\cdot(\boldsymbol{\sigma} \cdot \vec{w}) = \frac{\partial}{\partial x_i}(\sigma_{ij}w_j). \tag{18.18}$$

Application of the product rule of differentiation yields

$$\begin{aligned}
\vec{\nabla}\cdot(\boldsymbol{\sigma} \cdot \vec{w}) &= \frac{\partial}{\partial x_i}(\sigma_{ij}w_j) \\
&= \frac{\partial\sigma_{ij}}{\partial x_i}w_j + \sigma_{ij}\frac{\partial w_j}{\partial x_i}.
\end{aligned} \tag{18.19}$$

With the identifications

$$(\vec{\nabla}\cdot\boldsymbol{\sigma})\cdot\vec{w} = \frac{\partial\sigma_{ij}}{\partial x_i}w_j, \tag{18.20}$$

and

$$(\vec{\nabla}\vec{w})^{\mathrm{T}} : \boldsymbol{\sigma} = \frac{\partial w_j}{\partial x_i}\sigma_{ij}, \tag{18.21}$$

the product rule according to Eq. (18.15) is obtained.

Use of this result in Eq. (18.14) yields

$$\int_\Omega \vec{\nabla}\cdot(\boldsymbol{\sigma} \cdot \vec{w}) \, d\Omega - \int_\Omega (\vec{\nabla}\vec{w})^{\mathrm{T}} : \boldsymbol{\sigma} \, d\Omega + \int_\Omega \vec{w} \cdot \vec{f} \, d\Omega = 0. \tag{18.22}$$

The first integral may be transformed using the divergence theorem Eq. (16.5). This yields

$$\int_\Omega (\vec{\nabla}\vec{w})^{\mathrm{T}} : \boldsymbol{\sigma} \, d\Omega = \int_\Gamma \vec{w}\cdot(\boldsymbol{\sigma} \cdot \vec{n}) \, d\Gamma + \int_\Omega \vec{w} \cdot \vec{f} \, d\Omega. \tag{18.23}$$

18.4 Galerkin discretization

Within the context of the finite element method, the domain Ω is split into a number of non-overlapping subdomains (elements) Ω_e, such that this integral equation is rewritten as

$$\sum_{e=1}^{N_{el}} \int_{\Omega_e} (\vec{\nabla}\vec{w})^{\mathrm{T}} : \sigma \, d\Omega = \sum_{e=1}^{N_{el}} \left(\int_{\Gamma_e} \vec{w} \cdot (\sigma \cdot \vec{n}) \, d\Gamma + \int_{\Omega_e} \vec{w} \cdot \vec{f} \, d\Omega \right). \quad (18.24)$$

Clearly, the above $\int_{\Gamma_e} (.) \, d\Gamma$ denotes the integral along those element boundaries that coincide with the domain boundary Γ ($\Gamma_e = \Omega_e \cap \Gamma$). As an example of how to proceed based on Eq. (18.24) a **plane strain** problem is considered.

Step 1 The integrand of the integral in the left-hand side of Eq. (18.24) is elaborated first. If the double inner product of a symmetric tensor A and a skew-symmetric tensor B is taken, then

$$A : B = 0 \text{ if } A^{\mathrm{T}} = A \text{ and } B^{\mathrm{T}} = -B . \quad (18.25)$$

This can easily be proved. Using the properties of the double inner product it follows that

$$A : B = A^{\mathrm{T}} : B^{\mathrm{T}} = -A : B . \quad (18.26)$$

This equality can only hold if $A : B = 0$.

The dyadic product $(\vec{\nabla}\vec{w})^{\mathrm{T}}$ appearing in $(\vec{\nabla}\vec{w})^{\mathrm{T}} : \sigma$ may be split into a symmetric and a skew-symmetric part:

$$(\vec{\nabla}\vec{w})^{\mathrm{T}} = \frac{1}{2}\left[(\vec{\nabla}\vec{w}) + (\vec{\nabla}\vec{w})^{\mathrm{T}} \right] - \frac{1}{2}\left[(\vec{\nabla}\vec{w}) - (\vec{\nabla}\vec{w})^{\mathrm{T}} \right]. \quad (18.27)$$

Because of the symmetry of the Cauchy stress tensor it follows that

$$(\vec{\nabla}\vec{w})^{\mathrm{T}} : \sigma = \frac{1}{2}\left[(\vec{\nabla}\vec{w}) + (\vec{\nabla}\vec{w})^{\mathrm{T}} \right] : \sigma. \quad (18.28)$$

Notice that the expression $\frac{1}{2}\left[(\vec{\nabla}\vec{w}) + (\vec{\nabla}\vec{w})^{\mathrm{T}} \right]$ has a similar form to the infinitesimal strain tensor ε defined as

$$\varepsilon = \frac{1}{2}\left[(\vec{\nabla}\vec{u}) + (\vec{\nabla}\vec{u})^{\mathrm{T}} \right], \quad (18.29)$$

where $\vec{u}$ denotes the displacement field. This motivates the abbreviation:

$$\varepsilon^w = \frac{1}{2}\left[(\vec{\nabla}\vec{w}) + (\vec{\nabla}\vec{w})^{\mathrm{T}} \right]. \quad (18.30)$$

Consequently, the symmetry of the Cauchy stress tensor allows the double inner product $(\vec{\nabla}\vec{w})^{\mathrm{T}} : \sigma$ to be rewritten as

$$(\vec{\nabla}\vec{w})^{\mathrm{T}} : \sigma = \varepsilon^w : \sigma. \quad (18.31)$$

Step 2 To elaborate further, it is convenient to introduce a vector basis. Here, a Cartesian vector basis $\{\vec{e}_x, \vec{e}_y, \vec{e}_z\}$ is chosen. In the plane strain case it is assumed that $\varepsilon_{zz} = \varepsilon_{xz} = \varepsilon_{yz} = 0$ while the displacement field is written as

$$\vec{u} = u_x(x, y)\vec{e}_x + u_y(x, y)\vec{e}_y. \quad (18.32)$$

Likewise, the weighting function $\vec{w}$ is written as

$$\vec{w} = w_x(x,y)\,\vec{e}_x + w_y(x,y)\,\vec{e}_y. \tag{18.33}$$

In the plane strain case, the matrices associated with the tensors $\boldsymbol{\varepsilon}$ and $\boldsymbol{\varepsilon}^w$ are given by

$$\underline{\varepsilon} = \begin{bmatrix} \varepsilon_{xx} & \varepsilon_{xy} & 0 \\ \varepsilon_{xy} & \varepsilon_{yy} & 0 \\ 0 & 0 & 0 \end{bmatrix}, \qquad \underline{\varepsilon}^w = \begin{bmatrix} \varepsilon_{xx}^w & \varepsilon_{xy}^w & 0 \\ \varepsilon_{xy}^w & \varepsilon_{yy}^w & 0 \\ 0 & 0 & 0 \end{bmatrix}. \tag{18.34}$$

Consequently, the inner product $\boldsymbol{\varepsilon}^w : \boldsymbol{\sigma}$ equals

$$\boldsymbol{\varepsilon}^w : \boldsymbol{\sigma} = \varepsilon_{xx}^w \sigma_{xx} + \varepsilon_{yy}^w \sigma_{yy} + 2\varepsilon_{xy}^w \sigma_{xy}. \tag{18.35}$$

Notice the factor 2 in front of the last product on the right-hand side due to the symmetry of both $\boldsymbol{\varepsilon}^w$ and $\boldsymbol{\sigma}$. It is convenient to gather the relevant components of $\underline{\varepsilon}^w$, $\underline{\varepsilon}$ (for future purposes) and $\underline{\sigma}$ in a column:

$$(\underline{\varepsilon}^w)^{\mathrm{T}} = \begin{bmatrix} \varepsilon_{xx}^w & \varepsilon_{yy}^w & 2\varepsilon_{xy}^w \end{bmatrix}$$

$$\underline{\varepsilon}^{\mathrm{T}} = \begin{bmatrix} \varepsilon_{xx} & \varepsilon_{yy} & 2\varepsilon_{xy} \end{bmatrix}, \tag{18.36}$$

(notice the 2 in front of ε_{xy}^w and ε_{xy}) and

$$\underline{\sigma}^{\mathrm{T}} = \begin{bmatrix} \sigma_{xx} & \sigma_{yy} & \sigma_{xy} \end{bmatrix}. \tag{18.37}$$

This allows the inner product $\boldsymbol{\varepsilon}^w : \boldsymbol{\sigma}$ to be written as

$$\boldsymbol{\varepsilon}^w : \boldsymbol{\sigma} = (\underline{\varepsilon}^w)^{\mathrm{T}} \underline{\sigma}. \tag{18.38}$$

Step 3 The constitutive equation according to Eq. (18.4) may be recast in the form

$$\underline{\sigma} = \underline{H}\,\underline{\varepsilon}. \tag{18.39}$$

Dealing with the isotropic Hooke's law and plane strain conditions and after introduction of Eqs. (18.12) and (18.13) into Eq. (18.4), the matrix $\underline{H}$ can be written as

$$\underline{H} = K \begin{bmatrix} 1 & 1 & 0 \\ 1 & 1 & 0 \\ 0 & 0 & 0 \end{bmatrix} + \frac{G}{3} \begin{bmatrix} 4 & -2 & 0 \\ -2 & 4 & 0 \\ 0 & 0 & 3 \end{bmatrix}. \tag{18.40}$$

Consequently

$$\boldsymbol{\varepsilon}^w : \boldsymbol{\sigma} = (\underline{\varepsilon}^w)^{\mathrm{T}} \underline{H}\,\underline{\varepsilon}. \tag{18.41}$$

Step 4 With $\vec{u} = u_x \vec{e}_x + u_y \vec{e}_y$ and $\vec{x} = x\vec{e}_x + y\vec{e}_y$ the strain components may be written as

$$\underset{\sim}{\varepsilon} = \begin{bmatrix} \varepsilon_{xx} \\ \\ \varepsilon_{yy} \\ \\ 2\varepsilon_{xy} \end{bmatrix} = \begin{bmatrix} \frac{\partial u_x}{\partial x} \\ \\ \frac{\partial u_y}{\partial y} \\ \\ \frac{\partial u_x}{\partial y} + \frac{\partial u_y}{\partial x} \end{bmatrix}. \tag{18.42}$$

This is frequently rewritten as

$$\underset{\sim}{\varepsilon} = \underline{\hat{B}}\,\underset{\sim}{u}, \tag{18.43}$$

with $\underline{\hat{B}}$ an operator defined by

$$\underline{\hat{B}} = \begin{bmatrix} \frac{\partial}{\partial x} & 0 \\ \\ 0 & \frac{\partial}{\partial y} \\ \\ \frac{\partial}{\partial y} & \frac{\partial}{\partial x} \end{bmatrix}, \tag{18.44}$$

while $\underset{\sim}{u}$ represents the displacement field:

$$\underset{\sim}{u} = \begin{bmatrix} u_x \\ u_y \end{bmatrix}. \tag{18.45}$$

Step 5 Within each element the displacement field is interpolated according to

$$u_x|_{\Omega_e} = \sum_{i=1}^{n} N_i(x,y)\, u_{xei} = \underset{\sim}{N}^{\mathrm{T}} \underset{\sim}{u}_{xe}$$

$$u_y|_{\Omega_e} = \sum_{i=1}^{n} N_i(x,y)\, u_{yei} = \underset{\sim}{N}^{\mathrm{T}} \underset{\sim}{u}_{ye}, \tag{18.46}$$

where $\underset{\sim}{u}_{xe}$ and $\underset{\sim}{u}_{ye}$ denote the nodal displacements of element Ω_e in the x- and y-direction, respectively. Using this discretization the strain within an element can be written as

$$\underset{\sim}{\varepsilon} = \sum_{i=1}^{n} \begin{bmatrix} \frac{\partial N_i}{\partial x} u_{xei} \\ \\ \frac{\partial N_i}{\partial y} u_{yei} \\ \\ \frac{\partial N_i}{\partial y} u_{xei} + \frac{\partial N_i}{\partial x} u_{yei} \end{bmatrix}. \tag{18.47}$$

It is customary, and convenient, to gather all the nodal displacements u_{xei} and u_{yei} in one column, indicated by $\underset{\sim}{u}_e$, according to

$$
\underset{\sim}{u}_e =
\begin{bmatrix}
\left.\begin{array}{c} u_{xe1} \\ u_{ye1} \end{array}\right\} \text{node 1} \\[1em]
\left.\begin{array}{c} u_{xe2} \\ u_{ye2} \end{array}\right\} \text{node 2} \\[1em]
\vdots \\[1em]
\left.\begin{array}{c} u_{xen} \\ u_{yen} \end{array}\right\} \text{node } n
\end{bmatrix}.
\tag{18.48}
$$

Using this definition, the strain column for an element e can be rewritten as

$$
\underset{\sim}{\varepsilon} = \underline{B}\, \underset{\sim}{u}_e,
\tag{18.49}
$$

with $\underline{B}$ the so-called **strain displacement matrix**:

$$
\underline{B} =
\begin{bmatrix}
\frac{\partial N_1}{\partial x} & 0 & \frac{\partial N_2}{\partial x} & 0 & \cdots & \frac{\partial N_n}{\partial x} & 0 \\[0.8em]
0 & \frac{\partial N_1}{\partial y} & 0 & \frac{\partial N_2}{\partial y} & \cdots & 0 & \frac{\partial N_n}{\partial y} \\[0.8em]
\frac{\partial N_1}{\partial y} & \frac{\partial N_1}{\partial x} & \frac{\partial N_2}{\partial y} & \frac{\partial N_2}{\partial x} & \cdots & \frac{\partial N_n}{\partial y} & \frac{\partial N_n}{\partial x}
\end{bmatrix}.
\tag{18.50}
$$

Clearly, a similar expression holds for $\underset{\sim}{\varepsilon}^w$. So, patching everything together, the double inner product $\boldsymbol{\varepsilon}^w : \boldsymbol{\sigma}$ may be written as:

$$
\begin{aligned}
\boldsymbol{\varepsilon}^w : \boldsymbol{\sigma} &= (\underset{\sim}{\varepsilon}^w)^{\mathrm{T}} \underset{\sim}{\sigma} \\
&= (\underset{\sim}{\varepsilon}^w)^{\mathrm{T}} \underline{H}\, \underset{\sim}{\varepsilon} \\
&= \underset{\sim}{w}_e^{\mathrm{T}} \underline{B}^{\mathrm{T}} \underline{H}\, \underline{B}\, \underset{\sim}{u}_e ,
\end{aligned}
\tag{18.51}
$$

where $\underset{\sim}{w}_e$ stores the components of the weighting vector $\vec{w}$ structured in the same way as $\underset{\sim}{u}_e$. This result can be exploited to elaborate the left-hand side of Eq. (18.24):

$$
\int_{\Omega_e} (\vec{\nabla}\vec{w})^{\mathrm{T}} : \boldsymbol{\sigma}\, d\Omega = \underset{\sim}{w}_e^{\mathrm{T}} \int_{\Omega_e} \underline{B}^{\mathrm{T}} \underline{H} \underline{B}\, d\Omega\, \underset{\sim}{u}_e.
\tag{18.52}
$$

The element coefficient matrix, or stiffness matrix $\underline{K}_e$ is defined as

$$
\underline{K}_e = \int_{\Omega_e} \underline{B}^{\mathrm{T}} \underline{H} \underline{B}\, d\Omega.
\tag{18.53}
$$

Step 6 Writing the force per unit volume vector $\vec{f}$ as

$$\vec{f} = f_x\vec{e}_x + f_y\vec{e}_y, \tag{18.54}$$

and the weighting function $\vec{w}$ within element Ω_e as

$$\vec{w}|_{\Omega_e} = \underset{\sim}{N}^{\mathrm{T}}\underset{\sim}{w}_{xe}\vec{e}_x + \underset{\sim}{N}^{\mathrm{T}}\underset{\sim}{w}_{ye}\vec{e}_y, \tag{18.55}$$

the second integral on the right-hand side of Eq. (18.24) can be written as

$$\int_{\Omega_e} \vec{w} \cdot \vec{f} \, d\Omega = \int_{\Omega_e} (\underset{\sim}{N}^{\mathrm{T}}\underset{\sim}{w}_{xe}f_x + \underset{\sim}{N}^{\mathrm{T}}\underset{\sim}{w}_{ye}f_y) \, d\Omega$$

$$= \underset{\sim}{w}_e^{\mathrm{T}}\underset{\sim}{f}_e^v, \tag{18.56}$$

where

$$\underset{\sim}{f}_e^v = \int_{\Omega_e} \begin{bmatrix} N_1 f_{ex} \\ N_1 f_{ey} \\ \\ N_2 f_{ex} \\ N_2 f_{ey} \\ \vdots \\ N_n f_{ex} \\ N_n f_{ey} \end{bmatrix} d\Omega. \tag{18.57}$$

The first integral on the right-hand side of Eq. (18.24) may, if applicable, be elaborated in exactly the same manner. This results in

$$\int_{\Gamma_e} \vec{w} \cdot (\sigma \cdot \vec{n}) \, d\Gamma = \underset{\sim}{w}_e^{\mathrm{T}}\underset{\sim}{f}_e^p, \tag{18.58}$$

where $\underset{\sim}{f}_e^p$ is structured analogously to $\underset{\sim}{f}_e^v$. The contribution of the right-hand side triggers the abbreviation:

$$\underset{\sim}{f}_e = \underset{\sim}{f}_e^v + \underset{\sim}{f}_e^p, \tag{18.59}$$

often referred to as the element load contribution.

Step 7 Putting all the pieces together, the discrete weak formulation of Eq. (18.24) is written as:

$$\sum_{e=1}^{N_{el}} \underset{\sim}{w}_e^{\mathrm{T}}\underline{K}_e\underset{\sim}{u}_e = \sum_{e=1}^{N_{el}} \underset{\sim}{w}_e^{\mathrm{T}}\underset{\sim}{f}_e. \tag{18.60}$$

Following an equivalent assembling procedure as outlined in Chapter 14 the following result may be obtained:

$$\underset{\sim}{w}^{\mathrm{T}}\underline{K}\underset{\sim}{u} = \underset{\sim}{w}^{\mathrm{T}}\underset{\sim}{f}, \tag{18.61}$$

where w and u contain the global nodal weighting factors and displacements, respectively, and K is the global stiffness matrix. This equation should hold for all weighting factors, and thus

$$K\, u = f. \tag{18.62}$$

18.5 Solution

As outlined in Chapter 14 the nodal displacements u may be partitioned into two groups. The first displacement group consists of components of u that are prescribed: u_p. The remaining nodal displacements, which are initially unknown, are gathered in u_u. Hence:

$$u = \begin{bmatrix} u_u \\ u_p \end{bmatrix}. \tag{18.63}$$

In a similar fashion the stiffness matrix K and the load vector f are partitioned. As a result, Eq. (18.62) can be written as:

$$\begin{bmatrix} K_{uu} & K_{up} \\ K_{pu} & K_{pp} \end{bmatrix} \begin{bmatrix} u_u \\ u_p \end{bmatrix} = \begin{bmatrix} f_u \\ f_p \end{bmatrix}. \tag{18.64}$$

The force column f is split into two parts: f_u and f_p. The column f_u is known, since it stores the external loads applied to the body. The column f_p, on the other hand, is not known, since no external load may be applied to points at which the displacement is prescribed. In Eq. (18.64) f_u is known and f_p is unknown. The following set of equations results:

$$K_{uu} u_u = f_u - K_{up} u_p, \tag{18.65}$$

$$f_p = K_{pu} u_u + K_{pp} u_p. \tag{18.66}$$

The first equation is used to calculate the unknown displacements u_u. The result is substituted into the second equation to calculate the unknown forces f_p.

18.6 Example

Consider the bending of a beam subjected to a concentrated force (Fig. 18.2). Let the beam be clamped at $x = 0$ and the point load F be applied at $x = L$. It is interesting to investigate the response of the bar for different kinds of elements using a similar element distribution. Four different elements are tested: the linear and the quadratic triangular element, and the bi-linear and bi-quadratic quadrilateral

Table 18.1 Comparison of the relative accuracy of different element types for the beam bending case.

Element type	u_c/u_L
Linear triangle	0.231
Bi-linear quadrilateral	0.697
Quadratic triangle	0.999
Bi-quadratic quadrilateral	1.003

Figure 18.2

A beam that is clamped on one side and loaded with a vertical concentrated force at the other side.

element. The meshes for the linear triangular and the bi-linear quadrilateral element are shown in Fig. 18.3. The displacement at $x = L$ can be computed using standard beam theory, giving

$$u_L = \frac{FL^3}{3EI},\qquad(18.67)$$

with

$$I = \frac{bh^3}{12},\qquad(18.68)$$

where b is the width of the beam and h the height of the beam and E the Young's modulus of the material. The ratio h/L of height over length equal to 0.1 is chosen. Using the meshes depicted in Fig. 18.3, the results of Table 18.1 are obtained, which presents the ratio of u_L and the computed displacement u_c at $x = L$.

It is clear that the displacement of the beam, using a mesh of linear triangles is much too small. The poor performance of the linear triangle can easily be understood; because of the linear interpolation of the displacement field $\vec{u}$, the associated strains computed from

Figure 18.3

Triangular and quadrilateral element distribution for the bar problem.

$$\varepsilon = \underline{B}\,u \qquad\qquad (18.69)$$

are constant per element. The bi-linear quadrilateral element, on the other hand, is clearly enhanced. A typical shape function, of for example the first node in an element, is given by (with respect to the local coordinate system)

$$N_1 = \frac{1}{4}(1 - \xi)(1 - \eta). \qquad\qquad (18.70)$$

Hence

$$N_1 = \frac{1}{4}(1 - \xi - \eta + \xi\eta), \qquad\qquad (18.71)$$

which means that an additional non-linear term is present in the shape functions. Therefore a *linear* variation of the stress field within an element is represented.

Two remarks have to be made at this point:

- The numerical analysis in this example was based on plane stress theory, while in the chapter the equations were elaborated for a plane strain problem. How this elaboration is done for a plane stress problem is discussed in Exercise 18.1.
- Strictly speaking, a concentrated force in one node of the mesh is not correct. Mesh refinement in this case would lead to an infinite displacement of the node where the force is acting. This can be avoided by applying a distributed load over a small part of the beam.

Exercises

18.1   For Hooke's law the Cauchy stress tensor σ is related to the infinitesimal strain tensor ε via

$$\sigma = K\mathrm{tr}(\varepsilon)I + 2G\varepsilon^d.$$

(a)   What is $\mathrm{tr}(\varepsilon)$ for the plane strain case?

(b)   What is $\mathrm{tr}(\varepsilon)$ for the plane stress case?

(c) What are the non-zero components of ε and σ for the plane stress case?

(d) What are the non-zero components of ε and σ for the plane strain case?

(e) Consider the plane stress case. Let

$$
\underset{\sim}{\sigma} = \begin{bmatrix} \sigma_{xx} \\ \sigma_{yy} \\ \sigma_{xy} \end{bmatrix} \quad \underset{\sim}{\varepsilon} = \begin{bmatrix} \varepsilon_{xx} \\ \varepsilon_{yy} \\ \varepsilon_{xy} \end{bmatrix}.
$$

What is the related $\underline{H}$ matrix for this case, using

$$
\underset{\sim}{\sigma} = \underline{H}\,\underset{\sim}{\varepsilon}?
$$

(f) For the plane strain case the matrix $\underline{H}$ is given by

$$
\underline{H} = K \begin{bmatrix} 1 & 1 & 0 \\ 1 & 1 & 0 \\ 0 & 0 & 0 \end{bmatrix} + \frac{G}{3} \begin{bmatrix} 4 & -2 & 0 \\ -2 & 4 & 0 \\ 0 & 0 & 3 \end{bmatrix}.
$$

Rewrite this matrix in terms of Young's modulus E and Poisson's ratio v. Is the resulting matrix linearly dependent on E?

18.2 Consider the mesh given in the figure below. The mesh consists of two linear triangular elements and a linear elasticity formulation applies to this element configuration. The solution $\underset{\sim}{u}$ is given by

$$
\underset{\sim}{u}^{\mathrm{T}} = [\, u_1 \; w_1 \; u_2 \; w_2 \; u_3 \; w_3 \; u_4 \; w_4 \,],
$$

where u and v denote the displacements in the x- and y-direction, respectively.

(a) What is a possible `top` array of this element configuration assuming equal material and type identifiers for both elements?

(b) What is the `pos` array for this element configuration?

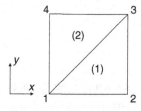

(c) What is the `dest` array?

(d) Based on the `pos` array the non-zero entries of the stiffness matrix can be identified. Visualize the non-zero entries of the stiffness matrix.

(e) Suppose that the boundary nodes in the array usercurves are stored as

```
usercurves=[
    1 3
    3 4
    2 4
    1 2];
```

The solution u is stored in the array sol. How are the displacements in the x-direction extracted from the sol array along the first usercurve?

(f) Let the solution array sol and the global stiffness matrix q be given. Suppose that both displacements at the nodes 1 and 2 are suppressed. Compute the reaction forces in these nodes. Compute the total reaction force in the y-direction along the boundary containing the nodes 1 and 2.

18.3 Consider the bi-linear element of the figure below. It covers exactly the domain $-1 \leq x \leq 1$ and $-1 \leq y \leq 1$. Assume that the plane strain condition applies.

(a) Compute the strain displacement matrix $\underline{B}$ for this element in point $(x, y) = (\frac{1}{2}, \frac{1}{2})$.

(b) Let the nodal displacements for this element be given by

$$\underline{u}_e^T = \begin{bmatrix} 0 & 0 & 0 & 0 & 1 & 0 & 1 & 0 \end{bmatrix}.$$

What are the strains in $(x, y) = (\frac{1}{2}, \frac{1}{2})$?

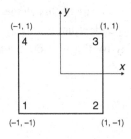

(c) If $G = 1$ and $K = 2$ what are the stress components σ_{xx}, σ_{yy} and σ_{xy} at $(x, y) = (0, 0)$? What is the value of σ_{zz}?

18.4 Given the solution vector sol and the pos array,

(a) Extract the solution vector for a given element ielem.

(b) Compute the strain at an integration point.

(c) Compute the stress at an integration point.

18.5　The m-file demo_bend in the directory twode analyses the pure bending of a single element (for quadrilaterals) or two elements (for triangles). The analysis is based on plane stress theory. The geometry is a simple square domain of dimensions 1×1. Along the left edge the displacements in the x-direction are set to zero, while at the lower left corner the displacement in the y-direction is set to zero to prevent rigid body motions. The two nodes at the right edge are loaded with a force F of opposite sign, to represent a pure bending moment. Investigate the stress field for various elements: linear triangle, bi-linear quadrilateral, and their quadratic equivalents. Explain the observed differences. To make these choices use itype and norder:

```
itype = 1  : quadrilateral element
itype = 20 : triangular element
norder = 1 :(bi-)linear element
norder = 2 :(bi-)quadratic element
```

18.6　In a shearing test a rectangular piece of material is clamped between a top and bottom plate, as schematically represented in the figure. This experiment is generally set-up to represent the so-called 'simple-shear' configuration. In the simple-shear configuration the strain tensor is given as

$$\boldsymbol{\varepsilon} = \varepsilon_{xy}\vec{e}_x\vec{e}_y + \varepsilon_{xy}\vec{e}_y\vec{e}_x$$

using the symmetry of the strain tensor. As a consequence, the stress-strain relation according to Hooke's law reduces to

$$\sigma_{xy} = 2G\varepsilon_{xy}.$$

Hence, measuring the clamp forces and the shear displacement provides

a direct means to identify the shear modulus G. However, the 'simple-shear' state is difficult to realize experimentally, since the configuration of the figure does *not* exactly represent the simple-shear case. This may be analysed using the m-file demo_shear.

(a)　Analyse the shear and the simple-shear case using this m-file. What is the difference in boundary conditions for these two cases?

    (b) Why is the `simple_shear=0` case not equal to the exact simple-shear case?

    (c) What ratio ℓ/h is required to measure the modulus G within 10 % accuracy? How many elements did you use to obtain this result? Explain the way the m-file computes the modulus G.

References

[1]   Adams, R. A. (2003) *Calculus: A Complete Course* (Addison, Wesley, Longman).

[2]   Brekelmans, W. A. M., Poort, H. W. and Slooff, T. J. J. H. (1972) A new method to analyse the mechanical behaviour of skeletal parts. *Acta Orthop. Scandnav.* **43**, 301–17.

[3]   Brooks, A. N. and Hughes, T. J. R. (1990) Streamline upwind/Petrov-Galerkin formulations for convection dominated flows with particular emphasis on the incompressible Navier-Stokes equations. *Computer Methods in Applied Mechanics and Engineering Archive – Special edition*, 199–259.

[4]   Carslaw, H. S. and Jaeger, J. C. (1980) *Conduction of Heat in Solids* (Clarendon Press).

[5]   Cacciola, G. R. C. (1998) Design, simulation and manufacturing of fibre reinforced polymer heart valves. Ph.D. thesis, Eindhoven University of Technology.

[6]   Frijns, A. J. H., Huyghe, J. M. R. J. and Janssen, J. D. (1997) A validation of the quadriphasic mixture theory for intervertebral disc tissue. *Int. J. Engng. Sci.* **35**(15), 1419–29.

[7]   Fung, Y. C. (1990) *Biomechanics: Motion, Flow, Stress, and Growth* (Springer-Verlag).

[8]   Fung, Y. C. (1993) *Biomechanics: Mechanical Properties of Living Tissues, 2nd edition* (Springer-Verlag).

[9]   Hill, A. V. (1938) The heat of shortening and the dynamic constants in muscle. *Proc. Roy. Soc. London* **126**, 136–65.

[10]   Hughes, T. J. R. (1987) *The Finite Element Method* (Prentice Hall).

[11]   Huyghe, J. M. R. J., Arts, T. and Campen, D. H. van (1992) Porous medium finite element model of the beating left ventricle. *Am.J.Physiol.* **262** H1256–H1267.

[12]   Huxley, A. F. (1957). Muscle structure and theory of contraction. *Prog. Biochem. Biophysic. Chem.*, 255–318.

[13]   Mow, V. C., Kuei, S. C. and Lai, W. M. (1980) Biphasic creep and stress relaxation of articular cartilage in compression. *J. Biomech. Engng.* **102** 73–84.

[14]   Oomens, C. W. J. (1985) A mixture approach to the mechanics of skin and subcutis. Ph.D. thesis, Twente University of Technology.

[15]   Oomens, C. W. J., Maenhout, M., Oijen, C. H. van, Drost, M. R. and
       Baaijens, F. P. T. (2003) Finite element modelling of contracting skeletal muscle.
       Phil. Trans. R. Soc. London B **358**, 1453–60.

[16]   Sengers, B. G., Oomens, C. W. J., Donkelaar, C. C. van, and Baaijens, F. P. T.
       (2005) A computational study of culture conditions and nutrient supply in cartilage
       tissue engineering. *Biotechnology Progress* **21**, 1252–1261.

[17]   Sengers, B. G. (2005) Modeling the development of tissue engineered cartilage.
       Ph.D. thesis, Eindhoven University of Technology.

[18]   Zienkiewicz, O. C. (1989) *The Finite Element Method, 4th edition* (McGraw-Hill).

Index

Printed in the United States
By Bookmasters